STRUCTURAL STABILITY

**THEORY
AND
IMPLEMENTATION**

STRUCTURAL STABILITY

THEORY AND IMPLEMENTATION

W.F. Chen, Ph.D.
Professor and Head of Structural Engineering
School of Civil Engineering
Purdue University
West Lafayette, Indiana

E.M. Lui, Ph.D.
Assistant Professor of Civil Engineering
Department of Civil Engineering
Syracuse University
Syracuse, New York

P T R Prentice Hall
Upper Saddle River, New Jersey 07458

Library of Congress Cataloging in Publication Data

Chen, Wai-Fah
  Structural stability.

  Includes bibliographies and index.
  1. Structural stability. I. Lui, E.M.II. Title.
TA656.C44 1987   624.1'71   86-19931
ISBN 0-13-500539-6

Cover Design: Miriam Recio

© 1987 by Prentice-Hall, Inc.
A Pearson Education Company
Upper Saddle River, NJ 07458

All rights reserved. No part of this book may be
reproduced, in any form or by any means,
without permission in writing from the publisher.

Printed in the United States of America

ISBN 0-13-500539-6

Prentice-Hall International (UK) Limited, London
Prentice-Hall of Australia Pty. Limited, Sydney
Prentice-Hall Canada Inc., Toronto
Prentice-Hall Hispanoamericana, S.A., Mexico
Prentice-Hall of India Private Limited, New Delhi
Prentice-Hall of Japan, Inc., Tokyo
Pearson Education Asia Pte. Ltd., Singapore
Editoria Prentice-Hall do Brasil, Ltda., Rio De Janeiro

CONTENTS

Preface   ix
Notation   xiii

CHAPTER 1   **GENERAL PRINCIPLES**   **1**
    1.1 Concepts of Stability   1
    1.2 Types of Stability   4
    1.3 Methods of Analyses in Stability   11
    1.4 Illustrative Examples—Small Deflection Analysis   12
    1.5 Illustrative Examples—Large Deflection Analysis   24
    1.6 Illustrative Examples—Imperfect Systems   31
    1.7 Design Philosophies   37
    1.8 Summary   42
    Problems   42
    References   43

CHAPTER 2   **COLUMNS**   **45**
    2.1 Introduction   45
    2.2 Classical Column Theory   48
    2.3 End-Restrained Columns   61
    2.4 Fourth-Order Differential Equation   79
    2.5 Special Members   85
    2.6 Initially Crooked Columns   91
    2.7 Inelastic Columns   96
    2.8 Design Curves for Aluminum Columns   108
    2.9 Stub Column Stress–Strain Curve   111
    2.10 Column Curves of Idealized Steel I-Section   117

2.11 Design Curves for Steel Columns **122**
2.12 Summary **137**
Problems **140**
References **145**

CHAPTER 3 BEAM-COLUMNS **147**

3.1 Introduction **147**
3.2 Beam-Column with a Uniformly Distributed Lateral Load **149**
3.3 Beam-Column with a Concentrated Lateral Load **156**
3.4 Beam-Columns Subjected to End Moments **161**
3.5 Superposition of Solutions **170**
3.6 Basic Differential Equations **175**
3.7 Slope-Deflection Equations **182**
3.8 Modified Slope-Deflection Equations **187**
3.9 Inelastic Beam-Columns **193**
3.10 Design Interaction Equations **205**
3.11 An Illustrative Example **219**
3.12 Summary **228**
Problems **231**
References **234**

CHAPTER 4 RIGID FRAMES **236**

4.1 Introduction **236**
4.2 Elastic Critical Loads by Differential Equation Method **239**
4.3 Elastic Critical Loads by Slope-Deflection Equation Method **248**
4.4 Elastic Critical Loads by Matrix Stiffness Method **253**
4.5 Second-Order Elastic Analysis **266**
4.6 Plastic Collapse Loads **270**
4.7 Merchant–Rankine Interaction Equation **280**
4.8 Effective Length Factors of Framed Members **281**
4.9 Illustrative Examples **291**
4.10 Summary **302**
Problems **303**
References **305**

CHAPTER 5 BEAMS **307**

5.1 Introduction **307**
5.2 Uniform Torsion of Thin-Walled Open Sections **309**

Contents

5.3 Non-Uniform Torsion of Thin-Walled Open Cross-Sections 311
5.4 Lateral Buckling of Beams 317
5.5 Beams with Other Loading Conditions 325
5.6 Beams with Other Support Conditions 333
5.7 Initially Crooked Beams 348
5.8 Inelastic Beams 351
5.9 Design Curves for Steel Beams 355
5.10 Other Design Approaches 370
5.11 Summary 373
Problems 375
References 378

CHAPTER 6 ENERGY AND NUMERICAL METHODS 381

6.1 Introduction 381
6.2 Principle of Virtual Work 382
6.3 Principle of Stationary Total Potential Energy 388
6.4 Calculus of Variations 391
6.5 Rayleigh–Ritz Method 414
6.6 Galerkin's Method 438
6.7 Newmark's Method 443
6.8 Numerical Integration Procedure 460
6.9 Summary 464
Problems 466
References 470

ANSWERS TO SOME SELECTED PROBLEMS 471

INDEX 485

PREFACE

This book presents a simple, concise, and reasonably comprehensive introduction to the principles and theory of structural stability that are the basis for structural steel design and shows how they may be used in the solution of practical building frame design problems. It provides the necessary background for the transition for students of structural engineering from fundamental theories of structural stability of members and frames to practical design rules in AISC Specifications. It was written for upper level undergraduate or beginning graduate students in colleges and universities on the one hand, and those in engineering practice on the other.

The scope of the book is indicated by its contents. The concepts and principles of structural stability presented in Chapter 1 form the basis for the elastic and plastic theories of stability of members and frames which are discussed separately in Chapter 2 (Columns), Chapter 3 (Beam-Columns), Chapter 4 (Rigid Frames), and Chapter 5 (Beams). The energy and numerical methods of analyzing a structure for its stability limit load are described in Chapter 6.

Each of these later chapters sets out initially to state the basic principles of structural stability, followed by the derivation of the necessary basic governing differential equations based on idealized conditions. These classical solutions and their physical significance are then examined. The chapter goes on to show how these solutions are affected by the inelasticity of the material and imperfection of the structural member and system associated with a real structure, using both hand techniques and modern computer capabilities. It finally outlines some of the popularly used techniques by which this voluminous information may be utilized to provide design rules and calculation techniques suitable for design office use. In this way, the reader not only will obtain an understanding of the fundamental principles and theory of structural stability from an idealized elastic, perfect system, but also to an inelastic imperfect system that leads to the necessary links between the code rules, design office practice, and the actual structural system in the real world.

The continued rapid development in computer hardware and software in recent years has made it possible for engineers and designers to predict structural behavior quite accurately. The advancement in structural analysis techniques coupled with the increased understanding of structural behavior has made it possible for engineers to adopt the Limit States Design philosophy. A limit state is defined as a condition at which a structural system or its component ceases to perform its intended function under normal conditions (Serviceability Limit State) or failure under severe conditions (Ultimate Limit State). The recently published Load and Resistance Factor Design (LRFD) Specification by the American Institute of Steel Construction (AISC) is based on the limit states philosophy and thus represents a more rational approach to the design of steel structures.

This book is not therefore just another book that presents Timoshenko's basic elastic theory (S. P. Timoshenko and J. M. Gere, "Theory of Elastic Stability," McGraw-Hill, 1961), or Bleich's inelastic buckling theory (F. Bleich, "Buckling Strength of Metal Structures," McGraw-Hill, 1952), or Chen's numerical analysis (W. F. Chen and T. Atsuta, "Theory of Beam-Columns," two-volume, McGraw-Hill, 1976, 1977) in a new style. Instead it presents theory and principles of structural stability in its most up-to-date form. This volume includes not only the state-of-the-art methods in the analysis and design of columns as individual members and as members of a structure, but also an introduction to engineers as to how these new developments have been implemented as the stability design criteria for members and frames in AISC/LRFD Specification.

This book is based on a series of lectures that Professor Chen gave at Purdue University and Lehigh University under the general heading of "Structural Stability." The preparation of the 1985 T. R. Higgins Lectureship Award paper entitled "Columns with End Restraint and Bending in Load and Resistance Factor Design" for AISC Engineering Journal (3rd Quarter, Vol. 22, No. 3, 1985) inspired us to attempt to create a useful textbook for the undergraduate and beginning graduate students in structural engineering as well as practicing structural engineers who are less familiar with the stability design criteria of members and frames in the newly published LRFD Specification.

Professor Chen wishes to extend his thanks to AISC for the 1985 T. R. Higgins Lectureship Award, when the book began to take shape, to

Professor H. L. Michael of Purdue University for continuing support over many years, and to the graduate students, C. Cheng, L. Duan, and F. H. Wu, among others, for preparing the Answers to Some Selected Problems during their course work on Structural Stability in the spring semester of 1986 in the School of Civil Engineering at Purdue University.

December, 1986
West Lafayette, IN

W.F. Chen
E.M. Lui

NOTATION

LOAD AND MOMENT

P	=	axial load
P_e	=	$\dfrac{\pi^2 EI}{L^2}$ = Euler buckling load
P_{cr}	=	elastic buckling load
P_{ek}	=	$\dfrac{\pi^2 EI}{(KL)^2}$
	=	elastic buckling load considering column end conditions
P_f	=	failure load by the elastic–plastic analysis
P_p	=	plastic collapse load or limit load by the simple plastic analysis
P_r	=	$P_e \dfrac{E_r}{E}$ = reduced modulus load
P_t	=	$P_e \dfrac{E_t}{E}$ = tangent modulus load
P_u	=	ultimate strength considering geometric imperfections and material plasticity
P_y	=	AF_y = yield load
M_a	=	amplified (design) moment
M_{cr}	=	elastic buckling moment
M_{ocr}	=	$\dfrac{\pi}{L}\sqrt{EI_y GJ}\sqrt{1+W^2}$, where $W^2 = \dfrac{\pi^2}{L^2}\left(\dfrac{EC_w}{GJ}\right)$
	=	elastic buckling moment under uniform moment
M_{eq}	=	$C_m M_2$ = equivalent moment

M_{ext}	=	moment at a section due to externally applied loads
M_{int}	=	internal resisting moment of the section
M_m	=	transition moment (in Plastic Design)
M_n	=	nominal flexural strength
M_p	=	ZF_y = plastic moment
M_{pcx}	=	$1.18 M_{px}\left[1 - \left(\dfrac{P}{P_y}\right)\right] \leq M_{px}$
		for H-section about strong axis.
	=	plastic moment capacity about the strong axis considering the influence of axial load
M_{pcy}	=	$1.19 M_{py}\left[1 - \left(\dfrac{P}{P_y}\right)^2\right] \leq M_{py}$
		for H-section about weak axis.
	=	plastic moment capacity about the weak axis considering the influence of axial load
M_u	=	ultimate moment capacity considering geometric imperfections and material plasticity
M_y	=	SF_y = yield moment
T_{sv}	=	$GJ \dfrac{d\gamma}{dz}$ = St. Venant (or uniform) torsion
T_w	=	$-EC_w \dfrac{d^3 \gamma}{dz^3}$ = warping restraint (or non-uniform) torsion
σ	=	stress
σ_{ij}	=	stress tensor
ε	=	strain
ε_{ij}	=	strain tensor

ENERGY AND WORK

U	=	$\dfrac{1}{2}\int_v \sigma_{ij}\varepsilon_{ij}\, dv = U_a + U_b + U_{sv} + U_w = -W_{int}$
	=	strain energy of a linear elastic system
U_a	=	$\dfrac{1}{2}\int_0^L \dfrac{P^2}{EA} dz = \dfrac{1}{2}\int_0^L EA \left(\dfrac{du}{dz}\right)^2 dz$
	=	strain energy due to axial shortening
U_b	=	$\dfrac{1}{2}\int_0^L \dfrac{M^2}{EI} dz = \dfrac{1}{2}\int_0^L EI \left(\dfrac{d^2 v}{dz^2}\right)^2 dz$
	=	strain energy due to bending
U_{sv}	=	$\dfrac{1}{2}\int_0^L \dfrac{T_{sv}^2}{GJ} dz = \dfrac{1}{2}\int_0^L GJ \left(\dfrac{d\gamma}{dz}\right)^2 dz$
	=	strain energy due to St. Venant torsion

Notation

U_w = $\dfrac{1}{2}\displaystyle\int_0^L EC_w\left(\dfrac{d^2\gamma}{dz^2}\right)^2 dz$
    = strain energy due to warping restraint torsion
V = $-W_{ext}$ = potential energy
W_{int} = $-U$ = work done by the internal resisting forces
W_{ext} = $-V$ = work done by the external applied forces
Π = $U + V$ = total potential energy

GEOMETRY AND DIMENSIONS

A = cross sectional area
b_f = flange width
C_w = warping constant
    = $\dfrac{1}{2}I_f h^2$ for I section
d = depth
h = distance between centroid of flanges
I = Ar^2 = moment of inertia
I_f = moment of inertia of one flange
J = uniform torsional (or St. Venant) constant
    = $\displaystyle\sum_{i=1}^{n}\dfrac{1}{3}b_i t_i^3$ for a thin-walled open section
L = length
r = $\sqrt{\dfrac{I}{A}}$ = radius of gyration
S = elastic section modulus
t = thickness
u = displacement in the X-direction
v = displacement in the Y-direction
W = $\dfrac{\pi}{L}\sqrt{\dfrac{EC_w}{GJ}}$
Z = plastic section modulus
ϕ = curvature
λ_b = $\sqrt{\dfrac{M_p}{M_{cr}}}$ = beam slenderness parameter
λ_c = $\sqrt{\dfrac{P_y}{P_{ek}}} = \dfrac{KL}{\pi r}\sqrt{\dfrac{P_y}{E}}$ = column slenderness parameter
γ = angle of twist

MATERIAL PARAMETERS

E = Young's modulus
    = 29,000 ksi for steel

E_{eff}	=	effective modulus
E_r	=	reduced modulus
E_t	=	tangent modulus
F_y, σ_y	=	yield stress
G	=	shear modulus
	=	$\dfrac{E}{2(1+\nu)} = 11{,}200$ ksi for steel
ν	=	Poisson's ratio
	=	0.3 for steel

STABILITY RELATED FACTORS

A_F = amplification factor

B_1 = $P-\delta$ moment amplification factor for beam-columns in LRFD

$$= \dfrac{C_m}{1-\left(\dfrac{P}{P_{ek}}\right)} \geq 1.0$$

B_2 = $P-\Delta$ moment amplification factor for beam-columns in LRFD

$$= \dfrac{1}{1-\sum\left(\dfrac{P}{P_{ek}}\right)} \text{ or}$$

$$= \dfrac{1}{1-\sum\left(\dfrac{P\Delta_0}{HL}\right)}$$

C_b = $\dfrac{M_{cr}}{M_{ocr}}$ = equivalent moment factor for beams

$= 1.75 + 1.05\left(\dfrac{M_1}{M_2}\right) + 0.3\left(\dfrac{M_1}{M_2}\right)^2 \leq 2.3$ in AISC Specifications for end moment case

$= \dfrac{12}{3\dfrac{M_1}{M_{max}} + 4\dfrac{M_2}{M_{max}} + 3\dfrac{M_3}{M_{max}} + 2}$ for other loading conditions

(see Table 5.2b, p. 334)

C_m = equivalent moment factor for beam-columns

$= 0.6 - 0.4\left(\dfrac{M_1}{M_2}\right) \geq 0.4$ in ASD for end moment case

$= 0.6 - 0.4\left(\dfrac{M_1}{M_2}\right)$ in LRFD for end moment case

Notation

	=	$1 + \psi \dfrac{P}{P_{ek}}$ = effective length factor
K	=	$\sqrt{\dfrac{P_e}{P_{ek}}}$ = effective length factor
r_i	=	load factors
ϕ	=	resistance factor
ϕ_b	=	resistance factor for flexure = 0.90
ϕ_c	=	resistance factor for compression = 0.85

Chapter 1

GENERAL PRINCIPLES

1.1 CONCEPTS OF STABILITY

When a change in the geometry of a structure or structural component under compression will result in the loss of its ability to resist loadings, this condition is called *instability*. Because instability can lead to a catastrophic failure of a structure, it must be taken into account when one designs a structure. To help engineers to do this, among other types of failure, a new generation of designing codes have been developed based on the concept of *limit states*.

In *limit states design*, the structure or structural component is designed against all pertinent limit states that may affect the safety or performance of the structure. Basically, there are two types of limit states: The first type, *Strength limit states*, deals with the performance of structures at their maximum load-carrying capacities. Examples of strength limit states include structural failure due to either the formation of a plastic collapse mechanism or to member or frame instability. *Serviceability limit states*, on the other hand, are concerned with the performance of structures under normal service conditions. Hence, they pertain to the appearance, durability, and maintainability of a structure. Examples of serviceability limit states include deflections, drift, vibration, and corrosion.

Stability, an important constituent of the strength limit states, is dealt with explicitly in the present American Institute for Steel Construction (AISC) limit state specification.[1] Although the importance of considering stability in design is recognized by most practicing engineers, the subject still remains perplexing to some. The reason for this perplexity is that the use of *first-order structural analysis*, which is familiar to most engineers, is not permissible in a stability analysis. In a true *stability analysis*, the change in geometry of the structure must be taken into account; as a consequence, equilibrium equations must be written based on the geometry of a structure that becomes deformed under load. This is known as the *second-order analysis*. The second-order analysis is further complicated by the fact that the resulting equilibrium equations are differential equations instead of the usual algebraic equations. Consequently, a mastery of differential calculus is a must before any attempt to solve these equations.

In what follows, we will explain the nature of structural stability and ways to analyze it accurately.

The concept of stability is best illustrated by the well-known example of a ball on a curved surface (Fig. 1.1). For a ball initially in equilibrium, a slight disturbing force applied to the ball on a concave surface (Fig. 1.1a) will displace the ball by a small amount, but the ball will return to its initial equilibrium position once it is no longer being disturbed. In this case, the ball is said to be in a *stable equilibrium*. If the disturbing force is applied to a ball on a convex surface (Fig. 1.1b) and then removed, the ball will displace continuously from, and never return to, its initial equilibrium position, even if the disturbance was infinitesimal. The ball in this case is said to be in an *unstable equilibrium*. If the disturbing force is

FIGURE 1.1 Stable, unstable, and neutral equilibrium

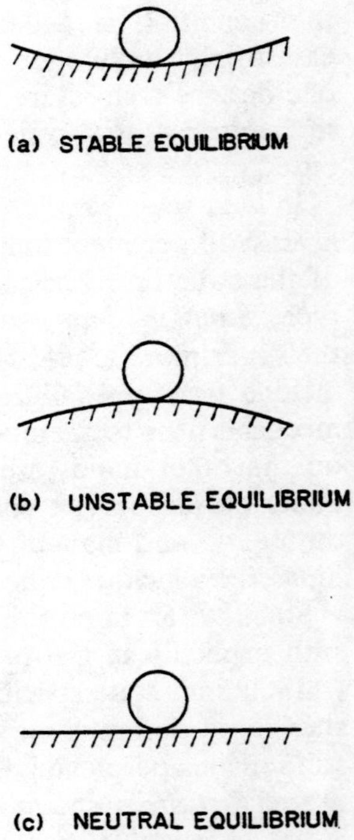

(a) STABLE EQUILIBRIUM

(b) UNSTABLE EQUILIBRIUM

(c) NEUTRAL EQUILIBRIUM

1.1 Concepts of Stability

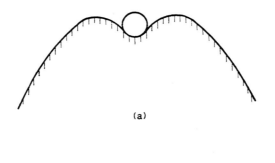

(a)

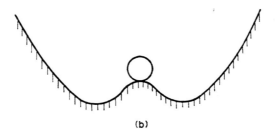

(b)

FIGURE 1.2 Effect of finite disturbance

applied to the ball on a flat surface (Fig. 1.1c), the ball will attain a new equilibrium position to which the disturbance has moved it and will stay there when the disturbance is removed. This ball is said to be in a *neutral equilibrium*.

Note that the definitions of stable and unstable equilibrium in the preceding paragraph apply only to cases in which the disturbing force is very small. These will be our *definitions of stability*. However, keep in mind that it is possible for a ball, under certain conditions (Fig. 1.2), to go from one equilibrium position to another; for example, a ball that is "stable" under a small disturbance may go to an unstable equilibrium under a large disturbance (Fig. 1.2a), or vice versa (Fig. 1.2b).

The concept of stability can also be explained by considering a system's stiffness. For an n-degrees-of-freedom system, the forces and displacements of the system are related by a stiffness matrix or function. If this stiffness matrix or function is *positive definite*, the system is said to be stable. The transition of the system from a state of stable to neutral equilibrium or from a state of stable to unstable equilibrium is marked by the *stability limit point*. The tangent stiffness of the system vanishes at just this point. We shall use the principle of vanishing tangent stiffness to calculate the buckling load of a system in subsequent sections and chapters.

Stability of an elastic system can also be interpreted by means of the concept of *minimum total potential energy*. In nature, an elastic system always tends to go to a state in which the total potential energy is at a minimum. The system is in a stable equilibrium if any deviation from its initial equilibrium state will result in an increase in the total potential energy of the system. The system is in an unstable equilibrium if any deviation from its initial equilibrium state will result in a decrease in total potential energy. Finally, the system is in a neutral equilibrium if any deviation from its initial equilibrium state will produce neither an increase nor a decrease in its total potential energy. Because of this principle, the energy concept can be used to find the buckling load of an elastic system. The elastic buckling analysis by energy method will be discussed in Chapter 6.

1.2 TYPES OF STABILITY

Stability of structures under compressive forces can be grouped into two categories: (1) instability that associates with a *bifurcation of equilibrium* (Fig. 1.3a); and (2) instability that associates with a *limit or maximum load* (Fig. 1.3b).

1.2.1 Bifurcation Instability

This type of instability is characterized by the fact that as the compressive load increases, the member or system that originally deflects in the direction of the applied loads suddenly deflects in a different direction. The point of transition from the usual deflection mode under loads to an alternative deflection mode is referred to as the *point of bifurcation of equilibrium*. The load at the point of bifurcation of equilibrium is called the *critical load*. The deflection path that exists before bifurcation is known as the *primary* or *fundamental path* and the deflection path that exists subsequent to bifurcation is known as the *secondary* or *postbuckling path* (Fig. 1.3a). Examples of this type of instability include the buckling of geometrically perfect columns loaded axially, buckling of thin plates subjected to in-plane compressive forces and buckling of rings subjected to radial compressive forces.

Depending on the nature of the postbuckling paths, two types of bifurcation can be identified: symmetric bifurcation and asymmetric bifurcation (Fig. 1.4).

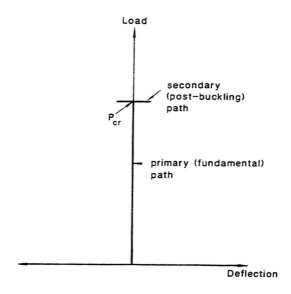

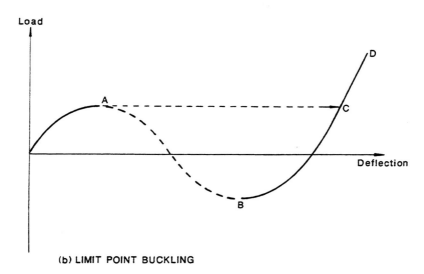

FIGURE 1.3 Bifurcation and limit point buckling

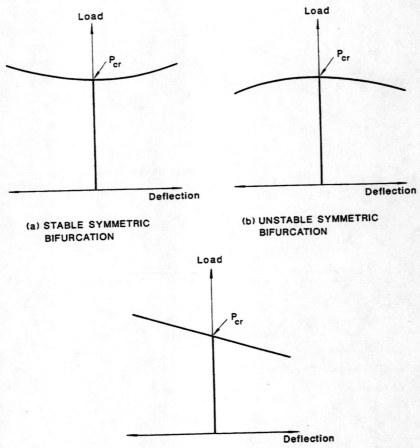

FIGURE 1.4 Postbuckling behavior

Symmetric Bifurcation

For symmetric bifurcation, the postbuckling paths are symmetric about the load axis. If the postbuckling paths rise above the critical load, the system is said to exhibit a *stable symmetric bifurcation*. For such a system, the load that is required to maintain equilibrium subsequent to buckling increases with increasing deformation, as shown in Fig. 1.4a. Examples of structures exhibiting stable postbuckling behavior include axially loaded elastic columns and in-plane loading of the thin elastic plates (Fig. 1.5). If the postbuckling paths drop below the critical load, the system is said to

1.2 Types of Stability

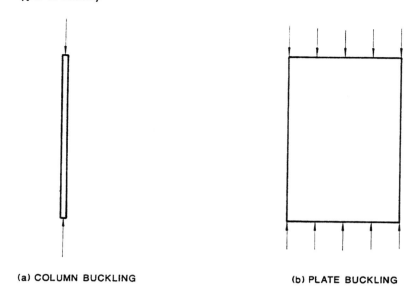

(a) COLUMN BUCKLING (b) PLATE BUCKLING

FIGURE 1.5 Examples of stable symmetric buckling

exhibit an *unstable symmetric bifurcation*. For such a system, the load that is required to maintain equilibrium subsequent to buckling decreases with increasing deflection as shown in Fig. 1.4b. The guyed tower shown in Fig. 1.6 is an example of a structure that exhibits an unstable postbuckling behavior. As the tower buckles and deflects, some of the cables are stretched, resulting in a detrimental pulling force on the tower.

Asymmetric Bifurcation

Figure 1.4c shows schematically the asymmetric bifurcation behavior of a system. For such a system, the load that is required to maintain equilibrium subsequent to buckling may increase or decrease with increasing deflection depending on the direction in which the structure deflects after buckling. The simple frame shown in Fig. 1.7 is an example of a structure that exhibits an *asymmetric postbuckling behavior*. If the frame buckles according to Mode 1, the shear force V induced in the beam will counteract the applied force P in the column. As a result, the load required to maintain equilibrium after buckling will increase with increasing deflection. On the other hand, if the frame buckles according to Mode 2, the shear force V induced in the beam will intensify the applied force P in the column. As a result, the load required to maintain equilibrium after buckling will decrease with increasing deflection.

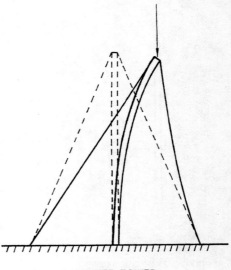

FIGURE 1.6 Example of unstable symmetric buckling

GUYED TOWER

1.2.2 Limit-Load Instability

This type of instability is characterized by the fact that there is only a single mode of deflection from the start of loading to the limit or maximum load (Fig. 1.3b). Examples of this type of instability are buckling of shallow arches and spherical caps subjected to uniform external pressure (Fig. 1.8). For this type of buckling, once the limit load is reached (Point A on the curve of Fig. 1.3b), the system will "*snap through*" from Point A to Point C, because the equilibrium path AB is an unstable one. This unstable equilibrium path will never be encountered under a load controlled testing condition, but it does exist and can be observed under a displacement controlled testing condition. The phenomenon in which a visible and sudden jump from one equilibrium configuration to another nonadjacent equilibrium configuration upon reaching the limit load is referred to as *snap-through buckling*.

Another type of buckling that is unique to shells under compressive forces (Fig. 1.9a) is referred to by Libove in reference 2 as *finite-disturbance buckling*. For this type of buckling the compressive force required to maintain equilibrium drops considerably as the structure buckles after reaching the critical load as shown in curve (i) of Fig. 1.9b. In fact, in reality the theoretical critical load will never be reached because of imperfections. The slightest imperfections in such structures will reduce the critical load tremendously and so curve (ii) in the figure will be more representative of the actual buckling behavior of the real structure.

1.2 Types of Stability

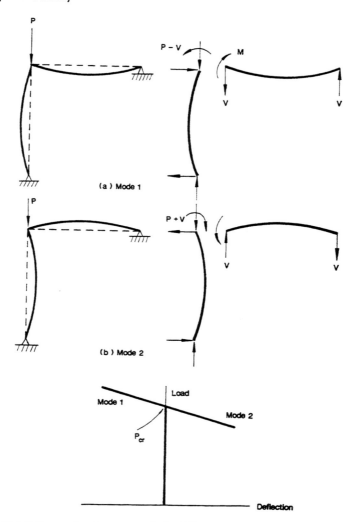

FIGURE 1.7 Example of asymmetric buckling

Finite disturbance buckling has the features of both bifurcation buckling and snap-through buckling. It resembles the former in that the shell deflects in one mode before the critical load is reached, but then deflects in a distinctly different mode after the critical load is reached. It resembles the latter in that a slight disturbance may trigger a jump from the original equilibrium configuration that exists before the critical load to a nonadjacent equilibrium configuration at finite deflections as indicated by the dotted line in the figure.

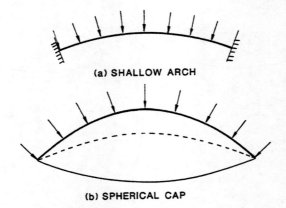

FIGURE 1.8 Examples of limit point buckling

FIGURE 1.9 Shell buckling

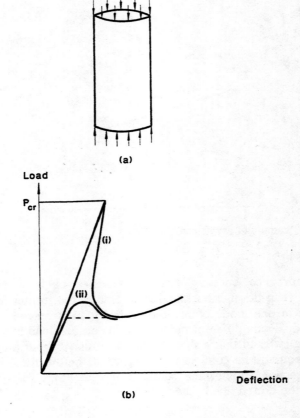

1.3 METHODS OF ANALYSES IN STABILITY

The concept of stability as described in Section 1.1 can be used to determine the critical conditions of an elastic system that is susceptible to instability.

Bifurcation Approach

The first approach is called the *bifurcation approach*. In this approach, the state at which two or more different but adjacent equilibrium configurations can exist is sought by an *eigenvalue analysis*. The lowest load that corresponds to this state is the critical load of the system. At the critical load, equilibrium can be maintained with alternative deflection modes that are infinitesimally close to one another.

To determine the critical load using the bifurcation approach, it is necessary to identify all possible equilibrium configurations the system can assume at the bifurcation load. This can best be accomplished by specifying a set of generalized displacements to describe all the possible displaced configurations of the system. If n parameters are required to describe the various modes of deflections, the system is said to have n degrees of freedom. For an n-degrees-of-freedom system, the determinant of the $n \times n$-system-stiffness matrix, which relates the generalized forces to the generalized displacements of the system, is a measure of the stiffness of the system. At the critical load, the tangent stiffness of the system vanishes. Thus, by setting the determinant of the system's-tangent-stiffness matrix equal to zero, the system's critical conditions can be identified.

The bifurcation approach is also known as the *eigenvalue approach*, because the technique used is identical to that used in the linear algebra for finding eigenvalues of a matrix. The critical conditions are represented by the eigenvalues of the system's stiffness matrix and the displaced configurations are represented by the eigenvectors. The lowest eigenvalue is the critical load of the system. The bifurcation or eigenvalue approach is an idealized mathematical approach to determine the critical conditions of a *geometrically perfect* system. If geometrical imperfections are present, deflection will commence at the beginning of loading. The problem then becomes a *load-deflection* rather than a bifurcation problem. For a load-deflection problem, the bifurcation approach cannot be applied.

Energy Approach

Another way to determine the critical conditions of a system is the *energy approach*. For an *elastic* system subjected to *conservative forces*, the total potential energy of the system can be expressed as a function of a set of generalized displacements and the external applied forces. The term "conservative forces" used here are those forces whose potential energy is dependent only on the final values of deflection, not the specific paths to reach these final values. If the system is in equilibrium, its total potential energy must be stationary. Thus, by setting the first derivative of the total potential energy function with respect to each generalized displacement equal to zero, we can identify the equilibrium conditions of the system. The critical load can then be calculated from the equilibrium equations.

Please note that by setting the first derivative of the total potential energy function equal to zero, we can only identify the equilibrium conditions of the system. To determine whether the equilibrium is stable or unstable, we must investigate higher order derivatives of the total potential energy function.

Dynamic Approach

The critical load of an elastic system can also be obtained by the *dynamic approach*. Here, a system of equations of motion governing the small free vibration of the system is written as a function of the generalized displacements and the external applied force. The critical load is obtained as the level of external force when the motion ceases to be bounded. The equilibrium is stable if a slight disturbance causes only a slight deviation of the system from its original equilibrium position and if the magnitude of the deviation decreases when the magnitude of the disturbance decreases. The equilibrium is unstable if the magnitude of motion increases without bound when subjected to a slight disturbance. The use of the dynamic approach requires a prerequisite of structural dynamics. This is beyond the scope of this book. However, the use of the other two approaches to determine the critical loads will be illustrated in the following sections. In the following examples, both the small and large deflection analyses will be used to demonstrate the significance and physical implications of each analysis.

1.4 ILLUSTRATIVE EXAMPLES—SMALL DEFLECTION ANALYSIS

In this section, the stability behavior of some simple structural models will be investigated in the context of a small deflection analysis by using both the bifurcation and energy approaches.

1.4.1 Rigid Bar Supported by a Rotational Spring

Consider the simple spring–bar system shown in Fig. 1.10a. The bar is assumed to be rigid, and the only possible mode of displacement for the system is the rigid body rotation of the bar about the pinned end as shown in Fig. 1.10b. The pinned end is supported by a linear rotational

1.4 Illustrative Examples—Small Deflection Analysis

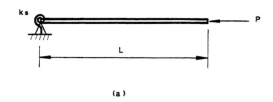

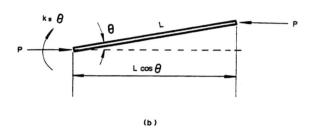

FIGURE 1.10 Rotational spring-supported rigid bar system (small deflection analysis)

spring of stiffness k_s. When the rigid bar is perfectly horizontal, the spring is in an unstrained state, and we shall denote any rotational displacement of the bar from this horizontal position by the angle θ.

The system will become unstable and buckle when P reaches its critical value, P_{cr}. We shall use the two methods already discussed to determine this initial value.

Bifurcation Approach

Assuming the rotational displacement θ is small, then the equilibrium condition of the bar at its displaced configuration can be written in the simple form as

$$k_s \theta - PL\theta = 0 \qquad (1.4.1)$$

This equation is always satisfied for $\theta = 0$. The horizontal position or $\theta = 0$ is therefore a trivial solution. For a nontrivial solution, we must have

$$P = P_{cr} = \frac{k_s}{L} \qquad (1.4.2)$$

Equation (1.4.2) indicates that when the applied force P reaches the value of k_s/L, the system will buckle. At this critical load, equilibrium for the bar is possible in both the original horizontal and slightly deflected positions.

Energy Approach

In using the energy approach to determine P_{cr}, one must write an expression for the total potential energy comprising the strain energy and the potential energy of the system. The strain energy stored in the system as the bar assumes its slightly deflected configuration is equal to the strain energy of the spring:

$$U = \tfrac{1}{2}k_s\theta^2 \tag{1.4.3}$$

The potential energy of the system is the potential energy of the external force and it is equal to the negative of the work done by the external force on the system:

$$V = -PL(1 - \cos\theta) \tag{1.4.4}$$

The term $L(1 - \cos\theta)$ represents the horizontal distance traveled by P as the bar rotates.

The total potential energy is the sum of the strain energy and the potential energy of the system:

$$\Pi = U + V = \tfrac{1}{2}k_s\theta^2 - PL(1 - \cos\theta) \tag{1.4.5}$$

For equilibrium, the total potential energy must assume a stationary value. Thus, we must have

$$\frac{d\Pi}{d\theta} = 0$$

or

$$k_s\theta - PL\sin\theta = 0 \tag{1.4.6}$$

For small θ, $\sin\theta \approx \theta$, therefore, we have

$$k_s\theta - PL\theta = 0 \tag{1.4.7}$$

Equation (1.4.7) is the equilibrium equation of the bar-spring system. The same equilibrium equation [Eq. (1.4.1)] has been obtained by considering equilibrium of a free body of the bar. The nontrivial solution for Eq. (1.4.7) is thus the critical load of the system.

$$P = P_{cr} = \frac{k_s}{L} \tag{1.4.8}$$

This value of the critical load is the same as that determined earlier by the bifurcation approach. Note that in the energy approach we can

1.4 Illustrative Examples—Small Deflection Analysis

determine not only the critical load, P_{cr}, but also the nature of the equilibrium of this system. That is, we can further determine whether the equilibrium of the system is stable or unstable at various stages of loading.

To determine whether the equilibrium is stable or unstable for the system in its original ($\theta = 0$) position, we need to investigate the positiveness or negativeness of higher derivatives of the total potential energy function. For this problem, if we take the second derivative of the total potential energy function given in Eq. (1.4.5), we have

$$\frac{d^2\Pi}{d\theta^2} = k_s - PL\cos\theta \approx k_s - PL \tag{1.4.9}$$

Thus, for $P < P_{cr}$, $d^2\Pi/d\theta^2$ is positive. This indicates that the equilibrium is *stable*. For $P > P_{cr}$, $d^2\Pi/d\theta^2$ is negative, and the equilibrium condition is therefore *unstable*. This behavior is illustrated in Fig. 1.11, in which the applied force P is plotted as a function of θ for *small* values of θ. The solid line represents a stable equilibrium loading path and the dotted line represents an unstable equilibrium path. The bar is in a stable equilibrium in the horizontal ($\theta = 0$) position when $P < P_{cr}$, but becomes unstable in that position when $P > P_{cr}$. For $P = P_{cr}$, $d^2\Pi/d\theta^2 = 0$ in a

FIGURE 1.11 Stability behavior of spring-bar system (small deflection analysis)

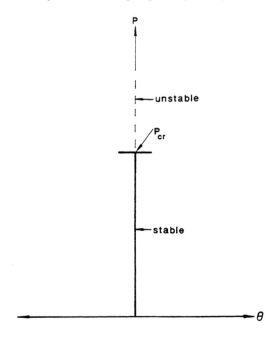

small θ analysis and so no conclusion can be drawn. However, as will be demonstrated in the next section, the stability condition at the critical load ($P = P_{cr}$) can be investigated using the energy approach with a large θ analysis.

1.4.2 Rigid Bar Supported by a Translational Spring

Figure 1.12a shows a rigid bar hinged at one end and supported by a linear translational spring of stiffness k_s at the free end. The spring is assumed to be able to move freely in the horizontal direction but retains its vertical orientation as the bar deflects. The bar is subjected to a concentrated force P at the free end. Assuming the system is geometrically perfect, we shall determine the critical load of the system.

Bifurcation Approach

In Fig. 1.12b we see a slightly deflected position of the bar. Consideration of equilibrium of the bar gives

$$k_s L^2 \theta - PL\theta = 0 \qquad (1.4.10)$$

This condition is always satisfied by the trivial solution $\theta = 0$. For a

FIGURE 1.12 Translational spring-supported rigid bar system (small deflection analysis)

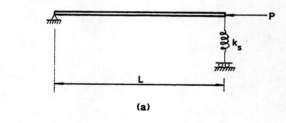

(a)

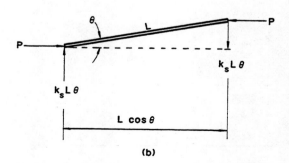

(b)

1.4 Illustrative Examples—Small Deflection Analysis

nontrivial solution, we must have

$$P = P_{cr} = k_s L \tag{1.4.11}$$

Energy Approach

The strain energy of the system at its deflected state is equal to the strain energy stored in the structural spring.

$$U = \tfrac{1}{2} k_s (L\theta)^2 \tag{1.4.12}$$

The potential energy of the system is equal to the negative of the work done by the applied force.

$$V = -PL(1 - \cos\theta) \tag{1.4.13}$$

The total potential energy is

$$\Pi = U + V = \tfrac{1}{2} k_s (L\theta)^2 - PL(1 - \cos\theta) \tag{1.4.14}$$

For equilibrium, the first derivative of Π with respect to θ must vanish.

$$\frac{d\Pi}{d\theta} = k_s L^2 \theta - PL \sin\theta = 0 \tag{1.4.15}$$

For small θ, Eq. (1.4.15) can be written as

$$k_s L^2 \theta - PL\theta = 0 \tag{1.4.16}$$

This equilibrium equation is identical to that of Eq. (1.4.10). Therefore, the nontrivial solution is

$$P = P_{cr} = k_s L \tag{1.4.17}$$

To investigate the nature of equilibrium of the system in its original ($\theta = 0$) position, we need to perform higher order derivatives of the total potential energy function. By taking the second derivative of the total potential energy function given in Eq. (1.4.14), we have

$$\frac{d^2\Pi}{d\theta^2} = k_s L^2 - PL \cos\theta \approx k_s L^2 - PL \tag{1.4.18}$$

Thus, for $P < P_{cr}$, $d^2\Pi/d\theta^2$ is positive, so the system is stable, but for $P > P_{cr}$, $d^2\Pi/d\theta^2$ is negative, so the system is unstable in its original ($\theta = 0$) position.

1.4.3 Two-Bar System

Consider the two-bar system shown in Fig. 1.13a. The two bars are linked together by a frictionless pin at C and the entire system is supported at three locations. The supports at B and D are hinged. The support at C is a spring with spring stiffness k_s. The bars are subjected to

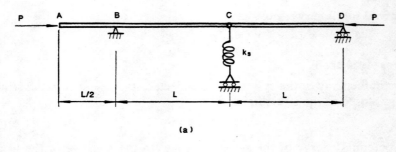

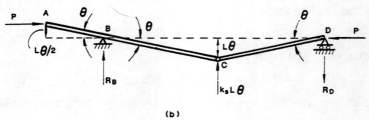

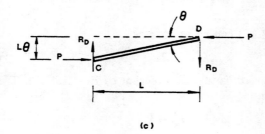

FIGURE 1.13 Two-bar system

a compressive force P at the ends. As P increases, a condition will be reached at which the bars will assume a slightly deflected position. This deflected position can be defined uniquely by a single parameter θ as shown in Fig. 1.13b.

Bifurcation Approach

Summing moments about B for the free body shown in Fig. 1.13b gives

$$P\left(\frac{L\theta}{2}\right) - k_s L\theta(L) + R_D(2L) = 0$$

from which, we solve for

$$R_D = \frac{2k_s L\theta - P\theta}{4} \qquad (1.4.19)$$

1.4 Illustrative Examples—Small Deflection Analysis

Summing moment about C for the free body in Fig. 1.13c gives

$$R_D L - PL\theta = 0 \tag{1.4.20}$$

Upon substitution of Eq. (1.4.19) into Eq. (1.4.20), we obtain

$$\tfrac{1}{2}k_s L\theta - \tfrac{5}{4}P\theta = 0 \tag{1.4.21}$$

The nontrivial solution for the equilibrium equation (1.4.21) gives the critical load of the system as

$$P = P_{cr} = \tfrac{2}{5}k_s L \tag{1.4.22}$$

Energy Approach

The strain energy of the system is

$$U = \tfrac{1}{2}k_s(L\theta)^2 \tag{1.4.23}$$

The potential energy of the system is

$$\begin{aligned} V &= -P[L(1-\cos\theta) + L(1-\cos\theta) \\ &\quad + \tfrac{1}{2}L(1-\cos\theta)] \\ &= -P[\tfrac{5}{2}L(1-\cos\theta)] \end{aligned} \tag{1.4.24}$$

The total potential energy of the system is

$$\Pi = U + V = \tfrac{1}{2}k_s(L\theta)^2 - P[\tfrac{5}{2}L(1-\cos\theta)] \tag{1.4.25}$$

For equilibrium, we must have

$$\frac{d\Pi}{d\theta} = k_s L^2\theta - \tfrac{5}{2}PL\sin\theta = 0 \tag{1.4.26}$$

and for small θ, we have

$$k_s L^2\theta - \tfrac{5}{2}PL\theta = k_s L\theta - \tfrac{5}{2}P\theta = 0 \tag{1.4.27}$$

The critical load is obtained as the nontrivial solution of Eq. (1.4.27).

$$P = P_{cr} = \tfrac{2}{5}k_s L \tag{1.4.28}$$

The nature of equilibrium of the system in its original ($\theta = 0$) position can be studied by observing higher derivatives of the total potential energy function. By taking the second derivative of the total potential energy function, we have

$$\frac{d^2\Pi}{d\theta^2} = k_s L^2 - \tfrac{5}{2}PL\cos\theta \approx k_s L^2 - \tfrac{5}{2}PL \tag{1.4.29}$$

For $P < P_{cr}$, $d^2\Pi/d\theta^2$ is positive, so the system is stable. For $P > P_{cr}$, $d^2\Pi/d\theta^2$ is negative, so the system is unstable in its original undeflected position.

1.4.4 Three-Bar System

Figure 1.14a shows a three-bar system supported at the ends A and D by frictionless hinges and connected to one another at B and C by linear rotational springs of spring stiffness k_s. The system is assumed geometrically perfect in that the springs are unstrained when all the bars are in their horizontal orientation. We shall now determine the critical load P_{cr} of the system.

Bifurcation Approach

In using the bifurcation approach, we are investigating the equilibrium conditions of the system in a slightly deflected configuration. A configuration of the system displaced by an arbitrary kinematic admissible displacement is shown in Fig. 1.14b. A *kinematic admissible displacement* is a displacement that does not violate the constraints of the system. Note that the two parameters (θ_1 and θ_2) are necessary to describe fully the displaced configuration of the bars. Thus, the system is said to have two degrees of freedom.

It is clear from equilibrium consideration that the support reactions at

FIGURE 1.14 Three-bar system

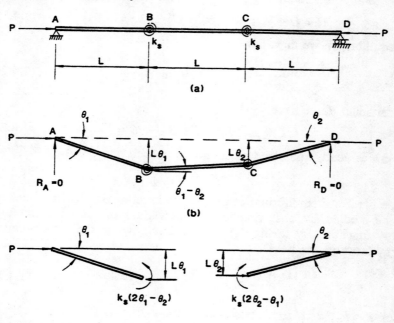

1.4 Illustrative Examples—Small Deflection Analysis

A and D are zeros. By using the free body diagrams of the left and right bars, respectively (Fig. 1.14c), we can write the following equilibrium equations:

$$k_s(2\theta_1 - \theta_2) - P(L\theta_1) = 0, \quad (1.4.30a)$$

$$k_s(2\theta_2 - \theta_1) - P(L\theta_2) = 0. \quad (1.4.30b)$$

In matrix form, we have

$$\begin{bmatrix} 2k_s - PL & -k_s \\ -k_s & 2k_s - PL \end{bmatrix} \begin{pmatrix} \theta_1 \\ \theta_2 \end{pmatrix} = \begin{pmatrix} 0 \\ 0 \end{pmatrix}. \quad (1.4.31)$$

For a nontrivial solution, we must have the determinant of the coefficient matrix equal to zero

$$\det \begin{vmatrix} 2k_s - PL & -k_s \\ -k_s & 2k_s - PL \end{vmatrix} = (2k_s - PL)^2 - k_s^2 = 0 \quad (1.4.32)$$

Equation (1.4.32) is the characteristic equation of the system. The two eigenvalues are

$$P = \frac{k_s}{L} \quad (1.4.33a)$$

and

$$P = \frac{3k_s}{L}. \quad (1.4.33b)$$

The corresponding eigenvectors are

$$\begin{pmatrix} \theta_1 \\ \theta_2 \end{pmatrix} = \begin{pmatrix} 1 \\ 1 \end{pmatrix}, \quad (1.4.34a)$$

and

$$\begin{pmatrix} \theta_1 \\ \theta_2 \end{pmatrix} = \begin{pmatrix} 1 \\ -1 \end{pmatrix}. \quad (1.4.34b)$$

The two deflected configurations of the system that correspond to the eigenvectors Eqs. (1.4.34a, b) are sketched in Fig. 1.15. Since the lowest value of the eigenvalue is the critical load of the system, we therefore have

$$P = P_{cr} = \frac{k_s}{L} \quad (1.4.35)$$

and the symmetric mode (Fig. 1.15a) is the buckling mode of the system.

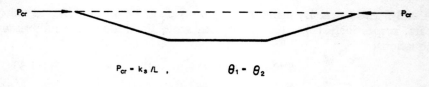

$P_{cr} = k_s/L$, $\theta_1 = \theta_2$

(a) Symmetric Buckling Mode

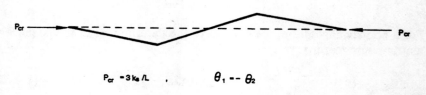

$P_{cr} = 3k_s/L$, $\theta_1 = -\theta_2$

(b) Antisymmetric Buckling Mode

FIGURE 1.15 Buckling modes of the three-bar system

Energy Approach

For the system shown in Fig. 1.14, the strain energy is equal to the strain energy stored in the two springs.

$$U = \tfrac{1}{2}k_s(2\theta_1 - \theta_2)^2 + \tfrac{1}{2}k_s(2\theta_2 - \theta_1)^2 \tag{1.4.36}$$

The potential energy is equal to the negative of the work done by the external forces.

$$\begin{aligned}V &= -PL[(1 - \cos\theta_1) + (1 - \cos\theta_2) + 1 - \cos(\theta_1 - \theta_2)] \\ &= -PL[3 - \cos\theta_1 - \cos\theta_2 - \cos(\theta_1 - \theta_2)]\end{aligned} \tag{1.4.37}$$

The total potential energy of the system is equal to the sum of the strain energy and the potential energy.

$$\begin{aligned}\Pi &= U + V \\ &= \tfrac{1}{2}k_s(2\theta_1 - \theta_2)^2 + \tfrac{1}{2}k_s(2\theta_2 - \theta_1)^2 \\ &\quad - PL[3 - \cos\theta_1 - \cos\theta_2 - \cos(\theta_1 - \theta_2)]\end{aligned} \tag{1.4.38}$$

For equilibrium, the total potential energy of the system must be stationary. In mathematical terms, this requires

$$\begin{aligned}\frac{\partial \Pi}{\partial \theta_1} &= 2k_s(2\theta_1 - \theta_2) - k_s(2\theta_2 - \theta_1) \\ &\quad - PL[\sin\theta_1 + \sin(\theta_1 - \theta_2)] = 0\end{aligned} \tag{1.4.39a}$$

1.4 Illustrative Examples—Small Deflection Analysis

$$\frac{\partial \Pi}{\partial \theta_2} = -k_s(2\theta_1 - \theta_2) + 2k_s(2\theta_2 - \theta_1)$$
$$- PL[\sin \theta_2 - \sin(\theta_1 - \theta_2)] = 0 \quad (1.4.39b)$$

Upon simplification and using small angle approximation, we can write Eqs. (1.4.39a and b) in matrix form as

$$\begin{bmatrix} 5k_s - 2PL & -4k_s + PL \\ -4k_s + PL & 5k_s - 2PL \end{bmatrix} \begin{pmatrix} \theta_1 \\ \theta_2 \end{pmatrix} = \begin{pmatrix} 0 \\ 0 \end{pmatrix} \quad (1.4.40)$$

For nontrivial solution, we must have

$$\det \begin{vmatrix} 5k_s - 2PL & -4k_s + PL \\ -4k_s + PL & 5k_s - 2PL \end{vmatrix} = 0 \quad (1.4.41)$$

from which the two eigenvalues are

$$P = \frac{k_s}{L} \quad (1.4.42a)$$

or

$$P = \frac{3k_s}{L} \quad (1.4.42b)$$

The smallest eigenvalue is the critical load of the system, therefore

$$P = P_{cr} = \frac{k_s}{L} \quad (1.4.43)$$

To study the nature of equilibrium for the system in its undeflected ($\theta_1 = \theta_2 = 0$) position, we need to investigate higher order derivatives of the total potential energy function. By taking the second derivative of the total potential energy function, we have

$$\frac{\partial^2 \Pi}{\partial \theta_1^2} = 5k_s - PL[\cos \theta_1 + \cos(\theta_1 - \theta_2)]$$
$$\approx 5k_s - 2PL \quad (1.4.44a)$$

$$\frac{\partial^2 \Pi}{\partial \theta_2^2} = 5k_s - PL[\cos \theta_2 + \cos(\theta_1 - \theta_2)]$$
$$\approx 5k_s - 2PL \quad (1.4.44b)$$

$$\frac{\partial^2 \Pi}{\partial \theta_1 \partial \theta_2} = -4k_s + PL \cos(\theta_1 - \theta_2)$$
$$\approx -4k_s + PL \quad (1.4.44c)$$

The equilibrium is stable if all of the following conditions are satisfied:

$$\frac{\partial^2 \Pi}{\partial \theta_1^2} > 0 \tag{1.4.45a}$$

$$\frac{\partial^2 \Pi}{\partial \theta_2^2} > 0 \tag{1.4.45b}$$

$$\left(\frac{\partial^2 \Pi}{\partial \theta_1^2}\right)\left(\frac{\partial^2 \Pi}{\partial \theta_2^2}\right) > \left(\frac{\partial^2 \Pi}{\partial \theta_1 \partial \theta_2}\right)^2 \tag{1.4.45c}$$

In view of Eqs. (1.4.44a–c), Equations (1.4.45a–c) become

$$P < \frac{5}{2}\left(\frac{k_s}{L}\right) \tag{1.4.46a}$$

$$P < \frac{5}{2}\left(\frac{k_s}{L}\right) \tag{1.4.46b}$$

$$(k_s - PL)(9k_s - 3PL) > 0 \tag{1.4.46c}$$

For $P < k_s/L$, all the inequalities expressed in Eqs. (1.4.46a–c) will be satisfied. Therefore for $P < k_s/L$, the equilibrium position $\theta_1 = \theta_2 = 0$ is stable, whereas for $P > k_s/L$ it is unstable.

1.5 ILLUSTRATIVE EXAMPLES—LARGE DEFLECTION ANALYSIS

In the foregoing analyses of the simple bar-spring models, the assumption of small deflection has been used because in these examples we are only interested in identifying the critical conditions and finding the critical loads of the system. Such an analysis is known as a *linear eigenvalue analysis*. Although, in addition to determining the critical loads, it is possible for us to investigate the nature of the equilibrium conditions of the systems in their *undeflected* configurations by studying the second derivatives of the total potential energy functions, a linear eigenvalue analysis can provide us with no information about the behavior of the systems after the critical loads have been reached. In other words, if the analysis is performed using the small displacement assumption, it is not possible to study the *postbuckling behavior* of the system. To study the postbuckling behavior of a system, we must use large displacement analysis. This is illustrated in the following examples.

1.5.1 Rigid Bar Supported by a Rotational Spring

Consider the simple one degree of freedom spring-bar system shown in Fig. 1.16. This simple model has been analyzed earlier using the small displacement assumption. The critical load was found to be $P_{cr} = k_s L$ and

1.5 Illustrative Examples—Large Deflection Analysis

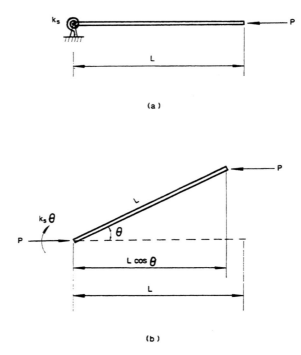

FIGURE 1.16 Rotational spring-supported rigid bar system (large deflection analysis)

by studying the nature of the second derivative of the total potential energy function, we concluded that the equilibrium position that corresponds to the initial (straight) configuration of the bar was stable if $P < P_{cr}$, but it became unstable if $P > P_{cr}$ (Fig. 1.11). However, no information about the nature of equilibrium can be obtained when $P = P_{cr}$ nor do we have any knowledge about the postbuckling behavior of the system. To obtain such information, it is necessary to perform a *large displacement analysis* as shown in the following.

Energy Approach

Although we could readily use the bifurcation approach to determine both the equilibrium paths of a system in a large displacement analysis and the critical load obtained at the point of intersection of the fundamental and postbuckling path(s), we will use the energy approach instead because, in addition to obtaining the critical load, with this second approach we can also investigate the stability of the postbuckling equilibrium paths of the system by examining the characteristic of the higher order derivatives of the total potential energy function.

The strain energy of the system is equal to the strain energy of the spring

$$U = \tfrac{1}{2} k_s \theta^2 \tag{1.5.1}$$

The potential energy of the system is the potential energy of the external force

$$V = -PL(1 - \cos \theta) \tag{1.5.2}$$

The total potential energy of the system is then

$$\Pi = U + V = \tfrac{1}{2} k_s \theta^2 - PL(1 - \cos \theta) \tag{1.5.3}$$

For equilibrium, we must have

$$\frac{d\Pi}{d\theta} = k_s \theta - PL \sin \theta = 0 \tag{1.5.4}$$

This equilibrium equation is satisfied for all values of P if $\theta = 0$. This trivial equilibrium path is the fundamental equilibrium path. It is plotted in Fig. 1.17. Note that this path is coincident with the load axis. The

FIGURE 1.17 Equilibrium paths of the spring-bar system shown in Fig. 1.16

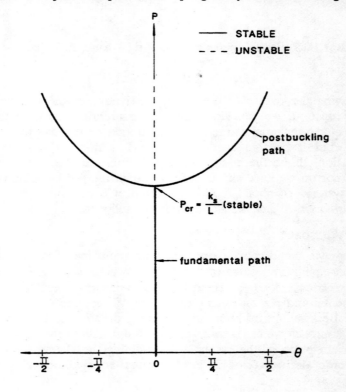

1.5 Illustrative Examples—Large Deflection Analysis

postbuckling path is given by

$$P = \frac{k_s \theta}{L \sin \theta} \quad (1.5.5)$$

This path is also plotted in Fig. 1.17 for the range $-\pi/2 \leq \theta \leq \pi/2$. It intersects the fundamental path at $P_{cr} = k_s/L$.

To study the stability of the equilibrium paths, we need to examine the higher order derivatives of the total potential energy function. By taking the second derivative of Π, we have

$$\frac{d^2\Pi}{d\theta^2} = k_s - PL \cos \theta \quad (1.5.6)$$

For the fundamental path, $\theta = 0$, therefore Eq. (1.5.6) becomes

$$\frac{d^2\Pi}{d\theta^2} = k_s - PL \quad (1.5.7)$$

The quantity $d^2\Pi/d\theta^2$ changes from positive to negative at $P = P_{cr} = k_s/L$, indicating that the initial horizontal position of the bar is stable if $P < P_{cr}$ but unstable if $P > P_{cr}$.

For the postbuckling path, $P = k_s\theta/L \sin \theta$, therefore Eq. (1.5.6) becomes

$$\frac{d^2\Pi}{d\theta^2} = k_s - \left(\frac{k_s \theta}{L \sin \theta}\right) L \cos \theta$$

$$= k_s\left(1 - \frac{\theta}{\tan \theta}\right) \quad (1.5.8)$$

The quantity $d^2\Pi/d\theta^2$ is always positive since the quantity in parenthesis is always greater than zero, indicating that the postbuckling path is always stable.

At the critical point ($P = P_{cr} = k_s/L$), $d^2\Pi/d\theta^2$ is zero according to Eq. (1.5.7), so no information concerning the stability of the system can be obtained. To investigate the stability of this critical equilibrium state, we need to examine the first nonzero term in a Taylor series expansion for Π. Using a Taylor series expansion for the total potential energy function about $\theta = 0$, we obtain

$$\Pi = \Pi\bigg|_{\theta=0} + \frac{d\Pi}{d\theta}\bigg|_{\theta=0} \theta + \frac{1}{2}\frac{d^2\Pi}{d\theta^2}\bigg|_{\theta=0} \theta^2$$

$$+ \frac{1}{6}\frac{d^3\Pi}{d\theta^3}\bigg|_{\theta=0} \theta^3 + \frac{1}{24}\frac{d^4\Pi}{d\theta^4}\bigg|_{\theta=0} \theta^4 + \cdots \quad (1.5.9)$$

It can easily be shown that the first four terms in Eq. (1.5.9) are zeros, thus the first nonzero term of the series is the fifth term

$$\frac{1}{24}\frac{d^4\Pi}{d\theta^4}\bigg|_{\theta=0}\theta^4 = \frac{1}{24}PL\theta^4 \qquad (1.5.10)$$

At $P = P_{cr} = k_s/L$, we have

$$\frac{1}{24}PL\theta^4 = \frac{1}{24}k_s\theta^4 \qquad (1.5.11)$$

which is positive, indicating that Π is a local minimum and so the equilibrium state at the critical point is stable.

From the above discussion it can be seen that as P increases gradually to P_{cr}, the stable fundamental equilibrium path bifurcates into an unstable equilibrium path corresponding to the original horizontal position of the bar and a stable postbuckling equilibrium path corresponding to the deflected configuration of the bar. The equilibrium state at the critical point is stable. The stable postbuckling equilibrium path is symmetric about the load axis, indicating that the bar can deflect either upward or downward with no particular preference.

1.5.2 Rigid Bar Supported by a Translational Spring

Consider now the one degree of freedom bar-spring system shown in Fig. 1.18. This model has been analyzed earlier using the small displacement assumption. The critical load was found to be $P_{cr} = k_s L$. To study the postbuckling behavior, we must use the large deflection analysis.

Energy Approach

Here, as in the previous example, the energy approach is used. The strain energy of the system is

$$U = \tfrac{1}{2}k_s(L\sin\theta)^2 \qquad (1.5.12)$$

The potential energy of the system is

$$V = -PL(1 - \cos\theta) \qquad (1.5.13)$$

The total potential energy of the system is

$$\Pi = U + V = \tfrac{1}{2}k_s(L\sin\theta)^2 - PL(1-\cos\theta) \qquad (1.5.14)$$

For equilibrium, we must have

$$\frac{d\Pi}{d\theta} = k_s L^2 \sin\theta\cos\theta - PL\sin\theta$$

$$= (k_s L^2 \cos\theta - PL)\sin\theta = 0 \qquad (1.5.15)$$

This equilibrium equation is satisfied for all values of P when $\theta = 0$,

1.5 Illustrative Examples—Large Deflection Analysis

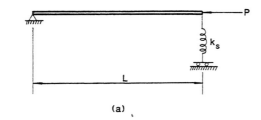

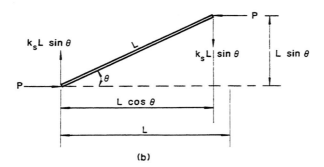

FIGURE 1.18 Translational spring-supported rigid bar system (large deflection analysis)

which is the fundamental equilibrium path (Fig. 1.19). The postbuckling path is given by

$$P = k_s L \cos \theta \tag{1.5.16}$$

The fundamental and postbuckling paths are plotted in Fig. 1.19 for the range $-\pi/2 \leqslant \theta \leqslant \pi/2$. They intersect at $P = P_{cr} = k_s L$.

To study the stability of the equilibrium paths, we form

$$\frac{d^2\Pi}{d\theta^2} = k_s L^2 (\cos^2 \theta - \sin^2 \theta) - PL \cos \theta \tag{1.5.17}$$

For the fundamental path, $\theta = 0$, Eq. (1.5.17) thus becomes

$$\frac{d^2\Pi}{d\theta^2} = k_s L^2 - PL \tag{1.5.18}$$

The quantity $d^2\Pi/d\theta^2$ changes from positive to negative at $P = P_{cr} = k_s L$, which indicates that the initial horizontal position of the bar is stable if $P < P_{cr}$ but unstable if $P > P_{cr}$.

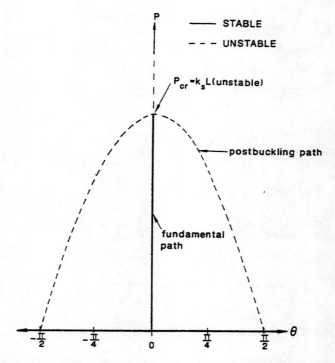

FIGURE 1.19 Equilibrium paths of the spring-bar system shown in Fig. 1.18

For the postbuckling path, $P = k_s L \cos \theta$, Eq. (1.5.17) thus becomes

$$\frac{d^2\Pi}{d\theta^2} = k_s L^2(\cos^2 \theta - \sin^2 \theta) - (k_s L \cos \theta)L \cos \theta$$

$$= -k_s L^2 \sin^2 \theta \qquad (1.5.19)$$

The quantity $d^2\Pi/d\theta^2$ is always negative, indicating that the postbuckling path is unstable.

At the critical point ($P = P_{cr} = k_s L$), $d^2\Pi/d\theta^2$ is zero according to Eq. (1.5.18). As a result, we need to expand the total potential energy function in a Taylor series and examine the first nonzero term in the series. If we substitute the expression for Π and its derivatives into the Taylor series, Eq. (1.5.9), it can be shown that the first nonzero term is

$$\frac{1}{24}\frac{d^4\Pi}{d\theta^4}\bigg|_{\theta=0} \theta^4 = \frac{1}{24}(-4k_s L^2 + PL)\theta^4 \qquad (1.5.20)$$

1.6 Illustrative Examples—Imperfect Systems

At $P = P_{cr} = k_s L$, we have

$$\frac{1}{24}(-4k_s L^2 + PL)\theta^4 = -\tfrac{1}{8}k_s L^2 \theta^4 \tag{1.5.21}$$

which is negative, indicating that Π is a local maximum and so the equilibrium state at the critical point is unstable.

From the above discussion, it can be seen that as P increases gradually to P_{cr}, the stable fundamental equilibrium path bifurcates into an unstable equilibrium path corresponding to the original horizontal position of the bar and an unstable equilibrium path corresponding to the deflected configuration of the bar. The equilibrium state at the critical point is also unstable.

1.6 ILLUSTRATIVE EXAMPLES—IMPERFECT SYSTEMS

Note that for all the examples presented in the preceding sections it has been assumed that the systems are geometrically *perfect*. When the bars are in their horizontal positions, the springs are unstrained at the commencement of the loadings. The systems will therefore remain undeflected until the values of P reach their critical values, P_{cr}. Suppose now that the systems are *imperfect* in the sense that the bars are slightly tilted when the springs are unstrained. The bar will deflect as soon as the load is applied. The problem then becomes a *load-deflection problem*. The following examples will illustrate the effect of this imperfection on the response of the systems.

1.6.1 Rigid Bar Supported by a Rotational Spring

Consider the one-degree-of-freedom-imperfect-bar-spring system shown in Fig. 1.20. The system is imperfect in that the bar is tilted slightly by an angle θ_0, and at this tilted position the spring is unstretched. We shall now study the behavior of the system using the energy approach.

Energy Approach

The strain energy of the system is equal to the strain energy stored in the spring

$$U = \tfrac{1}{2} k_s (\theta - \theta_0)^2 \tag{1.6.1}$$

The potential energy of the system is equal to the potential energy of the external force

$$V = -PL(\cos\theta_0 - \cos\theta) \tag{1.6.2}$$

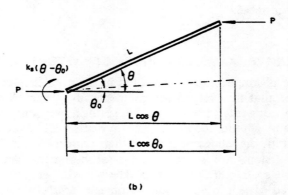

FIGURE 1.20 Rotational spring-supported imperfect rigid bar system

The total potential energy of the system is

$$\Pi = U + V = \tfrac{1}{2}k_s(\theta - \theta_0)^2 - PL(\cos\theta_0 - \cos\theta) \quad (1.6.3)$$

For equilibrium, the total potential energy of the system must be stationary

$$\frac{d\Pi}{d\theta} = k_s(\theta - \theta_0) - PL\sin\theta = 0 \quad (1.6.4)$$

from which, we obtain

$$P = \frac{k_s(\theta - \theta_0)}{L\sin\theta} \quad (1.6.5)$$

The equilibrium paths given in Eq. (1.6.5) for $\theta_0 = 0.1$ and 0.3 are plotted in Fig. 1.21. The figure also shows the equilibrium paths for the corresponding perfect system that was discussed earlier (Fig. 1.17). For the *imperfect* system, deflection commences as soon as the load is

1.6 Illustrative Examples—Imperfect Systems

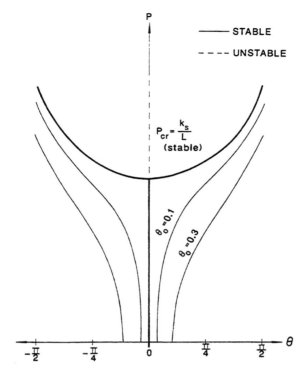

FIGURE 1.21 Equilibrium paths of the imperfect spring-bar system shown in Fig. 1.20

applied. The smaller the imperfection, the closer the equilibrium paths of the imperfect system is to that of the perfect system. In fact, if the imperfection vanishes, the equilibrium paths of the imperfect system will collapse onto the equilibrium paths of the perfect system.

To study the stability of the equilibrium paths of the imperfect system, we need to examine the second derivative of the total potential energy function

$$\frac{d^2\Pi}{d\theta^2} = k_s - PL \cos\theta \qquad (1.6.6)$$

Therefore, the equilibrium paths are stable if

$$P < \frac{k_s}{L \cos\theta} \qquad (1.6.7)$$

and they are unstable if

$$P > \frac{k_s}{L \cos \theta} \tag{1.6.8}$$

It is quite obvious that Eq. (1.6.7) always controls for $-\pi/2 \le \theta \le \pi/2$. Therefore, the equilibrium paths described by Eq. (1.6.5) are always stable, indicating that as P increases the deflections increase, as shown by the curves with the initial angle $\theta_0 = 0$, 0.1 and 0.3, and no instability will occur. The maximum load that the system can carry is greater than the critical load, P_{cr}.

1.6.2 Rigid Bar Supported by a Translation Spring

Consider now the imperfect system shown in Fig. 1.22. The system is imperfect, being tilted by an angle θ_0 when the spring is unstretched. We shall now use the energy approach to study the response of this imperfect system.

FIGURE 1.22 Translational spring-supported imperfect rigid bar system

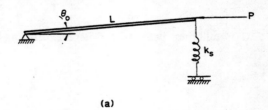

(a)

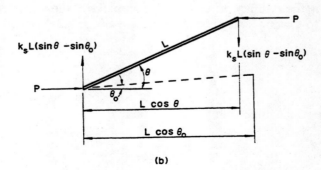

(b)

1.6 Illustrative Examples—Imperfect Systems

Energy Approach

The strain energy of the system is equal to the strain energy stored in the spring

$$U = \tfrac{1}{2}k_s L^2(\sin\theta - \sin\theta_0)^2 \tag{1.6.9}$$

The potential energy of the system is equal to the potential energy of the external force

$$V = -PL(\cos\theta_0 - \cos\theta) \tag{1.6.10}$$

The total potential energy of the system is

$$\Pi = U + V = \tfrac{1}{2}k_s L^2(\sin\theta - \sin\theta_0)^2 \\ - PL(\cos\theta_0 - \cos\theta) \tag{1.6.11}$$

For equilibrium, we must have

$$\frac{d\Pi}{d\theta} = k_s L^2(\sin\theta - \sin\theta_0)\cos\theta - PL\sin\theta = 0 \tag{1.6.12}$$

from which, we obtain

$$P = k_s L \cos\theta\left(1 - \frac{\sin\theta_0}{\sin\theta}\right) \tag{1.6.13}$$

The equilibrium paths described by Eq. (1.6.13) for $\theta_0 = 0.1$ and 0.3 are plotted in Fig. 1.23. The equilibrium paths for the corresponding perfect system are also shown in the figure. Again, as in the preceding example, as the imperfection θ_0 approaches zero, the equilibrium paths of the imperfect system collapse onto the equilibrium paths of the perfect system. The maximum loads P_{max} (the peak points of the load-deflection curves) that the imperfect system can carry are less than the critical load P_{cr} of the corresponding perfect system. These maximum loads can be evaluated by setting

$$\frac{dP}{d\theta} = k_s L\left(-\sin\theta + \frac{\sin\theta_0}{\sin^2\theta}\right) = 0 \tag{1.6.14}$$

from which we obtain the condition

$$\sin\theta_0 = \sin^3\theta \tag{1.6.15}$$

Substitution of Eq. (1.6.15) into Eq. (1.6.13) gives

$$P = P_{max} = k_s L \cos^3\theta \tag{1.6.16}$$

The locus of the maximum loads as described by Eq. (1.6.16) is plotted in Fig. 1.23 as a dash-and-dotted line. Note that P_{max} is always less than P_{cr}, except at $\theta = 0$, when P_{max} becomes P_{cr}. The larger the imperfection the smaller P_{max} will be.

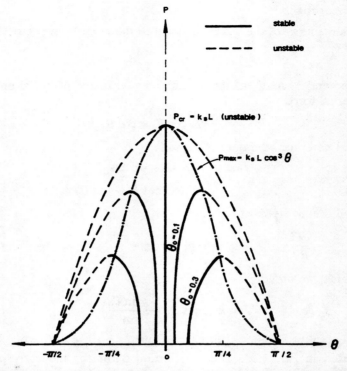

FIGURE 1.23 Equilibrium paths of the imperfect spring-bar system shown in Fig. 1.22

To study the stability of the equilibrium paths described by Eq. (1.6.13), we need to examine the second derivative of the total potential energy function.

$$\frac{d^2\Pi}{d\theta^2} = k_s L^2(\cos^2\theta - \sin^2\theta + \sin\theta_0 \sin\theta)$$
$$- PL\cos\theta \qquad (1.6.17)$$

Upon substitution of Eq. (1.6.13) for P into the above equation and simplifying, we obtain

$$\frac{d^2\Pi}{d\theta^2} = k_s L^2\left(\frac{\sin\theta_0 - \sin^3\theta}{\sin\theta}\right) \qquad (1.6.18)$$

Thus, for the range $-\pi/2 \le \theta \le \pi/2$, the equilibrium paths given by Eq. (1.6.13) will be stable if

$$\sin\theta_0 > \sin^3\theta \qquad (1.6.19)$$

Design Philosophies

and they will be unstable if

$$\sin\theta_0 < \sin^3\theta \qquad (1.6.20)$$

So, it can be seen that the rising equilibrium paths are stable and the falling equilibrium paths are unstable. As the load P is increased from zero, the load-deflection behavior of the imperfect system will follow the stable rising equilibrium paths as shown in Fig. 1.23, until P_{max} is reached, after which the system will become unstable.

From the foregoing examples of imperfect systems, we can conclude that for a system with stable postbuckling equilibrium paths, a small imperfection will not significantly affect the system's behavior. The maximum load the imperfect system can carry is larger than the critical load of the perfect system. On the other hand, for a system with unstable postbuckling equilibrium paths, a small imperfection may have a noticeable effect on the system's behavior. The maximum load that the imperfect system can carry is then smaller than the critical load of the perfect system, and the magnitude of this maximum load decreases with increasing imperfection.

1.7 DESIGN PHILOSOPHIES

To implement the mathematical theory of stability into engineering practice, it is necessary to review the various design philosophies and safety concepts upon which current design practice is based. Details of this implementation on various specific subjects will be given in the chapters that follow.

At present, design practice is based on one of these three design philosophies: Allowable Stress Design, Plastic Design, and Load and Resistance Factor Design. A brief discussion of these design philosophies will be given in the following sections.

1.7.1 Allowable Stress Design

The purpose of allowable stress design (ASD) is to ensure that the stresses computed under the action of the working, i.e., service loads, of a structure do not exceed some predesignated *allowable* values. The allowable stresses are usually expressed as a function of the yield stress or ultimate stress of the material. The general format for an allowable stress design is thus

$$\frac{R_n}{F.S.} \geq \sum_{i=1}^{m} Q_{ni} \qquad (1.7.1)$$

where

R_n = nominal resistance of the structural member expressed in unit of stress
Q_n = nominal working, or service stresses computed under working load conditions
$F.S.$ = factor of safety
i = type of load (i.e., dead load, live load, wind load, etc.)
m = number of load type

The left hand side of Eq. (1.7.1) represents the allowable stress of the structural member or component under various loading conditions (for example, tension, compression, bending, shear, etc.). The right hand side of the equation represents the combined stress produced by various load combinations (for example, dead load, live load, wind load, etc.). Formulas for the allowable stresses for various types of structural members under various types of loadings are specified in the AISC Specification.[3] A satisfactory design is when the stresses in the member computed using a first-order analysis under working load conditions do not exceed their allowable values. One should realize that in allowable stress design, the factor of safety is applied to the resistance term, and safety is evaluated in the service load range.

1.7.2 Plastic Design

The purpose of plastic design (PD) is to ensure that the maximum plastic strength of the structural member or component does not exceed that of the factored load combinations. It has the format

$$R_n \geq \gamma \sum_{i=1}^{m} Q_{ni} \qquad (1.7.2)$$

where

R_n = nominal plastic strength of the member
Q_n = nominal load effect (e.g., axial force, shear force, bending moment, etc.)
γ = load factor
i = type of load
m = number of load types

In steel design, the load factor is designated by the AISC Specification[3] as 1.7 if Q_n consists of only dead and live gravity loads, or as 1.3 if Q_n consists of dead and live gravity loads plus wind or earthquake loads. The use of a smaller load factor for the latter case is justified by the fact that the simultaneous occurrence of all these load effects is not very likely.

Design Philosophies

Note that in plastic design, safety is incorporated in the load term and is evaluated at the ultimate (plastic strength) limit state.

1.7.3 Load and Resistance Factor Design

The purpose of load and resistance factor design (LRFD) is to ensure that the nominal resistance of the structural member or component exceeds that of the load effects. Two safety factors are used, one applied to the loads, and the other to the resistance of the materials. Thus, the load and resistance factor design has the format

$$\phi R_n \geq \sum_{i=1}^{m} \gamma_i Q_{ni} \qquad (1.7.3)$$

where

R_n = nominal resistance of the structural member
Q_n = load effect (e.g., axial force, shear force, bending moment, etc.)
ϕ = resistance factor (≤ 1.0)
γ_i = load factor (usually >1.0) corresponding to Q_{ni}
i = type of load
m = number of load types

In the 1986 LRFD Specification,[1] the resistance factors were developed mainly through calibration,[4] whereas the load factors were developed based on statistical analysis.[5,6] In particular, the first-order probability theory is used. The load and resistance factors for various types of loadings and various load combinations are summarized in Tables 1.1 and 1.2, respectively. A satisfactory design is the one in which the probability of the structural member exceeding a limit state (for example, yielding, fracture, buckling, etc.) is minimal. Based on the first-order second-moment probabilistic analysis,[7] the safety of the structural member is measured by a *reliability* or *safety index*[4] defined as

$$\beta = \frac{\ln(\bar{R}_n/\bar{Q}_n)}{\sqrt{V_R^2 + V_Q^2}} \qquad (1.7.4)$$

where

$\bar{R}$ = mean resistance
$\bar{Q}$ = mean load effect
V_R = coefficient of variation of resistance
$\quad = \dfrac{\sigma_R}{\bar{R}}$
V_Q = coefficient of variation of load effect
$\quad = \dfrac{\sigma_Q}{\bar{Q}}$

in which σ equals the standard deviation.

Table 1.1 Load Factors and Load Combinations[a]

$$1.4\,D$$
$$1.2\,D + 1.6\,L + 0.5\,(L_r \text{ or } S \text{ or } R)$$
$$1.2\,D + 1.6\,(L_r \text{ or } S \text{ or } R) + (0.5\,L \text{ or } 0.8\,W)$$
$$1.2\,D + 1.3\,W + 0.5\,L + 0.5\,(L_r \text{ or } S \text{ or } R)$$
$$1.2\,D + 1.5\,E + (0.5\,L \text{ or } 0.2\,S)$$
$$0.9\,D - 1.3\,W \text{ or } 1.5\,E$$

where

D = dead load
L = live load
L_r = roof live load
W = wind load
S = snow load
E = earthquake load
R = nominal load due to initial rainwater or ice exclusive of the ponding contribution

[a] The load factor on L in the third, fourth and fifth load combinations shown above shall equal 1.0 for garages, areas occupied as places of public assembly and all areas where the live load is greater than 100 psf.

The physical interpretation of the reliability index β is shown in Fig. 1.24. It is the multiplier of the standard deviation $\sqrt{V_R^2 + V_Q^2}$ between the mean of the $\ln(R/Q)$ distribution and the ordinate. Note that both the resistance R and the load Q are treated as random parameters in LRFD, and so $\ln(R/Q)$ does not have a single value but follows a distribution. The shaded area in the figure represents the probability in which $\ln(R/Q) < 1$, i.e., the probability that the resistance will be smaller than

Table 1.2 Resistance Factors

Member type and limit state	ϕ
Tension member, limit state: yielding	0.90
Tension member, limit state: fracture	0.75
Pin-connected member, limit state: tension	0.75
Pin-connected member, limit state: shear	0.75
Pin-connected member, limit state: bearing	1.00
Columns, all limit states	0.85
Beams, all limit states	0.90
High-strength bolts, limit state: tension	0.75
High-strength bolts, limit state: shear	
A307 bolts	0.60
Others	0.65

Design Philosophies

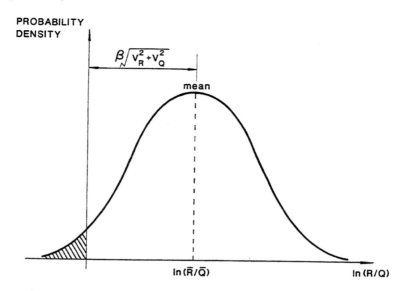

FIGURE 1.24 Reliability index

the load effect, indicating that a limit state has been exceeded. The larger the value of β, the smaller the area of the shaded area, and so it becomes more improbable that a limit state may be exceeded. Thus, the magnitude of β reflects the safety of the member.

In the development of the present LRFD Specification,[1] the following target values for β were selected:

1. $\beta = 3.0$ for members and $\beta = 4.5$ for connectors under dead plus live and/or snow loading;
2. $\beta = 2.5$ for members under dead plus live load acting in conjunction with wind loading, and;
3. $\beta = 1.75$ for members under dead plus live load acting in conjunction with earthquake loading.

A higher value of β for connectors ensures that the connections designed are stronger than their adjoining members. A lower value of β for members under the action of a combination of dead, live, wind, or earthquake loading reflects the improbability that these loadings will act simultaneously.

From the above discussion, it can be seen that in allowable stress design, the safety of the structural member is evaluated on the basis of service load conditions, whereas in plastic or load and resistance factor design, safety is evaluated on the basis of the ultimate or limit load conditions. In addition to strength, the designer must also pay attention to stiffness of the structure. One important criterion related to stiffness is that the structure or structural component must not deflect excessively under service load conditions. Thus, regardless of the design method, one should always investigate such *serviceability requirements* as deflection and vibration at service load conditions.

1.8 SCOPE

In this book, the discussion of *stability theory* will be limited to conservative systems under static or quasistatic loads. A *conservative system* is a system that is subjected only to conservative forces, that is, to forces whose potential energy is dependent only on the final values of deflection. In addition, most of the material presented in this book's later chapters will be based on what is called *small displacement theory*. Hence, the critical load but not the postbuckling behavior of the member or structure will be studied. Only the stability behavior of structural members and frames will be presented; the stability of plates and shells will not be discussed.

For a discussion of the stability behavior of elastic systems under nonconservative and dynamic forces, interested readers should refer to books by Bolotin.[8,9] For a discussion of the buckling behavior of plates and shells, please refer to books by Timoshenko and Gere[10] and Brush and Almroth.[11]

PROBLEMS

1.1 Find the critical load P_{cr} of the bar-spring systems shown in Fig. P1.1a–c using the bifurcation approach. Assume that all the bars are rigid.

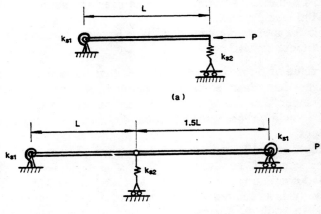

(a)

(b)

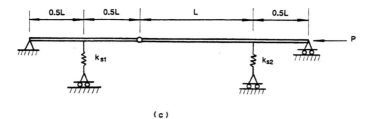

FIGURE P1.1

1.2 Repeat Prob. 1.1, using the energy approach.

1.3 Investigate the stability behavior of the asymmetric spring-bar model shown in Fig. P1.3.

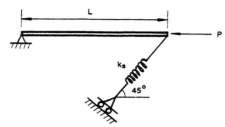

FIGURE P1.3

1.4 Investigate the stability behavior of the snap-through spring-bar model shown in Fig. P.1.4.

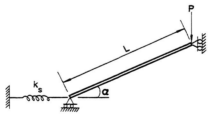

FIGURE P1.4

REFERENCES

1. Load and Resistance Factor Design Specification for Structural Steel Buildings, AISC, Chicago, IL, November, 1986.
2. Flügge, W. Handbook of Engineering Mechanics, McGraw-Hill, New York, 1962, Chapters 44 and 45.
3. Specification for the Design, Fabrication and Erection of Structural Steel for Buildings, AISC, Chicago, IL, November 1978.
4. Ravindra, M. K., and Galambos, T. V. Load and Resistance Factor Design for Steel. Journal of the Structural Division. ASCE, Vol. 104, No. ST. 9, September 1978, pp. 1337–1354.

5. Ellingwood, B., MacGregor, J. G., Galambos, T. V., and Cornell, C. A. Probability-Based Load Criteria Load Factors and Load Combinations. Journal of the Structural Division. ASCE, Vol. 108, No. ST5, May 1982, pp. 978–997.
6. ANSI, Building Code Requirements for Minimum Design Loads in Buildings and Other Structures, ANSI A58.1-1982, American National Standards Institute, New York.
7. Ang, A. H.-S., and Cornell, C. A. Reliability Bases of Structural Safety and Design. Journal of the Structural Division. ASCE, Vol. 100, No. ST9, September 1974, pp. 1755–1769.
8. Bolotin, V. V. Nonconservative Problems of the Theory of Elastic Stability, G. Herrmann Ed., Macmillan, New York, 1963 (translated from Russian).
9. Bolotin, V. V. The Dynamic Stability of Elastic Systems. Holden-Day, San Francisco, 1964.
10. Timoshenko, S. P., and Gere, J. M. Theory of Elastic Stability, 2nd Edition, McGraw-Hill, New York, 1961.
11. Brush, D. O., and Almroth, B. O. Buckling of Bars, Plates and Shells. McGraw-Hill, New York, 1975.

General References

Simitses, G. J. An Introduction to the Elastic Stability of Structures, Prentice-Hall, NJ, 1976.

Thompson, J.M.T., and Hunt, G.W. A General Theory of Elastic Stability. John Wiley & Sons, London, U.K., 1973.

Ziegler, H. Principles of Structural Stability. Blaisdell Publishing Co., Waltham, MA, 1968.

Chapter 2

COLUMNS

2.1 INTRODUCTION

In the preceding chapter, we investigated the critical conditions of several simple bar-spring models. Since the bars were assumed to be rigid, they did not deform as the system reached its critical state. As a result, the equations of equilibrium were algebraic rather than differential in form. In this and subsequent chapters, we shall study the buckling behavior of deformable systems. For such systems, internal forces will develop as the system deforms under the action of the applied external forces. Since internal forces are usually expressed as a function of the derivatives of the generalized coordinates, it follows that the resulting equilibrium equations that relate the external and internal forces will be differential in form. Therefore, to proceed with the calculation, a knowledge of differential calculus is indispensable.

We shall begin our discussion of a deformable system by studying the buckling behavior of columns. In particular, we will use the bifurcation approach to stability analysis in this chapter. The energy approach for the stability analysis of elastic columns will be deferred until Chapter 6. Shown in Fig. 2.1a is a perfectly straight elastic column loaded concentrically by an axial force P. If P is small, the column will remain in a straight position and undergoes only axial deformation. The column at this state is said to be in *stable equilibrium* since any lateral displacement produced by a slight disturbing lateral force will disappear when the lateral force is removed. As P is increased, a condition is reached in which equilibrium in a straight position of the column ceases to be stable. Under this condition, a very small lateral force will produce a very large lateral deflection that does not disappear when the lateral force is

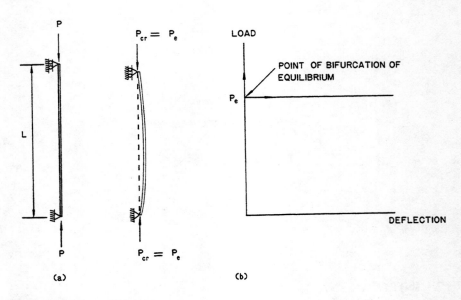

FIGURE 2.1 Euler load

removed. The axial load that demarcates the stable and *unstable equilibrium* of the straight column is referred to as the *critical load* (P_{cr}) or *Euler load* (P_e) (Fig. 2.1b). At the critical load, there also exists another equilibrium position in a slightly deflected configuration. This deflected position is favored when the straight column is disturbed by a small lateral force, and the column will not return to its straight position when the disturbing lateral force is removed. This slightly displaced configuration is a stable equilibrium position. The transition from the (unstable) straight configuration to the (stable) deflected configuration corresponds to a state of neutral equilibrium of the column. In the following section, we shall evaluate the critical load of this perfectly straight column by reference to this neutral equilibrium position. This technique for determining the critical load of a column is known as the *method of neutral equilibrium.*

The critical load also marks the point of *bifurcation of equilibrium* of the perfectly straight elastic column. It is at this point when the theoretical load-deflection curve of the column bifurcates into stable and unstable equilibrium branches that correspond to the deflected and straight configurations of the column, respectively [Fig. 2.2, curve (i)].

The bifurcation point exists only for a perfectly straight column. In reality, columns are rarely perfectly straight. Geometrical imperfection and/or load eccentricity, which are unavoidably present in an actual column, will cause the column to deflect laterally at the onset of loading.

2.1 Introduction

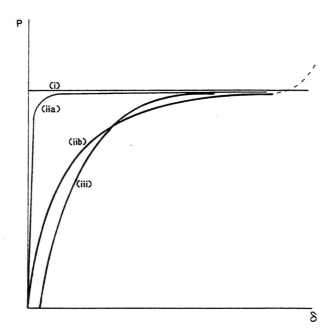

FIGURE 2.2 Load-deflection curves of (i) a perfectly straight column, (iia) a column with small initial crookedness, (iib) a column with large initial crookedness, (iii) a column with eccentrically applied load

Consequently, the load-deflection curve of an imperfect column is a smooth curve. Curves (iia), (iib), and (iii) of Fig. 2.2 show schematically the load-deflection behavior of a column with small geometric imperfection, large geometric imperfection, and load eccentricity, respectively. Initially crooked and eccentrically loaded columns will be discussed in subsequent sections of this chapter.

It should be mentioned that the behavior of a long column is quite different from that of a short column. For a *long* or *slender column*, buckling may occur when all fibers of the cross section are still elastic and so the Euler load (P_e) will govern the limit state of a slender column. For a *short* or *stocky column*, yielding of fibers over the entire cross section usually occurs when the yield stress of the material is reached before buckling can occur, and so, for a stocky column, the *yield load* P_y will govern the limit state of the column. For a *medium length column*, some of the fibers of the cross section may yield under the action of the applied force while some fibers still remain elastic. For this case, the limit load is denoted by P_u, the ultimate strength of the column. For a perfectly straight column, P_u can be represented by the *tangent modulus load* (P_t)

or the *reduced modulus load* (P_r). These loads can be obtained by making certain assumptions regarding the strain and stress distributions in the cross section of the column. The effect of this inelasticity can be taken into account by modifying the elastic modulus according to the two inelastic column theories: the *tangent modulus* and the *reduced modulus theories* for a perfectly straight column. These two inelastic column theories will be discussed in Section 2.7.

If the column is not perfectly straight, or if bending exists at the onset of loading, the tangent or reduced modulus theory is not applicable anymore. For such members, the ultimate load P_u must be determined numerically. Two commonly used numerical procedures to determine P_u will be discussed in Sections 6.7 and 6.8 of Chapter 6.

2.2 CLASSICAL COLUMN THEORY

A *column* is defined here as a member that sustains only axial load. If lateral loads are present in addition to the axial load, the member is referred to as a *beam-column* and will be treated separately in Chapter 3. Although a column can be considered as a limiting case of a beam-column when the lateral loads in a beam-column vanish, in this chapter we will treat the column problem independently.

2.2.1 Pinned-Ended Column

In deriving the basic differential equation of a pinned-ended column, the following assumptions regarding the geometry, kinematics, and material of the column are used:

1. The column is perfectly straight.
2. The axial load is applied along the centroidal axis of the column.
3. Plane sections before deformation remain plane after deformation.
4. Deflection of the member is due only to bending (i.e., shear deformation is ignored).
5. The material obeys Hooke's Law (i.e., the stress and strain are related linearly).
6. The deflection of the member is small. As a result, the curvature can be approximated by the second derivative of the lateral displacement.

With the above assumptions in mind, the governing differential equation of the column is derived as follows:

In Fig. 2.3a a column, pinned at both ends with the upper end free to move vertically, is loaded by an axial force P applied along its centroidal axis. To calculate the critical load of this column, one uses the *method of neutral equilibrium*. At the critical load, the column can be in equilibrium in both a straight and a slightly bent configuration. The critical load can

2.2 Classical Column Theory

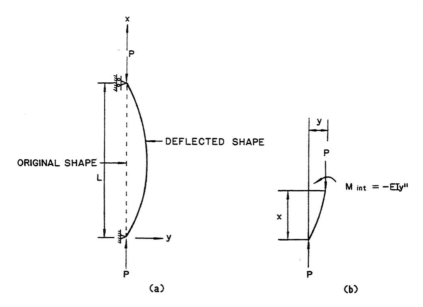

FIGURE 2.3 Pinned-ended column

be obtained from the governing differential equation written for the slightly bent configuration of the column using an *eigenvalue analysis*. In an eigenvalue analysis, only the deflected shape and not the magnitude of deflection of the buckled column can be determined. The critical load is the eigenvalue and the deflected shape is the eigenvector of the problem.

Figure 2.3b shows a free body diagram of a column segment of the column shown in Fig. 2.3a. Equilibrium of this free body requires that

$$-M_{int} + Py = 0 \qquad (2.2.1)$$

where M_{int} is the internal resisting moment and y is the lateral displacement of the cut section.

The internal moment M_{int} induced by the bending curvature Φ of the cross section is given by

$$M_{int} = EI\Phi \qquad (2.2.2)$$

where E is Young's modulus of the material and I is the moment of inertia of the cross section. The value EI here can be considered as the slope of the relation between moment M_{int} and curvature $\Phi = 1/R$ (Fig. 2.4). This linear moment-curvature relation can be derived directly from the kinematic and material assumptions 3, 4, and 5 given above in the following manner:

In Fig. 2.5, an infinitesimal segment of a column of length dx is shown

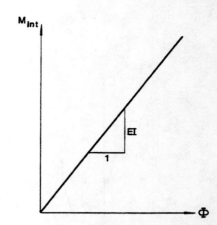

FIGURE 2.4 Linear moment-curvature relationship

with its undeformed and deformed positions. From similar triangles, we can write the kinematic relation as

$$\frac{\varepsilon_x \, dx}{y_1} = \frac{dx}{R} \qquad (2.2.3)$$

or

$$\varepsilon_x = \frac{y_1}{R} = y_1 \Phi \qquad (2.2.4)$$

where

ε_x = axial strain

R = radius of curvature

From Hooke's Law, the axial stress σ_x is related to the axial strain ε_x by the *linear relation*.

$$\sigma_x = E \varepsilon_x \qquad (2.2.5)$$

From statics, the internal moment M_{int} can be obtained by the

FIGURE 2.5 Kinematics of a column segment

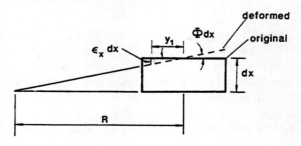

2.2 Classical Column Theory

integration of the moment induced by the stress σ_x over the cross section.

$$M_{\text{int}} = \int_A y_1 \sigma_x \, dA \qquad (2.2.6)$$

Substitution of the kinematic relation (2.2.4) and the stress–strain relation (2.2.5) into the equilibrium equation (2.2.6) gives

$$M_{\text{int}} = \frac{E}{R} \int_A y_1^2 \, dA \qquad (2.2.7)$$

By recognizing that $\int_A y_1^2 \, dA$ is the moment of inertia I of the cross section and $1/R$ is the curvature Φ, Eq. (2.2.7) can easily be reduced to Eq. (2.2.2).

If a small deflection is assumed, the curvature Φ can be approximated by the second derivative of the lateral displacement

$$\Phi = -\frac{d^2 y}{dx^2} = -y'' \qquad (2.2.8)$$

in which a prime indicates the derivative of y with respect to x. The negative sign in Eq. (2.2.8) indicates that the curvature Φ or the rate of change of the slope dy/dx of the deflected shape as sketched in Fig. 2.3b decreases with increasing x.

Using this approximation for curvature, the internal moment of Eq. (2.2.2) can be related to lateral displacement y by

$$M_{\text{int}} = -EI y'' \qquad (2.2.9)$$

Substitution of this expression for M_{int} into Eq. (2.2.1) gives

$$EI y'' + Py = 0 \qquad (2.2.10)$$

Introducing the notation

$$k^2 = \frac{P}{EI} \qquad (2.2.11)$$

Equation (2.2.10) can be written in the simple form

$$y'' + k^2 y = 0 \qquad (2.2.12)$$

Equation (2.2.12) is a *second-order linear* differential equation with constant coefficients. The general solution is

$$y = A \sin kx + B \cos kx \qquad (2.2.13)$$

Note that there are three unknowns, k, A, and B, in the above equation, but we have only two independent boundary conditions.

$$y(0) = 0 \qquad (2.2.14)$$
$$y(L) = 0 \qquad (2.2.15)$$

Therefore, we can only determine two unknowns. Substituting the first boundary condition (2.2.14) into Eq. (2.2.13), we have

$$B = 0 \tag{2.2.16}$$

Using the second boundary condition (2.2.15), we have

$$A \sin kL = 0 \tag{2.2.17}$$

Equation (2.2.17) is satisfied if

$$A = 0 \tag{2.2.18}$$

and/or

$$\sin kL = 0 \tag{2.2.19}$$

Equation (2.2.18) is a trivial solution that states that the straight configuration of the column is an equilibrium position. To obtain a nontrivial solution that describes the equilibrium position of the column in a slightly bent configuration, we must satisfy Eq. (2.2.19) with

$$kL = n\pi, \quad n = 1, 2, \ldots \tag{2.2.20}$$

or

$$k = \frac{n\pi}{L} \tag{2.2.21}$$

from which we can solve for P from Eq. (2.2.11)

$$P = \frac{n^2 \pi^2 EI}{L^2} \tag{2.2.22}$$

The value of P that corresponds to the smallest value of n (i.e., $n = 1$) is the critical load (P_{cr}) of the column. This load is also referred to as the *Euler load* (P_e), as Euler is the pioneer of this column-buckling problem.[1]

$$P_{cr} = P_e = \frac{\pi^2 EI}{L^2} \tag{2.2.23}$$

It will be seen in later sections and chapters that the Euler load constitutes an important reference load in the buckling and stability analysis of members and frames.

The deflected shape of the column at buckling can be found by substituting the constants B and k in Eqs. (2.2.16) and (2.2.21) with $n = 1$ into the deflection function (2.2.13). This gives

$$y = A \sin \frac{\pi x}{L} \tag{2.2.24}$$

Note that the constant A is still indeterminate. Thus, only the deflected

2.2 Classical Column Theory

shape and not the amplitude of the deflection can be determined. This is because we approximate the curvature Φ to the lateral displacement y of the column by the linear relation $\Phi = -y''$ using the small displacement assumption. This approximation leads to the *linear* differential equation [Eq. (2.2.12)]. If the small displacement assumption is obliterated, the resulting differential equation will be nonlinear. The use of formal mathematics for the solution to this nonlinear equation will result in not only the deflected shape but also the amplitude of the deflected column. The discussion of the large displacement behavior of an axially located column will be given in the later part of this section.

2.2.2 Eccentrically Loaded Column

Figure 2.6a shows the loading condition and the deflected shape of an eccentrically loaded column. The axial force P is loaded eccentrically at a distance e from the centroidal axis of the column.

Equilibrium of the free body of a column segment shown in Fig. 2.6b requires that

$$-M_{int} + P(e + y) = 0 \qquad (2.2.25)$$

Substituting the internal moment from Eq. (2.2.9) into the above equation gives

$$EIy'' + P(e + y) = 0 \qquad (2.2.26)$$

FIGURE 2.6 Eccentrically loaded column

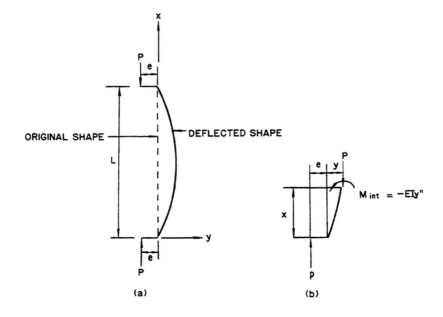

using $k^2 = P/EI$, we obtain

$$y'' + k^2 y = -k^2 e \qquad (2.2.27)$$

Equation (2.2.27) is the governing differential equation of an eccentrically loaded column. The general solution to this equation is

$$y = y_c + y_p \qquad (2.2.28)$$

where

y_c = complementary solution to the corresponding homogeneous differential equation (i.e., $y'' + k^2 y = 0$)
y_p = particular solution satisfying Eq. (2.2.27)

The homogeneous solution is given by

$$y_c = A \sin kx + B \cos kx \qquad (2.2.29)$$

The particular solution y_p can be obtained by either the *method of undetermined coefficient* or the *method of variation of parameters*. For this simple case, it can easily be shown that

$$y_p = -e \qquad (2.2.30)$$

Thus, the general solution is

$$y = A \sin kx + B \cos kx - e \qquad (2.2.31)$$

The constants A and B can be determined by the two boundary conditions

$$y(0) = 0 \qquad (2.2.32)$$
$$y(L) = 0 \qquad (2.2.33)$$

The first boundary condition leads to

$$B = e \qquad (2.2.34)$$

and the second boundary condition together with $B = e$ leads to

$$A = \left(\frac{1 - \cos kL}{\sin kL}\right) e \qquad (2.2.35)$$

Substituting these constants into the deflection equation (2.2.31), we obtain the deflected shape of the eccentrically loaded column as

$$y = \left(\frac{1 - \cos kL}{\sin kL} \sin kx + \cos kx - 1\right) e \qquad (2.2.36)$$

The corresponding moment is given by $M = -EIy''$ or

$$M = -EIk^2 e \left(\frac{\cos kL - 1}{\sin kL} \sin kx - \cos kx\right) \qquad (2.2.37)$$

2.2 Classical Column Theory

The maximum deflection and moment occur at midheight of the column. Therefore, by setting $x = L/2$, y_{max} and M_{max} are given respectively as

$$y_{max} = y(L/2) = \left(\frac{1 - \cos kL}{\sin kL} \sin \frac{kL}{2} + \cos \frac{kL}{2} - 1\right)e \qquad (2.2.38)$$

$$M_{max} = M(L/2) = EIk^2 e\left(\frac{1 - \cos kL}{\sin kL} \sin \frac{kL}{2} + \cos \frac{kL}{2}\right) \qquad (2.2.39)$$

Equations (2.2.38) and (2.2.39) can be simplified by using the trigonometric identities

$$\cos kL = 1 - 2\sin^2 \frac{kL}{2} \qquad (2.2.40)$$

$$\sin kL = 2\sin \frac{kL}{2} \cos \frac{kL}{2} \qquad (2.2.41)$$

and the resulting deflection and moment at the midheight are

$$y_{max} = \left(\sec \frac{kL}{2} - 1\right)e \qquad (2.2.42)$$

$$M_{max} = EIk^2 e \sec \frac{kL}{2} \qquad (2.2.43)$$

The maximum deflection (2.2.42) is measured from the original undeformed centroidal axis of the column. The total maximum deflection measured from the line of application of P is therefore

$$\delta_{max} = y_{max} + e = e\left(\sec \frac{kL}{2}\right) \qquad (2.2.44)$$

Defining

$$A_F = \sec \frac{kL}{2} = \sec\left(\frac{\pi}{2}\sqrt{\frac{P}{P_e}}\right) \qquad (2.2.45)$$

as the *amplification factor*, Eq. (2.2.44) can be written as

$$\delta_{max} = A_F e \qquad (2.2.46)$$

and similarly, we can write

$$M_{max} = A_F(EIk^2 e) \qquad (2.2.47)$$

since $k^2 = P/EI$, we have

$$M_{max} = A_F(Pe) \qquad (2.2.48)$$

Since e and Pe are, respectively, the end eccentricity and end

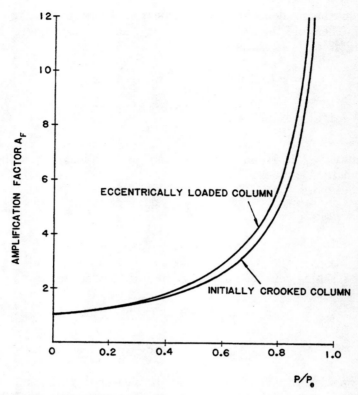

FIGURE 2.7 Amplification factors

moment of the eccentrically loaded column, Eqs. (2.2.46) and (2.2.48) indicate that the maximum displacement and maximum moment in the column can be obtained by simply multiplying the end eccentricity and end moment by the amplification factor A_F [Eq. (2.2.45)]. Note that this amplification factor depends on the axial force P. A plot of A_F as a function of P/P_e is shown in Fig. 2.7.

The normalized midheight deflection δ_{max}/L plotted as a function of P/P_e [Eq. (2.2.44)] for two end eccentricity ratios $e/L = 0.001$ and 0.005 is shown in Fig. 2.8. For an eccentrically loaded column, deflection begins as soon as the load is applied. The larger the end eccentricity, the more the column will deflect at the same load level. Deflection is relatively small at the commencement of loading, but increases progressively and rapidly as the load increases. At or near the Euler load, deflection increases drastically and the load-deflection curves approach asymptotically to the Euler load, P_e. Thus, the maximum load that a perfectly elastic eccentrically loaded column can carry is the Euler load.

2.2 Classical Column Theory

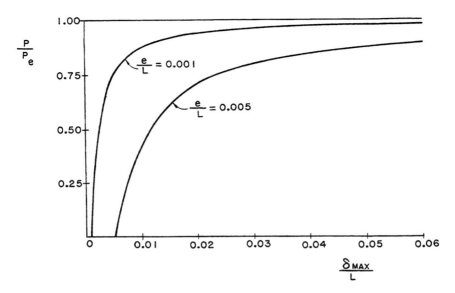

FIGURE 2.8 Load-deflection behavior of eccentrically loaded columns

In reality, however, because of material yielding, the Euler load is seldom reached and the maximum-load-carrying capacity of an eccentrically loaded column will fall far below the Euler load.

2.2.3 Secant Formula

The maximum stress in an eccentrically loaded elastic column is the sum of the axial stress and the maximum bending stress

$$\sigma_{max} = \frac{P}{A} + \frac{M_{max}c}{I} \tag{2.2.49}$$

where c is the distance from the neutral axis to extreme fiber of the cross section.

Substitution of M_{max} from Eq. (2.2.48) into Eq. (2.2.49) gives

$$\sigma_{max} = \frac{P}{A} + \frac{\left(Pe \sec \frac{\pi}{2}\sqrt{P/P_e}\right)c}{I} \tag{2.2.50}$$

Since

$$I = Ar^2 \tag{2.2.51}$$

where r is the radius of gyration of the cross section, we can write Eq.

(2.2.50) as

$$\sigma_{max} = \frac{P}{A}\left[1 + \frac{ec}{r^2}\sec\frac{\pi}{2}\sqrt{\frac{P}{P_e}}\right] \quad (2.2.52)$$

This *secant formula* enables us to calculate in a direct manner the maximum stress in an eccentrically loaded column. It will be used in what follows to develop a column strength equation for the design of axially loaded columns with imperfections.

If the first yield of the material is used as the criterion for failure, i.e., if the limit state of the column is defined as the state at which the maximum fiber stress just reaches the yield stress σ_y, the corresponding critical load (P_{cr}) can be calculated from Eq. (2.2.52) by setting $\sigma_{max} = \sigma_y$.

$$\sigma_y = \frac{P_{cr}}{A}\left[1 + \frac{ec}{r^2}\sec\frac{\pi}{2}\sqrt{\frac{P_{cr}}{P_e}}\right] \quad (2.2.53)$$

In actual design implementation, the secant formula is developed in conjunction with reference to experimental data. The *eccentricity factor* ec/r^2 is treated as an *imperfection factor* for an axially loaded column and is determined by calibration so that the formula will best fit the given experimental data.

2.2.4 Linear vs. Nonlinear Theory

In the preceding sections, the assumption of small displacement is used. As a result, the curvature Φ is approximated by the second derivative of the lateral displacement with respect to x, i.e.,

$$\Phi = -y'' \quad \text{(same as 2.2.8)}$$

If the small displacement assumption is obliterated, a more exact expression for the curvature must be used.

$$\Phi = \frac{-y''}{[1 + (y')^2]^{3/2}} \quad (2.2.54)$$

in which the second-order term $(y')^2$ in the denominator can not be neglected in the curvature-displacement relation. As a result, the governing differential equation (2.2.10) derived previously for a pinned-pinned column on the basis of small displacement assumption must be modified to include this term (Fig. 2.9a). For the column segment shown in Fig. 2.9b, the equilibrium equation is

$$-EI\Phi + Py = 0 \quad (2.2.55)$$

Upon substitution of the exact curvature expression (2.2.54) into the

2.2 Classical Column Theory

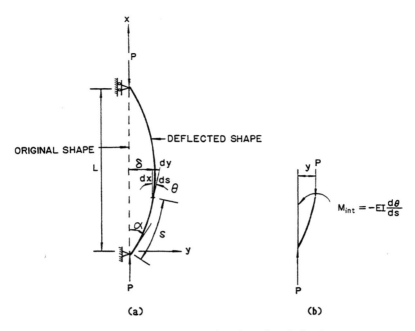

FIGURE 2.9 Large deflection analysis of a pinned-ended column

equilibrium equation (2.2.55), we have

$$\frac{EIy''}{[1+(y')^2]^{3/2}} + Py = 0 \tag{2.2.56}$$

This is a *nonlinear* differential equation. To simplify the equation, we express the curvature Φ in terms of the rate of change in slope along the deflected coordinate s of the member (Fig. 2.9a) by

$$\Phi = -\frac{d\theta}{ds} \tag{2.2.57}$$

and, hence, the governing differential equation (2.5.55) becomes

$$EI\frac{d\theta}{ds} + Py = 0 \tag{2.2.58}$$

Taking derivatives with respect to s and realizing that

$$\frac{dy}{ds} = \sin\theta \tag{2.2.59}$$

we have

$$EI\frac{d^2\theta}{ds^2} + P\sin\theta = 0 \qquad (2.2.60)$$

Equation (2.2.60) can be solved by using elliptical integrals. Details of these calculations are given in reference 2. In the following we shall discuss the result of these calculations.

The expression for the midheight deflection of this column as a function of P/P_e is given as (see reference 2)

$$\frac{\delta}{L} = \frac{2\sin\frac{\alpha}{2}}{\pi\sqrt{\frac{P}{P_e}}} \qquad (2.2.61)$$

where α is the end slope of the column (Fig. 2.9a).

Figure 2.10 shows a plot of Eq. (2.2.61). For $P/P_e < 1$, the straight configuration is the stable equilibrium position of the column. When $P/P_e = 1$, bifurcation of equilibrium takes place. The original straight configuration of the column will become unstable, in Fig. 2.10, this unstable equilibrium is represented by the line AB. A bent configuration

FIGURE 2.10 Large displacement load-deflection behavior of a pinned-ended column

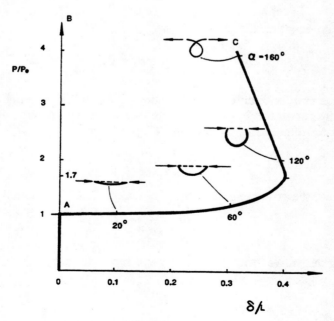

2.3 End-Restrained Columns

will be favored; in Fig. 2.10, this stable equilibrium position is represented by curve AC. The deflection modes at various stages of loading on the curve are represented by the inset diagrams. Both the end slope α and the midheight deflection δ increase rapidly as P/P_e rises slightly above unity. The midheight displacement is the greatest when P/P_e is about 1.7. After that, further increase on load will result in a decrease in midheight deflection due to the fact that the column has now turned inside out, and the applied force P will now act as a tensile rather than a compressive force to close the loop.

Some observations and conclusions can be made from the present large displacement analysis:

1. Both the linear and nonlinear theories give the same prediction of the critical load $(P_{cr}) = P_e$.
2. When $P/P_e > 1$, a slight increase in P will result in a large increase in displacement.
3. The postbuckling behavior of the column is stable because the buckled column can carry additional axial load beyond the Euler load (P_e).
4. Unlike the linear theory in which only the deflected shape, not the amplitude of deflection, can be determined, the nonlinear theory gives both the shape and amplitude of the buckled column.
5. The increase in load above the Euler load can only be achieved at a very large lateral deflection. At such a large deflection, inelastic behavior of material must be considered in the analysis. In the plastic or nonlinear range, the second-order differential equation becomes highly nonlinear and is often intractable. Recourse must then be had to numerical methods to obtain solutions. This is beyond the scope of this book. (Interested readers are referred to the two volume, comprehensive book, *Theory of Beam-Columns*, by Chen and Atsuta [1976, 1977].)

The consideration of material nonlinearity and yielding in the behavior of columns under the small displacement assumption will be given in Sections 2.7 to 2.9 of this chapter.

2.3 END-RESTRAINED COLUMNS

So far, we have considered the behavior of columns whose ends are pinned. In this section, we investigate the behavior of columns with other end conditions, and then compare them to the pinned-ended case in order to introduce the concept of effective length. The *effective length* of an end-restrained column is defined as the length of an *equivalent* pinned-ended column that will give the same critical load as the end-restrained column. Physically, the effective length can be visualized as the distance between the two inflection points (real or imaginary) of

2.3.1 Both Ends Fixed

Figure 2.11a shows a column built in at both ends. The forces acting on the column as it buckles are also shown in the figure. From Fig. 2.11b, it can be seen that equilibrium requires that

$$-M_{int} + Py + Vx - M_A = 0 \tag{2.3.1}$$

Since the internal resisting moment is

$$M_{int} = -EIy'' \tag{2.3.2}$$

The differential equation for equilibrium of this column can be written as

$$y'' + k^2 y = -\frac{V}{EI}x + \frac{M_A}{EI} \tag{2.3.3}$$

where $k^2 = P/EI$.

The general solution consists of a *complementary* solution satisfying the homogeneous equation and a *particular* solution satisfying the entire equation. The complementary solution is given by Eq. (2.2.13); and the particular solution can be obtained by inspection as $-Vx/P + M_A/P$. Thus, the general solution is

$$y = A \sin kx + B \cos kx - Vx/P + M_A/P \tag{2.3.4}$$

The boundary conditions are

$$y(0) = 0, \quad y'(0) = 0 \tag{2.3.5}$$

$$y(L) = 0, \quad y'(L) = 0 \tag{2.3.6}$$

Using the conditions at $x = 0$, we have

$$B = -M_A/P, \quad A = V/Pk \tag{2.3.7}$$

Substitution of Eq. (2.3.7) into Eq. (2.3.4) gives

$$y = \frac{V}{Pk} \sin kx - \frac{M_A}{P} \cos kx - \frac{Vx}{P} + \frac{M_A}{P} \tag{2.3.8}$$

Using $y(L) = 0$, we have

$$\frac{V}{Pk} \sin kL - \frac{M_A}{P} \cos kL - \frac{VL}{P} + \frac{M_A}{P} = 0 \tag{2.3.9}$$

Using $y'(L) = 0$, we have

$$\frac{V}{P} \cos kL + \frac{M_A k}{P} \sin kL - \frac{V}{P} = 0 \tag{2.3.10}$$

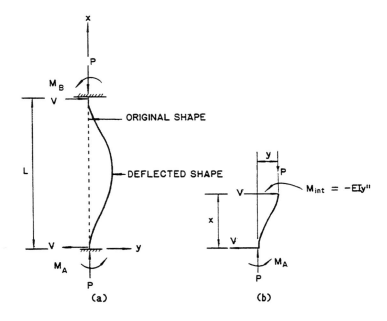

FIGURE 2.11 Fixed-fixed column

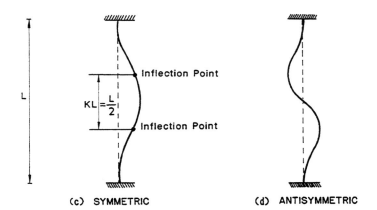

FIGURE 2.11 (*continued*) Symmetric and antisymmetric buckling modes of a fixed-fixed column

From Eqs. (2.3.9) and (2.3.10), for a nontrivial solution of V and M_a, we must have

$$kL \sin kL + 2 \cos kL - 2 = 0 \tag{2.3.11}$$

By using the trigonometrical identity $\sin kL = 2 \sin(kL/2) \cos(kL/2)$ and $\cos kL = 1 - 2 \sin^2(kL/2)$, we can write Eq. (2.3.11) as

$$\sin \frac{kL}{2} \left(\frac{kL}{2} \cos \frac{kL}{2} - \sin \frac{kL}{2} \right) = 0 \tag{2.3.12}$$

Equation (2.3.12) can be satisfied if either the first term $\sin(kL/2)$ or the terms in the parenthesis vanish.

If the first term vanishes, the solution is $kL = 2n\pi$ where, $n = 1, 2, 3, \ldots$, from which the critical load is obtained by setting $n = 1$, i.e.,

$$P_{cr} = \frac{4\pi^2 EI}{L^2} \tag{2.3.13}$$

If the terms in the parenthesis vanish, the lowest value that satisfies the equation $\frac{kL}{2} \cos \frac{kL}{2} - \sin \frac{kL}{2} = 0$ or $\tan \frac{kL}{2} = \frac{kL}{2}$ is $kL = 8.987$ from which

$$P_{cr} = \frac{80.766 EI}{L^2} \tag{2.3.14}$$

The values of Eqs. (2.3.13) and (2.3.14) correspond to the critical loads of the symmetric (Fig. 2.11c) and antisymmetric (Fig. 2.11d) buckling modes of the column, respectively. Since the critical load of the symmetric buckling mode is less than that of the antisymmetric buckling mode. The column will buckle in the symmetric mode. Unless the midheight of the column is braced against lateral movement, Eq. (2.3.14) will have little significance to us. The deflected shape of the symmetric buckling mode can be obtained by substituting $V = 0$ (because of symmetry) and $k = 2\pi/L$ into Eq. (2.3.8):

$$y = \frac{M_A}{P} \left(1 - \cos \frac{2\pi x}{L} \right) \tag{2.3.15}$$

If we define KL as the *effective length* of this fixed-fixed column, the equivalent pinned-pinned column with length KL (Fig. 2.11c) that will carry the same critical load as the fixed-fixed column with length L can be obtained by solving the following equation

$$\frac{\pi^2 EI}{(KL)^2} = \frac{4\pi^2 EI}{L^2} \tag{2.3.16}$$

which gives

$$KL = \tfrac{1}{2} L \tag{2.3.17}$$

2.3 End-Restrained Columns

In other words, the length of the equivalent pinned-pinned column is half that of the fixed-fixed column. Equivalently, the inflection points of the fixed-fixed column are at a distance of $L/2$ apart (Fig. 2.11c). The factor K is called the *effective length factor* of the fixed-fixed column. To verify that the inflection points are indeed as shown in Fig. 2.11c, we first write the moment expression along the length of the column and set it equal to zero to calculate the distances x that give the locations of the inflection points. By differentiating Eq. (2.3.15) twice, we can write the moment expression as

$$M = -EIy'' = EI\frac{M_A}{P}\frac{4\pi^2}{L^2}\cos\frac{2\pi x}{L} = 0 \qquad (2.3.18)$$

from which

$$\cos\frac{2\pi x}{L} = 0$$

or

$$x = \frac{nL}{4}, \quad n = 1, 3, 5, \ldots \qquad (2.3.19)$$

Using $n = 1$ and 3, give $x = L/4$ and $3L/4$. Hence, inflection points are located at $x = L/4$ and $3L/4$ and so the distance between them is $3L/4 - L/4 = L/2$, as shown in Fig. 2.11c.

In general, for a centrally loaded, and end-restrained column, the effective length factor K can be evaluated directly by the following equation

$$K = \sqrt{\frac{P_e}{P_{cr}}} \qquad (2.3.20)$$

where

P_{cr} = critical load of the end-restrained column
P_e = Euler load of the pinned-pinned column having the *same* length as the end-restrained column

Equation (2.3.20) can easily be derived from the definition of effective length factor similar to that of Eq. (2.3.16).

2.3.2 One End Fixed and One End Free

Figure 2.12a shows a column built in at one end and free at the other end. The corresponding free-body diagram of a short segment of the column is shown in Fig. 2.12b. The equilibrium equation for the free body is

$$M_{\text{int}} - P\Delta + Py = 0 \qquad (2.3.21)$$

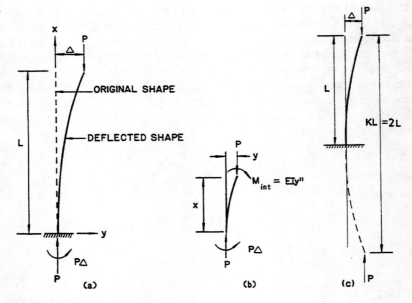

FIGURE 2.12 Fixed-free column

where Δ is the lateral deflection at the free end of the column. The internal moment is

$$M_{int} = EIy'' \tag{2.3.22}$$

The internal moment is related to the second derivative of the deflection d^2y/dx^2 positively because the curvature Φ (or the rate of change of the slope dy'/dx of the deflected curve as sketched in Fig. 2.12b, that corresponds to the positive moment M_{int}) increases with increasing x.

$$y'' + k^2 y = k^2 \Delta \tag{2.3.23}$$

where $k^2 = P/EI$.

The general solution is

$$y = A \sin kx + B \cos kx + \Delta \tag{2.3.24}$$

Using the boundary conditions

$$y(0) = 0 \tag{2.3.25}$$
$$y'(0) = 0 \tag{2.3.26}$$

we obtain

$$B = -\Delta \tag{2.3.27}$$
$$A = 0 \tag{2.3.28}$$

2.3 End-Restrained Columns

which, upon substitution into Eq. (2.3.24), gives

$$y = \Delta(1 - \cos kx) \qquad (2.3.29)$$

Using the condition

$$y(L) = \Delta \qquad (2.3.30)$$

in Eq. (2.3.29) will give

$$\cos kL = 0 \qquad (2.3.31)$$

from which

$$kL = \frac{n}{2}\pi, \quad n = 1, 3, \ldots \qquad (2.3.32)$$

The critical load is the load corresponding to $n = 1$ or

$$P_{cr} = \frac{\pi^2 EI}{4L^2} \qquad (2.3.33)$$

The deflected shape corresponding to P_{cr} is

$$y = \Delta\left(1 - \cos\frac{\pi x}{2L}\right) \qquad (2.3.34)$$

The effective length factor is

$$K = \sqrt{\frac{P_e}{P_{cr}}} = 2 \qquad (2.3.35)$$

The deflected shape of the equivalent pinned-pinned column with length $KL = 2L$ is shown in Fig. 2.12(c).

2.3.3 One End Hinged and One End Fixed

Shown in Fig. 2.13a,b are the diagrams of a hinged-fixed column and the free body of a short segment of the same column cut at a distance x from the hinged support. Note that for moment equilibrium, a shear force of M_F/L must be present at both ends of the column to balance the fixed end moment M_F, which is induced in the built-in end as the column buckles.

The equilibrium equation for the column segment shown in Fig. 2.13b is

$$-M_{int} + Py - \frac{M_F}{L}x = 0 \qquad (2.3.36)$$

and since

$$M_{int} = -EIy'' \qquad (2.3.37)$$

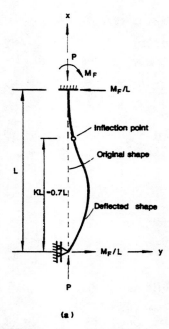
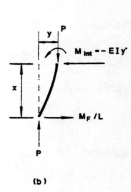

FIGURE 2.13 Hinged-fixed column

we can write the equilibrium equation (2.3.36) as

$$y'' + k^2 y - \frac{M_F}{EIL} x = 0 \qquad (2.3.38)$$

where $k^2 = P/EI$.

The general solution is

$$y = A \sin kx + B \cos kx + \frac{M_F}{PL} x \qquad (2.3.39)$$

The boundary conditions are

$$y(0) = 0 \qquad (2.3.40)$$
$$y(L) = 0 \qquad (2.3.41)$$
$$y'(L) = 0 \qquad (2.3.42)$$

Using the first two boundary conditions, we obtain

$$B = 0 \qquad (2.3.43)$$
$$A = -\frac{M_F}{P \sin kL} \qquad (2.3.44)$$

2.3 End-Restrained Columns

and the deflection function (2.3.39) becomes

$$y = \frac{M_F}{P}\left(\frac{x}{L} - \frac{\sin kx}{\sin kL}\right) \quad (2.3.45)$$

Using the third boundary condition in Eq. (2.3.45) gives

$$\tan kL = kL \quad (2.3.46)$$

from which kL can be solved by trial and error or by graphical means. The lowest value of kL that satisfies Eq. (2.3.46) is

$$kL = 4.4934 \quad (2.3.47)$$

which gives

$$P_{cr} = \frac{20.19 EI}{L^2} \quad (2.3.48)$$

The deflected shape of the column at buckling is

$$y = \frac{M_F}{P}\left[\frac{x}{L} + 1.0245 \sin\left(4.4934\frac{x}{L}\right)\right] \quad (2.3.49)$$

and the effective length factor is

$$K = \sqrt{\frac{P_e}{P_{cr}}} = \sqrt{\frac{\pi^2}{20.19}} = 0.7 \quad (2.3.50)$$

The effective length $KL = 0.7L$ is shown in Fig. 2.13a.

Note that the term M_F/P in Eq. (2.3.49) represents the displacement at $x = 0.2 L$ of the buckled member. This displacement is indeterminate as for the other cases of end-restrained columns shown in this section because of the use of the linear theory based on the small displacement assumption.

2.3.4 One End Fixed and One End Guided

A column with one end fixed and the other end guided is shown in Fig. 2.14a. Note that the shear force is zero but the moment is not zero at the guided end. Because of antisymmetry, the moment at the fixed end has the same direction and magnitude as the moment at the guided end. As a result, if we denote Δ the relative horizontal displacement of the two ends of the column, it follows that the end moment from equilibrium consideration will be $P\Delta/2$.

The equilibrium equation for a segment of this column is (Fig. 2.14b)

$$M_{int} - \frac{P\Delta}{2} + Py = 0 \quad (2.3.51)$$

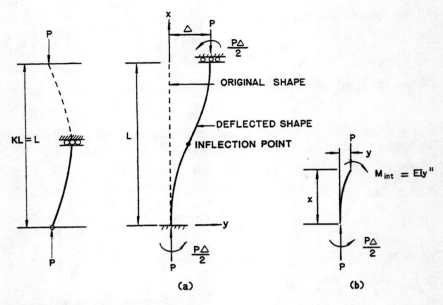

FIGURE 2.14 Fixed-guided column

and upon substitution of the expression for the internal moment

$$M_{int} = EIy'' \quad (2.3.52)$$

we obtain the differential equilibrium equation as

$$y'' + k^2 y = \frac{P\Delta}{2EI} \quad (2.3.53)$$

where $k^2 = P/EI$.

The general solution is

$$y = A \sin kx + B \cos kx + \frac{\Delta}{2} \quad (2.3.54)$$

Using the boundary conditions of

$$y(0) = 0, \quad y'(0) = 0 \quad (2.3.55)$$

respectively in the general solution (2.2.54) gives

$$B = -\frac{\Delta}{2}, \quad \text{and} \quad A = 0 \quad (2.3.56)$$

Thus

$$y = \frac{\Delta}{2}(1 - \cos kx) \quad (2.3.57)$$

2.3 End-Restrained Columns

Using the boundary condition
$$y(L) = \Delta$$
gives
$$\cos kL = -1 \quad (2.3.58)$$
or
$$kL = n\pi, \quad n = 1, 3, \ldots \quad (2.3.59)$$

The critical load is given by letting $n = 1$, hence
$$P_{cr} = \frac{\pi^2 EI}{L^2} \quad (2.3.60)$$

The deflected shape at P_{cr} is
$$y = \frac{\Delta}{2}\left(1 - \cos\frac{\pi x}{L}\right) \quad (2.3.61)$$

and the effective length factor is
$$K = \sqrt{\frac{P_e}{P_{cr}}} = 1 \quad (2.3.62)$$

The effective length KL of the equivalent pinned-pinned column is also shown in Fig. 2.14a.

2.3.5 One End Hinged and One End Guided

If a column is hinged at one end and guided at the other, as shown in Fig. 2.15a, the equilibrium equation for a segment of this column (Fig. 2.15b) can be written as
$$-M_{int} + Py = 0 \quad (2.3.63)$$

Since the internal moment is
$$M_{int} = -EIy'' \quad (2.3.64)$$

the differential equation governing the behavior of this column is
$$EIy'' + Py = 0 \quad (2.3.65)$$
or
$$y'' + k^2 y = 0 \quad (2.3.66)$$

The general solution is
$$y = A \sin kx + B \cos kx \quad (2.3.67)$$

The boundary conditions are
$$y(0) = 0 \quad (2.3.68)$$
$$y'(L) = 0 \quad (2.3.69)$$

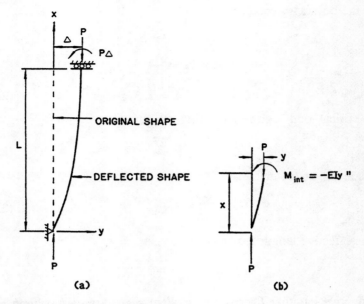

FIGURE 2.15 Hinged-guided column

Using the first boundary condition, we have
$$B = 0 \tag{2.3.71}$$
Therefore, Eq. (2.3.67) becomes
$$y = A \sin kx \tag{2.3.72}$$
Using the second boundary condition, we obtain
$$\cos kL = 0 \tag{2.3.73}$$
from which
$$kL = \frac{n\pi}{2}, \quad n = 1, 3, 5, \ldots \tag{2.3.74}$$
The lowest value of n gives the critical load of the column
$$P_{cr} = \frac{\pi^2 EI}{4L^2} \tag{2.3.75}$$
The deflected shape at buckling is
$$y = A \sin \frac{\pi x}{2L} \tag{2.3.76}$$

2.3 End-Restrained Columns

The effective length factor is

$$K = \sqrt{\frac{P_e}{P_{cr}}} = 2 \tag{2.3.77}$$

2.3.6 AISC Effective Length Factor and Column Curve

From the above discussion, it can be seen that for an isolated column, regardless of its end conditions, the critical load can be expressed in the general form as

$$P_{cr} = \frac{\pi^2 EI}{(KL)^2} \tag{2.3.78}$$

where K is the effective length factor that is dependent on the boundary conditions of the column. The theoretical values of K for various boundary conditions have been derived in the preceding sections and are summarized in Table 2.1. On the same table, the K-values recommended by the AISC[3] are also shown. The recommended K-values involving cases with fixed support are higher than their theoretical counterparts because full join fixity is seldom realized in actual columns.

For design purpose, it is more convenient to express Eq. (2.3.78) in graphical form. Realizing that $I = Ar^2$ and defining

$$P_y = A\sigma_y \tag{2.3.79}$$

as the *yield load* and

$$\lambda_c = \frac{KL}{r}\sqrt{\frac{\sigma_y}{\pi^2 E}} \tag{2.3.80}$$

as the *slenderness parameter*, Eq. (2.3.78) can be written as

$$\frac{P_{cr}}{P_y} = \lambda_c^{-2} \tag{2.3.81}$$

Equation (2.3.81) is plotted in Fig. 2.16. Note that the curve terminates at $P = 0.5P_y$, because Eq. (2.3.81) is only valid for perfectly elastic columns. For columns in the inelastic range ($P > 0.5P_y$ or $\lambda_c < \sqrt{2}$), a different column curve by AISC based on the *tangent modulus concept* (to be discussed in Section 2.7) is used for practical design. The demarcation point ($P = 0.5P_y$ or $\lambda_c = \sqrt{2}$) for elastic and inelastic column behavior is based on experimental observations that the maximum compressive residual stress of a hot-rolled–column section is approximately $0.3\sigma_y$. The use of the number 0.5 rather than 0.3 is for conservative purposes.

Table 2.1 Theoretical and Recommended K Values for Idealized Columns

	(a)	(b)	(c)	(d)	(e)	(f)
Buckled shape of column is shown by dashed line						
Theoretical K value	0.5	0.7	1.0	1.0	2.0	2.0
Recommended design value when ideal conditions are approximated	0.65	0.80	1.2	1.0	2.10	2.0
End condition code			Rotation fixed and translation fixed Rotation free and translation fixed Rotation fixed and translation free Rotation free and translation free			

Adapted from reference 3.

2.3.7 Elastically Restrained Ends

The discussion so far has been focused on axially loaded columns with rotational restraint at their ends that are either fully rigid (fixed-ended case) or nonexistent (pinned-ended case). In actual structures, columns usually do not exist alone but connected to other structural members that will provide accountable rotational restraint to the columns. Consequently, it is pertinent to investigate the behavior of columns with elastically restrained ends.

Figure 2.17a shows an end-restrained column acted on by an axial force P. Here, it is convenient to represent the effect of end restraint by a spring with rotational stiffnesses R_{kA} and R_{kB} at the A and B ends of the column, respectively. The rotational stiffness is defined as the moment the spring can sustain for a unit rotation.

Referring to Fig. 2.17b, the equilibrium equation for the column segment is expressed as

$$-M_{\text{int}} + Py + Vx - M_A = 0 \qquad (2.3.82)$$

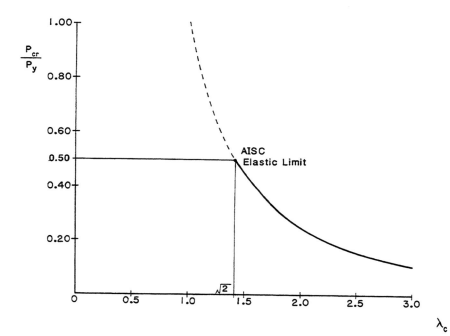

FIGURE 2.16 Column design curve for elastic column

FIGURE 2.17 Column with elastically restrained ends

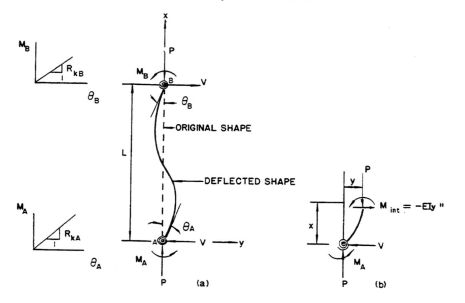

The internal resisting moment is

$$M_{int} = -EIy'' \tag{2.3.83}$$

Therefore, the differential equation of equilibrium can be written as

$$y'' + k^2 y = \frac{-V}{EI}x + \frac{M_A}{EI} \tag{2.3.84}$$

The general solution of the differential equation is

$$y = A \sin kx + B \cos kx - \frac{V}{P}x + \frac{M_A}{P} \tag{2.3.85}$$

Using the displacement and slope boundary conditions for the specific problem and recognizing that

$$M_A = R_{kA}\theta_A \tag{2.3.86}$$

$$M_B = R_{kB}\theta_B \tag{2.3.87}$$

the critical load of the end-restrained column can be obtained as shown in the forthcoming example.

As an illustrative example, the buckling load of the column in the simple frame shown in Fig. 2.18a will be determined.

At buckling, the forces that act on the column are shown in the free-body diagram of Fig. 2.18c. Because of symmetry there is no shear force acting on the column. Since the applied force P is generally much greater than the shear forces V_A and V_B that are induced as the beams bend during column buckling, the differential equation of the column can be written as (Fig. 2.18d)

$$y'' + k^2 y = \frac{M_A}{EI} \tag{2.3.88}$$

where M_A is the end moment induced at joint A as a result of buckling of the column.

Note that Eq. (2.3.88) is a special form of Eq. (2.3.84), with the column shear force V equal to zero. The general solution is

$$y = A \sin kx + B \cos kx + \frac{M_A}{P} \tag{2.3.89}$$

Using the boundary condition

$$y(0) = 0 \tag{2.3.90}$$

we obtain

$$B = -\frac{M_A}{P} \tag{2.3.91}$$

2.3 End-Restrained Columns

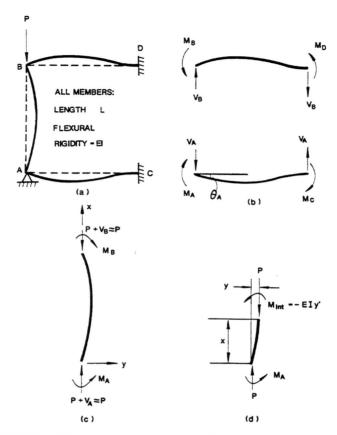

FIGURE 2.18 Buckling of an end-restrained column

Using the symmetry condition,

$$\left.\frac{dy}{dx}\right|_{x=L/2} = 0 \tag{2.3.92}$$

we have

$$A = -\frac{M_A}{P}\tan\frac{kL}{2} \tag{2.3.93}$$

Hence

$$y = \frac{M_A}{P}\left(1 - \tan\frac{kL}{2}\sin kx - \cos kx\right) \tag{2.3.94}$$

and

$$\frac{dy}{dx} = \frac{M_A k}{P}\left(-\tan\frac{kL}{2}\cos kx + \sin kx\right) \tag{2.3.95}$$

By referring to the free-body diagram of the lower beam (Fig. 2.18b) and setting $\theta_c = M_{FA} = M_{FC} = 0$ in the following slope-deflection equations

$$M_A = \frac{EI}{L}(4\theta_A + 2\theta_C) + M_{FA} \qquad (2.3.96a)$$

$$M_C = \frac{EI}{L}(4\theta_C + 2\theta_A) + M_{FC} \qquad (2.3.96b)$$

we can obtain

$$M_A = \frac{4EI}{L}\theta_A \qquad (2.3.97a)$$

$$M_C = \frac{2EI}{L}\theta_A = \tfrac{1}{2}M_A \qquad (2.3.97b)$$

from which we have

$$\theta_A = \frac{M_A L}{4EI} \qquad (2.3.98)$$

If rigid connection is assumed, the beam end rotation will be exactly equal to the column-end rotation, i.e.,

$$\theta_A = \frac{dy}{dx}\bigg|_{x=0} \qquad (2.3.99)$$

Thus, by equating Eq. (2.3.98) to Eq. (2.3.95) evaluated at $x = 0$, we obtain

$$\frac{M_A L}{4EI} = \frac{M_A k}{P}\left(-\tan\frac{kL}{2}\right) \qquad (2.3.100)$$

or

$$\tan\frac{kL}{2} + \frac{kL}{4} = 0 \qquad (2.3.101)$$

Equation (2.3.101) is the transcendental equation whose solution will give the value of the buckling load.

Using graphical method or by trial and error, it can be shown that the smallest value of kL satisfying Eq. (2.3.101) is

$$kL = 4.586 \qquad (2.3.102)$$

or

$$P_{cr} = \frac{21.03 EI}{L^2} \qquad (2.3.103)$$

Note that this critical load for the column restrained by the two beams falls between P_{cr} of a pinned-pinned column and that of a fixed-fixed column. The effective length factor for this elastically end-restrained column in the simple frame is

$$K = \sqrt{\frac{P_e}{P_{cr}}} = \sqrt{\frac{\pi^2}{21.03}} = 0.685 \qquad (2.3.104)$$

2.4 FOURTH-ORDER DIFFERENTIAL EQUATION

In the previous section, the governing differential equation describing the behavior of the column has been developed by consideration of equilibrium for a column segment of *finite* size. The resulting equilibrium equation is second order. Depending on the end conditions, this second-order differential equation may or may not be homogeneous. By enforcing proper geometric (or kinematic) boundary conditions, the critical load can be obtained as the eigenvalue of the characteristic or transcendental equation of the differential equation. In this section, a fourth-order differential equation (2.4.6), which is applicable to all columns with any boundary condition, will be developed.

Figure 2.19 shows the free-body diagram of an *infinitesimal* segment of

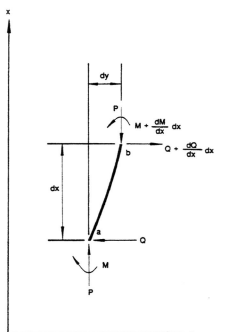

FIGURE 2.19 Free-body diagram of an infinitesimal segment of a column

the *column*. By summing the moment about point b, we obtain

$$Q\,dx + P\,dy + M - \left(M + \frac{dM}{dx}dx\right) = 0$$

or, upon simplification

$$Q = \frac{dM}{dx} - P\frac{dy}{dx} \qquad (2.4.1)$$

Summing force horizontally, we can write

$$-Q + \left(Q + \frac{dQ}{dx}dx\right) = 0$$

or, upon simplification

$$\frac{dQ}{dx} = 0 \qquad (2.4.2)$$

Differentiating Eq. (2.4.1) with respect to x, we obtain

$$\frac{dQ}{dx} = \frac{d^2M}{dx^2} - P\frac{d^2y}{dx^2} \qquad (2.4.3)$$

which, when compared with Eq. (2.4.2), gives

$$\frac{d^2M}{dx^2} - P\frac{d^2y}{dx^2} = 0 \qquad (2.4.4)$$

Since $M = -EI\frac{d^2y}{dx^2}$, Eq. (2.4.4) can be written as

$$EI\frac{d^4y}{dx^4} + P\frac{d^2y}{dx^2} = 0 \qquad (2.4.5)$$

or

$$y^{IV} + k^2 y'' = 0 \qquad (2.4.6)$$

Equation (2.4.6) is the general fourth-order differential equation that is valid for all support conditions. The general solution to this equation is

$$y = A \sin kx + B \cos kx + Cx + D \qquad (2.4.7)$$

To determine the critical load, we need to specify four boundary conditions: two at each end of the column. In most cases, mixed—i.e., both geometric and force—boundary conditions are needed to be specified.

To show how to obtain critical loads using the fourth-order differential equation, we will solve the cases of a pinned-pinned, a fixed-fixed, and a fixed-free column.

2.4 Fourth-Order Differential Equation

Pinned-Pinned Column

For a pinned-pinned column (Fig. 2.20), the four boundary conditions are:

$$y(0) = 0, \quad M(0) = 0 \quad (2.4.8)$$
$$y(L) = 0, \quad M(L) = 0 \quad (2.4.9)$$

Since $M = -EIy''$, the moment conditions can be written as

$$y''(0) = 0 \quad (2.4.10)$$
$$y''(L) = 0 \quad (2.4.11)$$

Using the conditions $y(0) = y''(0) = 0$, we obtain

$$B = D = 0 \quad (2.4.12)$$

The deflection function (2.4.7) reduces to

$$y = A \sin kx + Cx \quad (2.4.13)$$

Using the conditions $y(L) = y''(L) = 0$, Eq. (2.4.13) gives

$$A \sin kL + CL = 0 \quad (2.4.14)$$

and

$$-Ak^2 \sin kL = 0 \quad (2.4.15)$$

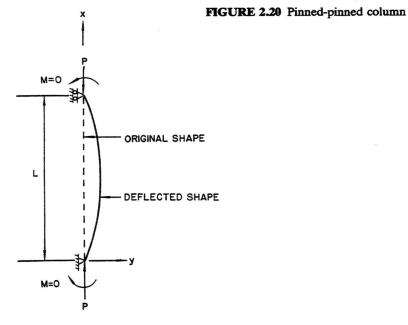

FIGURE 2.20 Pinned-pinned column

In matrix form

$$\begin{bmatrix} \sin kL & L \\ -k^2 \sin kL & 0 \end{bmatrix} \begin{bmatrix} A \\ C \end{bmatrix} = \begin{bmatrix} 0 \\ 0 \end{bmatrix} \quad (2.4.16)$$

If $A = C = 0$, the solution is trivial. Therefore, to obtain a nontrivial solution, the determinant of the coefficient matrix of Eq. (2.4.16) must vanish, i.e.,

$$\det \begin{vmatrix} \sin kL & L \\ -k^2 \sin kL & 0 \end{vmatrix} = 0 \quad (2.4.17)$$

or

$$k^2 L \sin kL = 0 \quad (2.4.18)$$

Since $k^2 \neq 0$, we must have

$$\sin kL = 0 \quad (2.4.19)$$

or $kL = n\pi$, $n = 1, 2, 3, \ldots$.

The critical load can be obtained by setting $n = 1$ to give

$$P_{cr} = \frac{\pi^2 EI}{L^2} \quad (2.4.20)$$

Fixed-Fixed Column

The four boundary conditions for this case are (Fig. 2.21)

$$y(0) = y'(0) = 0 \quad (2.4.21)$$
$$y(L) = y'(L) = 0 \quad (2.4.22)$$

Using the first two boundary conditions, we obtain

$$D = -B, \quad C = -Ak \quad (2.4.23)$$

The deflection function (2.4.7) becomes

$$y = A(\sin kx - kx) + B(\cos kx - 1) \quad (2.4.24)$$

Using the last two boundary conditions, we have

$$\begin{bmatrix} \sin kL - kL & \cos kL - 1 \\ \cos kL - 1 & -\sin kL \end{bmatrix} \begin{bmatrix} A \\ B \end{bmatrix} = \begin{bmatrix} 0 \\ 0 \end{bmatrix} \quad (2.4.25)$$

For a nontrivial solution, we must have

$$\det \begin{vmatrix} \sin kL - kL & \cos kL - 1 \\ \cos kL - 1 & -\sin kL \end{vmatrix} = 0 \quad (2.4.26)$$

or, after expanding

$$kL \sin kL + 2 \cos kL - 2 = 0 \quad (2.4.27)$$

2.4 Fourth-Order Differential Equation

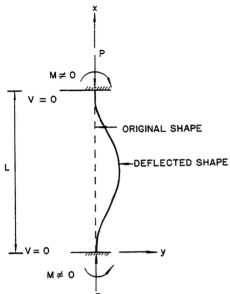

FIGURE 2.21 Symmetric buckling mode of a fixed-fixed column

Note that Eq. (2.4.27) is identical to Eq. (2.3.11) and thus the critical load for the symmetric buckling mode is $P_{cr} = 4\pi^2 EI/L^2$ and that for the antisymmetric buckling mode it is $P_{cr} = 80.766 EI/L^2$.

Fixed-Free Column

The boundary conditions for a fixed-free column are (Fig. 2.22).
At the fixed end

$$y(0) = y'(0) = 0 \tag{2.4.28}$$

and, at the free end, the moment $M = EIy''$ is equal to zero

$$y''(L) = 0 \tag{2.4.29}$$

and the shear force $V = -dM/dx = -EIy'''$ is equal to the transverse component of P acting at the free end of the column cross section Py' (Fig. 2.22).

$$V = -EIy''' = Py' \tag{2.4.30}$$

It follows that the shear force condition at the free end has the form

$$y''' + k^2 y' = 0 \tag{2.4.31}$$

Using the boundary conditions at the fixed end, we have

$$B + D = 0, \qquad Ak + C = 0 \tag{2.4.32}$$

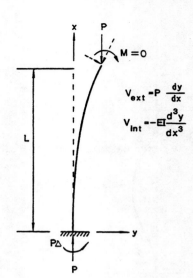

FIGURE 2.22 Fixed-free column

The boundary conditions at the free end gives

$$A \sin kL + B \cos kL = 0, \quad C = 0 \tag{2.4.33}$$

In matrix form, Eqs. (2.4.32) and (2.4.33) can be written as

$$\begin{bmatrix} 0 & 1 & 1 \\ k & 0 & 0 \\ \sin kL & \cos kL & 0 \end{bmatrix} \begin{bmatrix} A \\ B \\ D \end{bmatrix} = \begin{bmatrix} 0 \\ 0 \\ 0 \end{bmatrix} \tag{2.4.34}$$

For a nontrivial solution, we must have

$$\det \begin{vmatrix} 0 & 1 & 1 \\ k & 0 & 0 \\ \sin kL & \cos kL & 0 \end{vmatrix} = 0 \tag{2.4.35}$$

or, after expanding

$$k \cos kL = 0 \tag{2.4.36}$$

Since $k \neq 0$, we must have $\cos kL = 0$ or

$$kL = \frac{n\pi}{2} \quad n = 1, 3, 5, \ldots \tag{2.4.37}$$

The smallest root ($n = 1$) gives the critical load of the column

$$P_{cr} = \frac{\pi^2 EI}{4L^2} \tag{2.4.38}$$

2.5 Special Members

Table 2.2 Boundary Conditions for Various End Conditions

End conditions	Boundary conditions	
Pinned	$y = 0$,	$y'' = 0$
Fixed	$y = 0$,	$y' = 0$
Guided	$y' = 0$,	$y''' = 0$
Free	$y'' = 0$,	$y''' + k^2 y' = 0$

For the three cases studied, the solutions of the fourth-order differential equation are seen to lead to the same critical loads as the second-order equation. Note that the second-order equations for these cases studied previously are different because of different boundary conditions, but the fourth-order equation is the same for all cases. In determining the critical load using the fourth-order equation, four boundary conditions must be specified. The boundary conditions for various support cases are summarized in Table 2.2.

2.5 SPECIAL MEMBERS

The discussion so far has been restricted to columns for which the axial force P and flexural rigidity EI are constant along the length of the member. Furthermore, no intermediate support is present, so that restraint is provided only at the ends of the column under investigation. In this section, we shall extend the solution for evaluating critical loads of prismatic isolated columns with constant axial force to columns with a change in axial force, a change in flexural rigidity, or with intermediate support (Fig. 2.23).

2.5.1 Two-Axial-Force Column

As an illustration, consider the cantilever column shown in Fig. 2.24a. The column is subjected to two axial forces P: one applied at the free end and the other at midheight. As a result, the axial force along the entire length of the column is not a constant. The axial force is equal to $2P$ for the lower portion of the column from A to B (segment 1), but it is equal to P for the upper portion of the column from B to C (segment 2). To determine the critical load of this column, it is therefore necessary to write two differential equations, one for each segment of the column for which the axial force is a constant. For convenience, two sets of coordinates are established: $x_1 - y_1$ for column segment 1 and $x_2 - y_2$ for column segment 2.

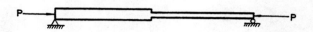

FIGURE 2.23 Special members

FIGURE 2.24 Cantilever column subjected to two axial forces

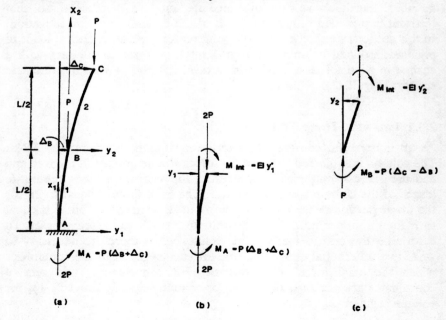

2.5 Special Members

With reference to Fig. 2.24b, the differential equation for segment 1 is

$$EI y_1'' + 2P y_1 = M_A \tag{2.5.1}$$

or

$$y_1'' + k_1^2 y_1 = \frac{M_A}{EI} \tag{2.5.2}$$

where

$$k_1^2 = \frac{2P}{EI} \tag{2.5.3}$$

$$M_A = P(\Delta_B + \Delta_C) \tag{2.5.4}$$

in which Δ_B and Δ_C are the lateral deflections with respect to the $x_1 - y_1$ axes for points B and C, respectively. The general solution of Eq. (2.5.2) is

$$y_1 = A \sin k_1 x + B \cos k_1 x + \frac{M_A}{EI k_1^2} \tag{2.5.5}$$

With reference to Fig. 2.24c, the differential equation for segment 2 is

$$EI y_2'' + P y_2 = M_B \tag{2.5.6}$$

or

$$y_2'' + k_2^2 y_2 = \frac{M_B}{EI} \tag{2.5.7}$$

where

$$k_2^2 = \frac{P}{EI} \tag{2.5.8}$$

$$M_B = P(\Delta_C - \Delta_B) \tag{2.5.9}$$

The general solution of Eq. (2.5.7) is

$$y_2 = C \sin k_2 x_2 + D \cos k_2 x_2 + \frac{M_B}{EI k_2^2} \tag{2.5.10}$$

The four constants A, B, C, and D in Eqs. (2.5.5) and (2.5.10) can be evaluated using the following boundary and continuity conditions

$$y_1(0) = 0 \tag{2.5.11}$$

$$y_1'(0) = 0 \tag{2.5.12}$$

$$y_1\left(\frac{L}{2}\right) = y_2(0) + \Delta_B \tag{2.5.13}$$

$$y_1'\left(\frac{L}{2}\right) = y_2'(0) \tag{2.5.14}$$

The first two boundary conditions give

$$B = -\frac{M_A}{EIk_1^2} \qquad (2.5.15)$$

$$A = 0 \qquad (2.5.16)$$

and the continuity conditions (2.5.13) and (2.5.14), give

$$D = -\frac{M_A}{EIk_1^2}\cos\frac{k_1 L}{2} + \frac{M_B}{EIk_1^2} - \frac{M_B}{EIk_2^2} \qquad (2.5.17)$$

$$C = \frac{M_A}{EIk_1 k_2}\sin\frac{k_1 L}{2} \qquad (2.5.18)$$

Substitution of these expressions into the deflection functions (2.5.5) and (2.5.10) gives

$$y_1 = \frac{M_A}{EIk_1^2}(1 - \cos k_1 x_1) \qquad (2.5.19)$$

and

$$y_2 = \frac{M_A}{EIk_1 k_2}\sin\frac{k_1 L}{2}\sin k_2 x_2$$
$$- \left(\frac{M_A}{EIk_1^2}\cos\frac{k_1 L}{2} - \frac{M_B}{EIk_1^2} + \frac{M_B}{EIk_2^2}\right)\cos k_2 x_2 + \frac{M_B}{EIk_2^2} \qquad (2.5.20)$$

Finally, using the conditions

$$y_1\left(\frac{L}{2}\right) = \Delta_B \qquad (2.5.21)$$

$$y_2\left(\frac{L}{2}\right) = \Delta_C - \Delta_B \qquad (2.5.22)$$

and realizing that $M_A = P(\Delta_B + \Delta_C)$ and $M_B = P(\Delta_C - \Delta_B)$, we obtain

$$\begin{bmatrix} k_{11} & k_{12} \\ k_{21} & k_{22} \end{bmatrix}\begin{bmatrix} \Delta_B \\ \Delta_C \end{bmatrix} = \begin{bmatrix} 0 \\ 0 \end{bmatrix} \qquad (2.5.23)$$

where

$$k_{11} = -1 - \cos\frac{k_1 L}{2} \qquad (2.5.24a)$$

$$k_{12} = 1 - \cos\frac{k_1 L}{2} \qquad (2.5.24b)$$

$$k_{21} = \frac{1}{\sqrt{2}}\sin\frac{k_1 L}{2}\sin\frac{k_2 L}{2} + \cos\frac{k_2 L}{2} \qquad (2.5.24c)$$

$$k_{22} = \frac{1}{\sqrt{2}}\sin\frac{k_1 L}{2}\sin\frac{k_2 L}{2} - \cos\frac{k_2 L}{2} \qquad (2.5.24d)$$

2.5 Special Members

For a nontrivial solution, we must have

$$\det \begin{vmatrix} k_{11} & k_{12} \\ k_{21} & k_{22} \end{vmatrix} = 0 \qquad (2.5.25)$$

By trial and error, and by recognizing that $k_1 = \sqrt{2}\, k_2$, we find that the lowest value of k_2 satisfying Eq. (2.5.25) is

$$k_2 = \frac{1.4378}{L} \qquad (2.5.26)$$

Using the definition of k_1 [Eq. (2.5.8)], we obtain

$$P_{cr} = 2.067 \frac{EI}{L^2} \qquad (2.5.27)$$

2.5.2 Continuous Member

In the preceding example, the second-order differential equation has been used in the solution procedure. The fourth-order differential equation is equally applicable, of course. To illustrate this, consider the two span continuous column shown in Fig. 2.25a. The column is subjected to an axial force P. It is desired to determine the critical load of this column. Here, as in the preceding example, the column is divided

FIGURE 2.25 Continuous member

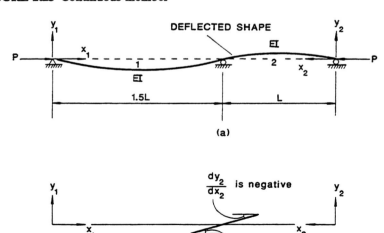

into two segments, 1 and 2, and a differential equation is written for each segment. For convenience, two sets of coordinate axes are again established: $x_1 - y_1$ for segment 1 and $x_2 - y_2$ for segment 2.

The fourth-order differential equation for segment 1 is

$$y_1^{IV} + k^2 y_1'' = 0 \qquad (2.5.28)$$

where $k^2 = P/EI$ and the general solution is

$$y_1 = A \sin kx_1 + B \cos kx_1 + Cx_1 + D \qquad (2.5.29)$$

Similarly, the fourth-order differential equation for segment 2 is

$$y_2^{IV} + k^2 y_2'' = 0 \qquad (2.5.30)$$

and the general solution is

$$y_2 = E \sin kx_2 + F \cos kx_2 + Gx_2 + H \qquad (2.5.31)$$

By using the boundary conditions

$$y_1(0) = 0, \qquad y_1''(0) = 0 \qquad (2.5.32)$$
$$y_1(\tfrac{3}{2}L) = 0 \qquad (2.5.33)$$
$$y_2(0) = 0, \qquad y_2''(0) = 0 \qquad (2.5.34)$$
$$y_2(L) = 0 \qquad (2.5.35)$$

and the continuity conditions

$$y_1'(\tfrac{3}{2}L) = -y_2'(L) \qquad (2.5.36)$$
$$y_1''(\tfrac{3}{2}L) = y_2''(L) \qquad (2.5.37)$$

it can be shown that

$$B = D = F = H = 0 \qquad (2.5.38)$$

and

$$\begin{bmatrix} \sin \tfrac{3}{2}kL & \tfrac{3}{2}L & 0 & 0 \\ 0 & 0 & \sin kL & L \\ k \cos \tfrac{3}{2}kL & 1 & k \cos kL & 1 \\ -\sin \tfrac{3}{2}kL & 0 & \sin kL & 0 \end{bmatrix} \begin{bmatrix} A \\ C \\ E \\ G \end{bmatrix} = \begin{bmatrix} 0 \\ 0 \\ 0 \\ 0 \end{bmatrix} \qquad (2.5.39)$$

The minus sign in Eq. (2.5.36) indicates that a positive slope at the intemediate support with respect to the $x_1 - y_1$ axes corresponds to a negative slope with respect to the $x_2 - y_2$ axes (Fig. 2.25b). For a nontrivial solution, the determinant of the coefficient matrix of Eq.

2.6 Initially Crooked Columns

(2.5.39) must vanish

$$\det \begin{vmatrix} \sin\frac{3}{2}kL & \frac{3}{2}L & 0 & 0 \\ 0 & 0 & \sin kL & L \\ k\cos\frac{3}{2}kL & 1 & k\cos kL & 1 \\ -\sin\frac{3}{2}kL & 0 & \sin kL & 0 \end{vmatrix} = 0 \qquad (2.5.40)$$

or, after simplification, we obtain the characteristic equation as

$$5\sin\tfrac{3}{2}kL \sin kL - 3kL \sin\tfrac{3}{2}kL = 0 \qquad (2.5.41)$$

By trial and error, the smallest value of k that satisfies the characteristic equation is

$$k = \frac{2.427}{L} \qquad (2.5.42)$$

using $k^2 = P/EI$, we have

$$P_{cr} = 5.89\frac{EI}{L^2} \qquad (2.5.43)$$

2.6 INITIALLY CROOKED COLUMNS

In reality, all columns are imperfect. There are two types of imperfections: geometrical imperfection and material imperfection. In this section, we investigate the behavior of geometrical imperfect column. The behavior of column with material imperfection will be discussed in the next section.

2.6.1 Pinned-Ended Column

Figure 2.26a shows a geometrical imperfect column. To begin with, let us assume that the initial out-of-straightness is in the form of a half sine curve described by

$$y_0 = \delta_0 \sin\frac{\pi x}{L} \qquad (2.6.1)$$

where δ_0 is the amplitude of the crookedness at midheight of the column.

If we consider equilibrium of a segment of column (Fig. 2.26b), the equilibrium equation is

$$-M_{int} + P(y + y_0) = 0 \qquad (2.6.2)$$

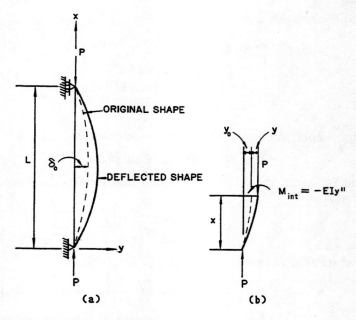

FIGURE 2.26 Initially crooked column

noting that y is the deflection of the column from its original crooked position.

The internal resisting moment is

$$M_{int} = -EIy'' \qquad (2.6.3)$$

This internal bending moment results from a change in curvature y'' and not from the total curvature $y'' + y_0''$, since it is tacitly assumed that the column is stress-free in its initially crooked position before the application of the load P.

Substituting the expression for the internal moment (2.6.3) into the equilibrium equation (2.6.2), the differential equation that describes the behavior of an initially crooked pinned-ended column takes the form

$$EIy'' + P(y + y_0) = 0$$

or using Eq. (2.6.1), we have

$$y'' + k^2 y = -k^2 \delta_0 \sin \frac{\pi x}{L} \qquad (2.6.4)$$

where $k^2 = P/EI$ and the complementary solution is

$$y_c = A \sin kx + B \cos kx \qquad (2.6.5)$$

2.6 Initially Crooked Columns

The particular solution can be obtained by using the method of undetermined coefficient. Since the right-hand side of the equation consists of a sine and/or cosine term, the particular solution is of the form

$$y_p = C \sin \frac{\pi x}{L} + D \cos \frac{\pi x}{L} \qquad (2.6.6)$$

in which C and D are the undetermined coefficients.

To determine C and D, we substitute Eq. (2.6.6) into the differential equation (2.6.4), and, after combining terms, we obtain

$$\left[C\left(k^2 - \frac{\pi^2}{L^2}\right) + k^2 \delta_0\right] \sin \frac{\pi x}{L} + \left[D\left(k^2 - \frac{\pi^2}{L^2}\right)\right] \cos \frac{\pi x}{L} = 0 \qquad (2.6.7)$$

This equation must be satisfied for all values of x. Thus, the terms in both square brackets must vanish. The vanishing of the first square brackets leads to

$$C = \frac{-k^2 \delta_0}{k^2 - \pi^2/L^2} = \frac{\delta_0 P/P_e}{1 - P/P_e} \qquad (2.6.8)$$

in which P_e is the Euler load. The vanishing of the second square brackets gives either

$$D = 0 \qquad (2.6.9)$$

or

$$k^2 = \frac{\pi^2}{L^2} \qquad (2.6.10)$$

If we use Eq. (2.6.10), we obtain $P_e = \pi^2 EI/L^2$, which is the Euler load. This is not the solution we are interested in here. Therefore, we must have $D = 0$. As a result, the general solution is

$$y = A \sin kx + B \cos kx + \frac{\delta_0 P/P_e}{1 - P/P_e} \sin \frac{\pi x}{L} \qquad (2.6.11)$$

To determine the two constants A and B, we use the boundary conditions

$$y(0) = 0 \qquad (2.6.12)$$
$$y(L) = 0 \qquad (2.6.13)$$

Using the first boundary condition, we have

$$B = 0 \qquad (2.6.14)$$

and the second boundary condition leads to

$$A \sin kL = 0 \qquad (2.6.15)$$

from which either

$$A = 0 \tag{2.6.16}$$

or

$$\sin kL = 0 \tag{2.6.17}$$

If we let $\sin kL = 0$, we again limit the solution to $P = P_e$. Therefore, we must have $A = 0$. With $A = B = 0$, Eq. (2.6.11) becomes

$$y = \frac{P/P_e}{1 - P/P_e} \delta_0 \sin \frac{\pi x}{L} \tag{2.6.18}$$

Equation (2.6.18) expresses the deflection from the initial crooked position of the column. To obtain the total deflection (i.e., deflection from the x-axis), we need to add Eq. (2.6.1) to Eq. (2.6.18).

$$y_{\text{total}} = y_0 + y$$

or

$$y_{\text{total}} = \left(\frac{1}{1 - P/P_e}\right) \delta_0 \sin \frac{\pi x}{L} \tag{2.6.19}$$

Equation (2.6.19) states that the total deflection (induced as a result of the applied compressive force) can be obtained simply by multiplying the initial deflection by a factor $1/(1 - P/P_e)$. The term in parenthesis in Eq. (2.6.19) is called the *amplification factor* (A_F)

$$A_F = \frac{1}{1 - P/P_e} \tag{2.6.20}$$

The moment in the column is

$$M = P(y + y_0) = P y_{\text{total}}$$

or

$$M = \left(\frac{1}{1 - P/P_e}\right) P \delta_0 \sin \frac{\pi x}{L} \tag{2.6.21}$$

If we denote M_I as the first-order moment, or the moment evaluated by considering equilibrium with respect to the initial geometry of the geometrical imperfect column, then we can write Eq. (2.6.21) as

$$M = A_F M_I \tag{2.6.22}$$

Equation (2.6.22) states that the moment evaluated based on the deformed geometry of the column (second-order moment) can be obtained from the moment evaluated based on the initial geometry of the column (first-order moment) simply by multiplying the latter by the *amplification factor*. The variation of this *moment amplification (or magnification) factor* as a function of P/P_e is shown in Fig. 2.7.

If the initial crookedness of the column is not a half sine wave but

2.6 Initially Crooked Columns

some general shape, it is advantageous to express this general shape in a Fourier sine series

$$y_0 = \delta_1 \sin\frac{\pi x}{L} + \delta_2 \sin\frac{2\pi x}{L} + \cdots \qquad (2.6.23)$$

If we proceed as before with each term of the series, and then sum the results together, we have

$$y_{\text{total}} = \frac{\delta_1}{1 - P/P_e}\sin\frac{\pi x}{L} + \frac{2^2\delta_2}{2^2 - P/P_e}\sin\frac{2\pi x}{L} + \cdots \qquad (2.6.24)$$

2.6.2 Perry–Robertson Formula

For an eccentrically loaded column, it has been shown that a formula (the Secant Formula) can be developed that relates the stress in the column to its slenderness ratio. For an initially crooked column, a similar approach can be taken to develop an equation called the *Perry–Robertson formula*,[5,6] which also relates the stress in the column to its slenderness ratio.

The maximum stress in an initially crooked column can be expressed as the sum of axial and bending stress.

$$\sigma_{\max} = \frac{P}{A} + \frac{M_{\max}c}{I} \qquad (2.6.25)$$

where

A = cross-section area
c = distance from neutral axis to extreme fiber
I = moment of inertia

The maximum moment for an initially crooked column occurs at midheight and is given by setting $x = L/2$ in Eq. (2.6.21)

$$M_{\max} = \frac{P\delta_0}{1 - P/P_e} \qquad (2.6.26)$$

Substitution of the $M_{\max}$ into the expression for $\sigma_{\max}$ and using $I = Ar^2$ gives

$$\sigma_{\max} = \frac{P}{A}\left[1 + \frac{\delta_0 c}{r^2}\frac{1}{1 - P/P_e}\right] \qquad (2.6.27)$$

This is called the *Perry–Robertson formula*. Again, using the first yielding of the material as the criterion of failure for the column, the ultimate or critical load can be determined from Eq. (2.6.27) by setting

$\sigma_{max} = \sigma_y$, i.e.,

$$\sigma_y = \frac{P_{cr}}{A}\left[1 + \frac{\delta_0 c}{r^2}\frac{1}{1 - P_{cr}/P_e}\right] \quad (2.6.28)$$

The Secant and Perry–Robertson formulas are, strictly speaking, valid only for very long or slender columns for which their slenderness ratios L/r are large, so that the stress in the column will remain in the elastic range at the buckling load. For shorter columns, material yielding is more important than geometrical imperfection. As a result, it is essential to consider the material imperfection in describing the behavior of shorter or stocky columns. The behavior of inelastic columns will be discussed in the next section.

2.7 INELASTIC COLUMNS

The discussion so far pertains to columns for which the material remains fully elastic and obeys Hooke's Law. This assumption is valid as long as the column is slender enough so that buckling occurs only at a stress level below the *proportional limit* of the stress-strain relationship of the material. For shorter columns, buckling will occur at a stress level above the proportional limit (Fig. 2.27). This type of buckling is referred to as the *inelastic buckling*. For columns that buckle inelastically, some of the fibers in the cross section have been yielded before buckling occurs. As a result, only the fibers that remain elastic are effective in resisting the

FIGURE 2.27 Critical stress above proportional limit

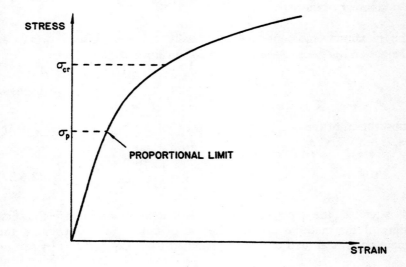

2.7 Inelastic Columns

additional applied force. Since only a portion of the cross section is effective in resisting the axial force at buckling, the elastic modulus E must be replaced by an *effective modulus* E_{eff} to describe the behavior of an inelastic column. In this section, we will discuss the buckling behavior of a perfectly straight column buckled in the inelastic range; in particular, we will discuss the *tangent modulus theory* and the *reduced modulus theory* proposed by Engesser[7] and the *inelastic column theory* of Shanley.[8]

2.7.1 Tangent Modulus Theory

The *tangent modulus theory* was proposed by Engesser[7] in 1889 to describe the buckling behavior of columns whose buckling stress is above the proportional limit of the material. The following assumptions are made in the tangent modulus theory:

1. The column is perfectly straight.
2. The ends of the column are pinned and the load is applied along the centroidal axis of the column.
3. The bending deformation of the column is small.
4. Plane sections before bending remain plane after bending.
5. During bending, no strain reversal (i.e., unloading of fibers) occurs across the cross section of the column.

For inelastic buckling, the stress in the fibers is above the proportional limit of the material (Fig. 2.27). In what follows we shall show that the *tangent modulus* E_t governs the behavior of the fibers during buckling of the column.

Figure 2.28a shows a pinned-pinned column buckling at the tangent modulus load P_t. The distributions of strain and stress across the cross section are shown in Fig. 2.28b. In the figure, σ_t and ε_t are, respectively, the stress and strain at the tangent modulus load before buckling. When the column buckles at the tangent modulus load, it is assumed that there is an increase in the axial force ΔP together with the bending moment ΔM. This increase in axial force ΔP combined with the incresing bending moment ΔM is such that it will cause an overall increase in axial strain across the section, so that no strain reversal will take place anywhere in the cross section. As a result, the tangent modulus E_t will govern the stress-strain behavior of all fibers of the cross section as shown in Fig. 2.28c.

The differential equation governing the behavior of this column can now be derived as follows:

For a column segment of length x from the support, the equation of equilibrium can be written as

$$-M_{int} + Py = 0 \quad (2.7.1)$$

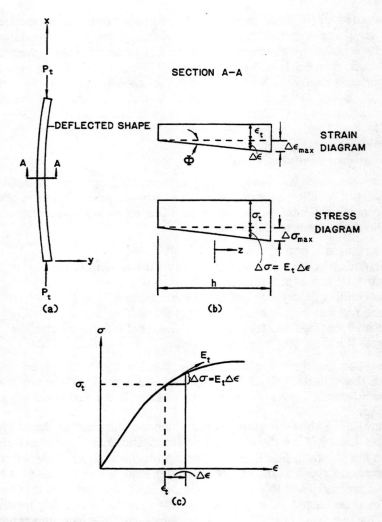

FIGURE 2.28 Tangent modulus theory

where P is the applied centroidal axial force and y is the distance from the line of action of the axial force to the centroidal axis of the cross section as the column bends.

The internal moment at the section due to bending has the general form

$$M_{\text{int}} = \int_A \sigma z \, dA \qquad (2.7.2)$$

2.7 Inelastic Columns

where σ is the longitudinal stress of a fiber in the cross section and z is the distance from that fiber to the centroidal axis of the cross section.

From the stress diagram in Fig. 2.28b, it can easily be seen that σ can be expressed as

$$\sigma = \sigma_t + \Delta\sigma = \sigma_t + \frac{\Delta\sigma_{max}}{h}\left(z + \frac{h}{2}\right) \qquad (2.7.3)$$

Substituting Eq. (2.7.3) into Eq. (2.7.2) gives

$$M_{int} = \int_A \left[\sigma_t + \frac{\Delta\sigma_{max}}{h}\left(z + \frac{h}{2}\right)\right] z \, dA \qquad (2.7.4)$$

or, expanding

$$M_{int} = \sigma_t \int_A z \, dA + \frac{\Delta\sigma_{max}}{h} \int_A z^2 \, dA + \frac{\Delta\sigma_{max}}{2} \int_A z \, dA \qquad (2.7.5)$$

Since the first moment of area about the centroidal axis is zero, i.e.,

$$\int_A z \, dA = 0 \qquad (2.7.6)$$

and realizing that

$$\int_A z^2 \, dA = I \qquad (2.7.7)$$

is the moment of inertia I of the cross section, we can write Eq. (2.7.5) as

$$M_{int} = \frac{\Delta\sigma_{max}}{h} I = \Delta M \qquad (2.7.8)$$

By substituting $\Delta\sigma_{max} = E_t \Delta\varepsilon_{max}$ in Eq. (2.7.8) and recognizing that $\Delta\varepsilon_{max}/h$ is the curvature Φ of the cross section (Fig. 2.28b), Eq. (2.7.8) becomes

$$M_{int} = E_t I \Phi \qquad (2.7.9)$$

Assuming a small deflection, the curvature Φ can be approximated by the second derivatives of the deflection or $\Phi = -y''$. As a result, Eq. (2.7.9) can be written in the usual form

$$M_{int} = -E_t I y'' \qquad (2.7.10)$$

Substituting the internal moment (2.7.10) into the equilibrium equation (2.7.1), we have

$$E_t I y'' + P y = 0 \qquad (2.7.11)$$

Equation (2.7.11) is the governing differential equation for an inelastic column developed on the basis of the tangent modulus theory.

The only difference between Eq. (2.7.11) and Eq. (2.2.10), is that E in Eq. (2.2.10) is replaced by E_t in Eq. (2.7.11). As a result, by following the same procedure as in Section 2.2, the critical load of this inelastic column based on the tangent modulus theory is

$$P_t = \frac{\pi^2 E_t I}{L^2} = \frac{E_t}{E} P_e \qquad (2.7.12)$$

The load expressed in Eq. (2.7.12) is called the *tangent modulus load*. We will show later that this load is the smallest load at which bifurcation of equilibrium of a perfectly straight column can take place in the inelastic range. The *effective modulus* in the tangent modulus theory is therefore the tangent modulus, E_t. The tangent modulus used in Eq. (2.7.12) depends only on the material property (i.e., stress-strain relationship of the material).

2.7.2 Double Modulus Theory

The *double modulus theory*, also referred to as the *reduced modulus theory*, was proposed by Engesser[7] in 1895, based on the concept given by Considere.[9]

The first four assumptions used for the development of tangent modulus theory are also used in the reduced modulus theory. However, the fifth assumption is different. In the reduced modulus theory, the axial force is assumed to remain constant during buckling; consequently, the bending deformation at buckling will cause strain reversal on the convex side of the column. The strain on the concave side of the column, on the other hand, continues to increase. As a result, the increments of stress and strain induced as a result of the bending of the column at the buckling load will be related by the elastic modulus on the convex side of the column, but the increments of stress and strain on the concave side of the column are related by the tangent modulus. Since two moduli, E and E_t, are necessary to describe the moment-curvature relationship of the cross section, the name *double modulus* was used. Because the double modulus is less than that of the elastic modulus that appeared in the Euler buckling formula, the double-modulus load will be less than that of the Euler buckling load. This will be demonstrated in what follows. Thus, the double-modulus load is also called the *reduced modulus load*.

Figure 2.29a shows the buckled shape of a centrally loaded pinned-ended inelastic column at the reduced modulus load, P_r. The corresponding strain and stress distributions are shown in Fig. 2.29b. The relationship between the increments of stress and strain as a result of bending deformation is shown in Fig. 2.29c. The governing differential equation describing the behavior of this column can be derived as follows.

2.7 Inelastic Columns

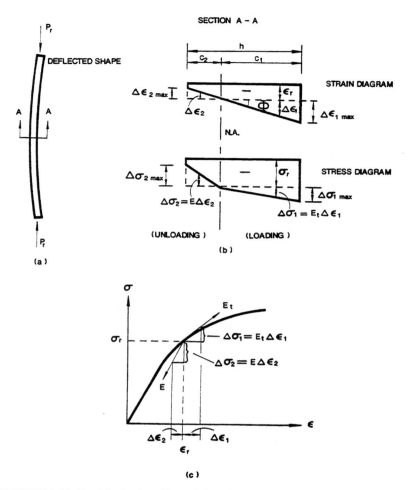

FIGURE 2.29 Double (reduced) modulus theory

For a column segment of length x from the support, the equation of equilibrium can be written as

$$-M_{int} + Py = 0 \qquad (2.7.13)$$

where P is the applied centroidal axial force and y is the distance from the line of action of the axial force to the centroidal axis of the cross section as the column bends.

The internal moment-curvature ($M_{int} - \Phi$) relationship for the cross section will be derived in the following. Because the axial force remains constant as the column buckles at the reduced (or double) modulus load,

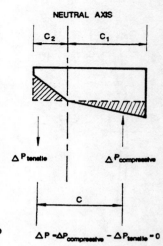

FIGURE 2.30 Forces due to bending according to the double modulus theory

the additional compressive force induced as a result of bending of the column must be equal to the additional tensile force such that the net increase in axial force ΔP is zero (Fig. 2.30), i.e.,

$$\Delta P_{\text{compressive}} = \Delta P_{\text{tensile}} \qquad (2.7.14)$$

This pair of forces $\Delta P_{\text{compressive}}$ and $\Delta P_{\text{tensile}}$ constitutes a couple and the internal resisting moment of the column is equal to this couple

$$M_{\text{int}} = \Delta P_{\text{compressive}}\, c = \Delta P_{\text{tensile}}\, c \qquad (2.7.15)$$

where c is the distance between this pair of forces.

Since the moment arm or the locations at which $\Delta P_{\text{compressive}}$ and $\Delta P_{\text{tensile}}$ act depends on the geometry of the cross section, we must specify the geometry of the cross section before we can proceed to evaluate the internal resisting moment.

Rectangular Cross Section

Considering a rectangular cross section with dimensions b and h as shown in Fig. 2.31, we can write

$$\Delta P_{\text{compressive}} = \tfrac{1}{2} b c_1 (\Delta \sigma_{1\max}) \qquad (2.7.16a)$$

$$\Delta P_{\text{tensile}} = \tfrac{1}{2} b c_2 (\Delta \sigma_{2\max}) \qquad (2.7.16b)$$

Equating these two forces,

$$\tfrac{1}{2} b c_1 \Delta \sigma_{1\max} = \tfrac{1}{2} b c_2 \Delta \sigma_{2\max} \qquad (2.7.17)$$

we obtain

$$\frac{c_1}{c_2} = \frac{\Delta \sigma_{2\max}}{\Delta \sigma_{1\max}} \qquad (2.7.18)$$

2.7 Inelastic Columns

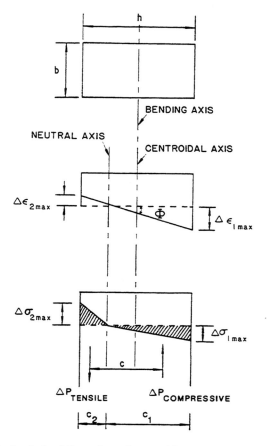

FIGURE 2.31 Inelastic buckling of a column with rectangular cross section

Since

$$\Delta\sigma_{1max} = E_t \Delta\varepsilon_{1max} = E_t c_1 \Phi \qquad (2.7.19)$$

$$\Delta\sigma_{2max} = E \Delta\varepsilon_{2max} = E c_2 \Phi \qquad (2.7.20)$$

where Φ is the curvature, we can write Eq. (2.7.18) as

$$\left(\frac{c_1}{c_2}\right)^2 = \frac{E}{E_t} \qquad (2.7.21)$$

Realizing that

$$c_1 + c_2 = h \qquad (2.7.22)$$

we can solve for c_1 and c_2 from Eqs. (2.7.21) and (2.7.22)

$$c_1 = \frac{h\sqrt{E}}{\sqrt{E}+\sqrt{E_t}} \tag{2.7.23}$$

$$c_2 = \frac{h\sqrt{E_t}}{\sqrt{E}+\sqrt{E_t}} \tag{2.7.24}$$

Using Eqs. (2.7.23) and (2.7.19), Eq. (2.7.16a) can be written as

$$\Delta P_{\text{compressive}} = \tfrac{1}{2}bh^2 \frac{EE_t}{(\sqrt{E}+\sqrt{E_t})^2}\Phi \tag{2.7.25}$$

Similarly, using Eq. (2.7.24) and (2.7.20), Eq. (2.7.16b) can be written as

$$\Delta P_{\text{tensile}} = \tfrac{1}{2}bh^2 \frac{EE_t}{(\sqrt{E}+\sqrt{E_t})^2}\Phi \tag{2.7.26}$$

Since the distance, c, between this pair of forces $\Delta P_{\text{compressive}}$ and $\Delta P_{\text{tensile}}$ for a rectangular section is

$$c = \tfrac{2}{3}h \tag{2.7.27}$$

the internal resistance moment, Eq. (2.7.15), has the value

$$M_{\text{int}} = \tfrac{1}{3}bh^3 \frac{EE_t}{(\sqrt{E}+\sqrt{E_t})^2}\Phi \tag{2.7.28}$$

or

$$M_{\text{int}} = E_r I \Phi \tag{2.7.29}$$

where

$$E_r = \frac{4EE_t}{(\sqrt{E}+\sqrt{E_t})^2} \tag{2.7.30}$$

is the *reduced modulus* for the *rectangular cross section*, and

$$I = \tfrac{1}{12}bh^3 \tag{2.7.31}$$

is the moment of inertia of the rectangular cross section.

Again, for a small deflection analysis, we can write Eq. (2.7.29) as

$$M_{\text{int}} = -E_r I y'' \tag{2.7.32}$$

Substitution of the above equation into the equilibrium equation (2.7.13)

$$E_r I y'' + Py = 0 \tag{2.7.33}$$

Equation (2.7.33) is the governing differential equation for the inelastic column derived on the basis of the reduced modulus theory.

Again, comparing Eq. (2.7.33) with Eq. (2.2.10), the only difference is that E in Eq. (2.2.11) is replaced by E_r in Eq. (2.7.33). Therefore, by

2.7 Inelastic Columns

following the same procedure as in Section 2.2, the critical load of the inelastic column is

$$P_r = \frac{\pi^2 E_t I}{L^2} = \frac{E_t}{E} P_e \qquad (2.7.34)$$

The load expressed in Eq. (2.7.34) is called the *reduced modulus load*. It is the largest load under which a real column can remain straight. This will be described further in the next section.

In the reduced modulus theory, the effective modulus of the column is of course the reduced modulus, E_r. Unlike the tangent modulus, the reduced modulus is a function of both material property and geometry of the cross section of a column. In other words, given the same material property, the reduced modulus will be different for different cross-sectional shapes. Thus, the expression for E_r given in Eq. (2.7.30) is only valid for rectangular cross sections.

Idealized I Section

For idealized symmetric I-sections (i.e., I-sections of equal flange areas connected by a web of negligible thickness), it can be shown that the reduces modulus is (see Problem 2.12)

$$E_r = \frac{2EE_t}{E + E_t} \qquad (2.7.35)$$

The reduced modulus, E_r, is always smaller than the elastic modulus, E, but larger than the tangent modulus, E_t, i.e.,

$$E_t < E_r < E \qquad (2.7.36)$$

hence

$$P_t < P_r < P_e \qquad (2.7.37)$$

2.7.3 Shanley's Inelastic Column Theory

Shanley's inelastic column theory[8] uses a simplified column model to explain the postbuckling behavior of an inelastic column. Recall in the tangent modulus theory, a slight increase in axial force is assumed at the onset of buckling, so that no *strain reversal* occurs in any cross section as the column bends at the tangent modulus load. On the other hand, in the reduced modulus theory, the axial force is assumed to remain constant at buckling, so that a complete strain reversal occurs at the convex side of the column as the column bends at the reduced modulus load. In Shanley's inelastic column theory, it is assumed that buckling is accompanied simultaneously by an increase in the axial force. Thus, at any instant as the column buckles, the net increase in axial force is not zero as postulated in the reduced modulus theory, but equal to a finite value

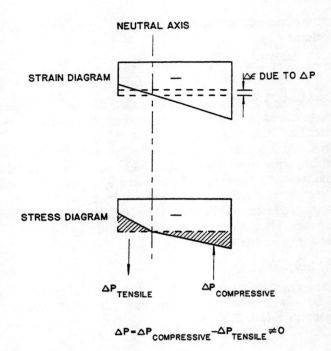

FIGURE 2.32 Shanley's inelastic column theory

given by the difference of $\Delta P_{\text{compressive}}$ and $\Delta P_{\text{tensile}}$ (Fig. 2.32). The magnitude of this increase in axial force ΔP is such that strain reversal may occur at the convex side of the column as shown in the stress diagram of Fig. 2.32.

By using a simple column model (Fig. 2.33) in which the column is represented by two rigid bars connected by a deformable cell consisting of two small longitudinal links, Shanley shows that the relationship between the applied load P and the midheight deflection d can be expressed as

$$P = P_t\left(1 + \frac{1}{\dfrac{b}{2d} + \dfrac{1+\tau}{1-\tau}}\right) \qquad (2.7.38)$$

where

P_t = tangent modulus load of the column
b = width of the column cross section
$\tau = E_t/E$, in which E_t is the tangent modulus

In deriving Eq. (2.7.38), Shanley assumes that the column begins to

2.7 Inelastic Columns

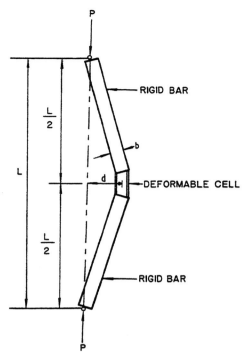

FIGURE 2.33 Shanley's inelastic column model

bend as soon as P_t is reached. Thus, Eq. (2.7.38) gives the postbuckling behavior of the column, i.e., the behavior of the column when $P > P_t$.

Figure 2.34 shows a plot of Eq. (2.7.38). It can be seen that as P increases above P_t, there is an increase in midheight deflection d. As d becomes large, P approaches P_r, the reduced modulus load. It should be remembered that Eq. (2.7.38) was developed on the basis of the simple column model (Fig. 2.33). For a real column in which E_t varies across the cross section and along the length of the column, the load-deflection behavior of the column will follow the dashed lines in Fig. 2.34. An important observation is that the maximum load of a really perfectly straight inelastic column lies somewhere between the tangent modulus load and the reduced modulus load. Hence, the *tangent modulus* load represents a *lower bound* and the *reduced modulus* load represents an *upper bound* to the strength of a concentrically loaded, perfectly straight inelastic column. Experiments on real columns show that their maximum strengths usually fall closer to the tangent modulus load than to the reduced modulus load. This is because unavoidable imperfections always exist in real columns along with accidental load eccentricities during the testings. Both of these effects tend to lower the strength of real columns.

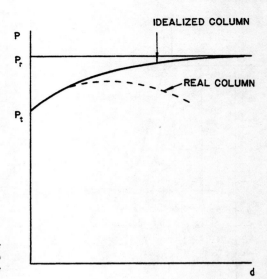

FIGURE 2.34 Postbuckling behavior of column according to Shanley's inelastic column theory

Because of this, and because of the ease with which the tangent modulus load can be obtained when compared to the reduced modulus load, the tangent modulus is usually adopted in practice to represent the ultimate strength of a centrally loaded real column.

In deriving the inelastic buckling loads of the column (P_r and P_t), the ends of the column are assumed to be pinned. If the end conditions of the column are not pinned, the concept of effective length should be applied with the term L on Eqs. (2.7.12) and (2.7.34) replaced by KL, the effective length of the column.

2.8 DESIGN CURVES FOR ALUMINUM COLUMNS

The tangent modulus concept discussed in the preceding section can be used directly to construct column curves for the design of aluminum columns. A *column curve* is a curve that gives the critical buckling load or critical buckling stress (σ_{cr}) of a column as a function of its slenderness ratio (KL/r).

If we divide both sides of the tangent modulus load equation by A, the cross-section area of the column, we obtain

$$\frac{P_t}{A} = \frac{\pi^2 E_t I}{A(KL)^2} \qquad (2.8.1)$$

Denoting

$$\sigma_{cr} = \frac{P_t}{A} \qquad (2.8.2)$$

2.8 Design Curves for Aluminum Columns

and recognizing that $I = Ar^2$, where r is the radius of gyration of the cross section, Eq. (2.8.1) can be written as

$$\sigma_{cr} = \frac{\pi^2 E_t}{(KL/r)^2} \tag{2.8.3}$$

To plot σ_{cr} versus KL/r using Eq. (2.8.3), we have to know E_t, which, in turn, is a function of σ_{cr}. Thus, to obtain a proper value of E_t, we must first know the stress–strain relationship of the material. The stress–strain relationship of aluminum alloys as obtained from experiments (coupon tests) can best be fitted by the Ramberg–Osgood equation.[10]

The Ramberg–Osgood equation has the form

$$\varepsilon = \frac{\sigma}{E} + 0.002 \left(\frac{\sigma}{\sigma_{0.2}}\right)^n \tag{2.8.4}$$

where

E = elastic Young modulus
$\sigma_{0.2}$ = 0.2% offset yield stress
n = hardening parameter

Since E_t is the slope of the $\sigma - \varepsilon$ curve, we can therefore determine E_t from Eq. (2.8.4) by a direct differentiation

$$E_t = \frac{d\sigma}{d\varepsilon} = \frac{E}{\left[1 + \frac{0.002nE}{\sigma_{0.2}}\left(\frac{\sigma}{\sigma_{0.2}}\right)^{n-1}\right]} \tag{2.8.5}$$

Figure 2.35 shows the nondimensional stress–strain curve of the aluminum alloy 6061-T6 (described in reference 11) with $E = 10,100$ ksi (69,640 MPa), $\sigma_{0.2} = 40.15$ ksi (277 MPa), and $n = 18.55$. Using Eq. (2.8.5), the values of E_t for any given stress σ can be determined. A plot of E_t/E versus $\sigma/\sigma_{0.2}$ is shown in Fig. 2.36. Using this figure, the column curve for this aluminum alloy can be obtained as follows:

1. Pick a value of σ, called σ_{cr}.
2. Obtain E_t from Eq. (2.8.5) or from Fig. 2.36.
3. Calculate the slenderness ratio from the equation

$$\frac{KL}{r} = \pi \sqrt{\frac{E_t}{\sigma_{cr}}} \tag{2.8.6}$$

which is simply a rearrangement of Eq. (2.8.3).

Following the procedures outlined above, for any given σ_{cr}, a corresponding value of KL/r can be obtained. The variation of σ_{cr} with KL/r is shown in Fig. 2.37.

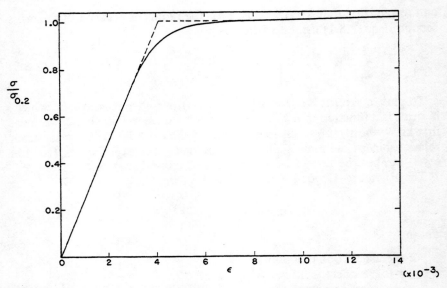

FIGURE 2.35 Stress–strain curve of aluminum alloy 6061-T6

This curve can be used directly for column design. To select a section to carry a specific load, the designer first picks a trial section, calculates the value KL/r, and reads the corresponding σ_{cr} from Fig. 2.37. If this σ_{cr} is greater than the actual stress $\sigma = P/A$ acting on the column, the section is considered satisfactory. Usually in design, a safety margin is established by either lowering the value of critical stress σ_{cr} and/or by increasing the value of applied stress σ so that σ_{cr} is guaranteed to exceed σ.

Note that the column curve in Fig. 2.37 is applicable only to columns of aluminum alloy 6061-T6. For other types of aluminum alloys, the stress–strain curve will be different and hence the column curve will be different. In fact, the column curve is extremely sensitive to the shape of the stress–strain curve. For instance, if we approximate the stress–strain curve of Fig. 2.35 (solid line) by two straight lines (dotted lines in figure), the corresponding column curve will be like the dotted lines in Fig. 2.37. As can be seen, there is a large discrepancy between the actual (solid line) and approximate (dotted lines) column curves. The apparent discontinuity of the approximate column curves in Fig. 2.37 is due to the sudden change in the value of E_t as the initial slope of the approximate stress–strain curve in Fig. 2.35 is replaced by a very much shallower slope.

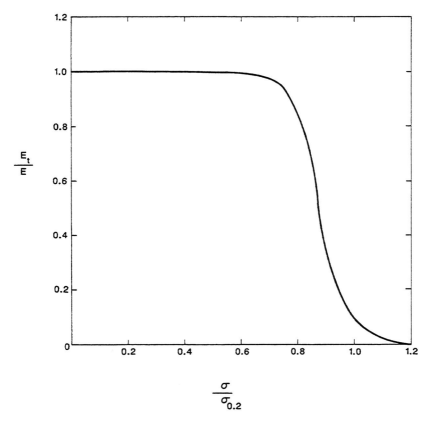

FIGURE 2.36 Variation of tangent modulus with stress for aluminum alloy 6061-T6

2.9 STUB COLUMN STRESS–STRAIN CURVE

The stub (or short) column stress–strain curve for a steel member is generally used to determine the critical load of the steel column directly from the tangent modulus equation (2.7.12). The value of E_t in the formula is evaluated from the slope of the *stub column stress–strain curve*. This curve can be obtained by one of the following two methods

1. Experimental
2. Numerical

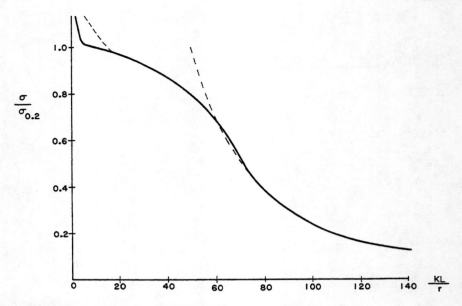

FIGURE 2.37 Column curve for aluminum alloy 6061-T6

2.9.1 Experimental Stub-Column Stress–Strain Curve

If a simple compression test is performed on a coupon cut from a steel member, the stress–strain relationship of the coupon will exhibit a linear elastic–perfectly plastic behavior as shown in Fig. 2.38 (dashed lines). However, if the same test is performed on a short length of column (*stub column test*) and the average stress $\sigma_{av} = P/A$ is plotted against the axial strain ε, the $\sigma - \varepsilon$ relationship of the stub-column test will deviate from that of the coupon as shown in Fig. 2.38 (solid line). In particular, after a certain average stress σ_{av} has been reached, the stress–strain relationship of the stub column follows the curve ABC instead of AEC. This phenomenon is attributed to the presence of "lock-in" or *residual stresses* in the steel column. Residual stresses are created in the column in the following way: Steel members are usually heated at some stage during the fabrication process. As they cool down, the part of the cross section for which the surface area to volume ratio is the largest will lose heat more rapidly than the part for which the ratio of surface area to volume is small. This uneven cooling creates a set of *self-equilibrating stresses* in the cross section. These are what is called the residual stresses.

For hot-rolled wide-flange shapes, which are used extensively in building construction, the toes of the flanges have a larger surface area to volume ratio than the regions where the web joins the flanges, hence the

2.9 Stub Column Stress–Strain Curve

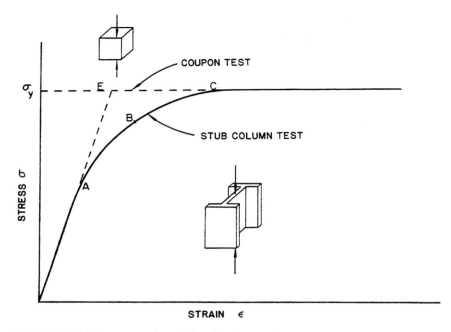

FIGURE 2.38 Stress–strain relationship for steel

toes of the flanges will cool faster. As the junctions of the web and flanges begin to cool and shrink, the toes of the flanges, which have been cooled and hardened already, will prevent the junctions from shrinking, with the result that the junctions of the web and flanges will be left in tension while the toes of the flanges will be left in compression. As for the web, if the height to thickness ratio is large, then the central portion will cool much faster than the portion where the web joins the flanges, and so a compressive residual stress will be induced at the central portion. On the other hand, if the height to thickness ratio is small, then cooling will be more uniform, so that the whole web will be in a state of tension.

Figure 2.39 shows schematically the residual stress distributions on the flanges and web of a W8 × 31 hot-rolled section.[12] As an axial force P is applied to the sections, the stress distribution over the cross section will change in several stages, as shown in Fig. 2.40. As the stress in any fiber equals or exceeds the yield stress, that particular fiber will yield, and any additional load will be carried by the fibers that are still elastic. From the figure, it is clear that yielding over the cross section is a gradual process. The fibers that have the highest value of compressive residual stresses will yield the soonest, followed by the fibers that have a lower value of

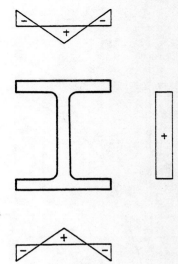

FIGURE 2.39 Residual stress distribution of a W8 × 31 section

compressive residual stresses. Finally, the fibers that have tensile residual stresses will yield as the applied load further increases. Because of this gradual yielding process (plastification) over the cross section, the stress–strain curve of a stub column follows the rather smooth curve ABC in Fig. 2.38.

The main difference between the aluminum and steel members is that for aluminum members the effect of residual stress is negligible and the nonlinear stress–strain behavior is due primarily to the material behavior. This nonlinearity shows up in a coupon test. For steel members, coupon tests show an elastic–perfectly plastic behavior, but the stub-column tests show a gradual yielding because of the presence of residual stresses.

If a stub column stress–strain curve (obtained either experimentally or numerically) is available, then the critical load of the steel column can be determined directly from the tangent modulus formula Eq. (2.7.12) with a simple modification for the value of E_t that is evaluated from the slope of the *stub column stress–strain curve*.

2.9.2 Numerical Stub Column Stress–Strain Curve

To generate the stub column stress–strain curve numerically, the cross section is first divided into a number of small elements as shown in Fig. 2.41. Denote A_j^e as the area of the j element and A_{el} as the remaining area of the cross section that is still elastic, the stub column stress–strain

2.9 Stub Column Stress–Strain Curve

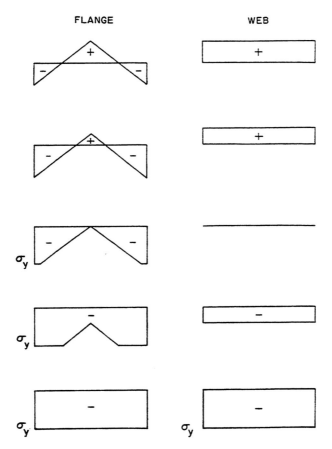

FIGURE 2.40 Change in stress distributions on the flanges and web as P increases

curve can be traced numerically as follows:

1. Specify a strain increment $\Delta \varepsilon^i$ (a negative quantity) at the i cycle.
2. Calculate the stress increment $\Delta \sigma^i$ from

$$\Delta \sigma^i = E \, \Delta \varepsilon^i \qquad (2.9.1)$$

   for every element that is still elastic.
3. Calculate the current state of stress σ^i from

$$\sigma^i = \sigma_r + \Sigma \, \Delta \sigma^i \qquad (2.9.2)$$

   for every element that is still elastic. The stress σ_r is the value of residual stress for that particular element and is taken as positive if the

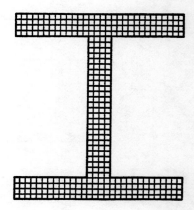

FIGURE 2.41 Discretization of cross section

residual stress is tensile and negative if the residual stress is compressive.
4. Check whether a particular element has been yielded.
   a. If $|\sigma^i| = \sigma_y$, the element has yielded. Subtract the area A_j^e of this j element from the area of the cross section that is still elastic, i.e.,

   $$A_{el}^i = A_{el}^{i-1} - \Sigma A_j^e \qquad (2.9.3)$$

   where

   A_{el}^i = area of the cross section that is still elastic at the i cycle
   A_{el}^{i-1} = area of the cross section that is still elastic at the end of the previous cycle

   Proceed to the next element.
   b. If $|\sigma^i| < \sigma_y$, the element is still elastic. Proceed to the next element.
   c. If $|\sigma^i| > \sigma_y$, the state of stress of the element is larger than the yield stress. We need to scale down the strain increment $\Delta \varepsilon^i$ by a factor r given by

   $$r = \frac{\sigma_y - \sigma^{i-1}}{\Delta \sigma^i} \qquad (2.9.4)$$

   in which σ^{i-1} is the state of stress of the element at the end of the previous cycle. Go back to Step 2 with $\Delta \varepsilon^i$ replaced by $r \Delta \varepsilon^i$.
5. Calculate the load increment ΔP^i corresponding to the strain increment $\Delta \varepsilon^i$ from

   $$\Delta P^i = \Delta \sigma^i A_{el}^i = E A_{el}^i \Delta \varepsilon^i \qquad (2.9.5)$$

6. Calculate the *average* stress increment $\Delta \sigma_{av}^i$ at the end of the i cycle by

   $$\Delta \sigma_{av}^i = \Delta P^i / A \qquad (2.9.6)$$

2.10 Column Curves of Idealized Steel I-Section

7. Evaluate the current state of average stress of the cross section

$$\sigma_{av}^i = \sigma_{av}^{i-1} + \Delta\sigma_{av}^i \qquad (2.9.7)$$

8. Finally, evaluate the current state of strain of the cross section from

$$\varepsilon^i = \varepsilon^{i-1} + \Delta\varepsilon^i \qquad (2.9.8)$$

Thus, for each value of strain calculated in Step 8, there is a corresponding value of average stress calculated in Step 7. By repeating the process, an average stress versus strain curve can be plotted. This numerically generated $\sigma - \varepsilon$ curve can be used in conjunction with Eq. (2.7.12) to obtain P_t for the column. E_t is obtained as the slope of this stress–strain curve.

E_t can also be represented by the expression EA_{el}/A. This is demonstrated as follows:
For a given load increment ΔP, we can write

$$\Delta P = A \, \Delta\sigma_{av} \qquad (2.9.9)$$

where A is the area of the cross section and $\Delta\sigma_{av}$ is the average stress increment of the cross section. On the other hand, ΔP is related to $\Delta\varepsilon$ by

$$\Delta P = EA_{el} \, \Delta\varepsilon \qquad (2.9.10)$$

where E is the elastic modulus and A_{el} is the area of the elastic cross section.
Equating Eqs. (2.9.9) and (2.9.10), we have

$$A \, \Delta\sigma_{av} = EA_{el} \, \Delta\varepsilon \qquad (2.9.11)$$

and

$$\frac{\Delta\sigma_{av}}{\Delta\varepsilon} = \frac{EA_{el}}{A} \qquad (2.9.12)$$

Since E_t is the slope of the average stress–strain curve, it follows that

$$E_t = \frac{d\sigma_{av}}{d\varepsilon} \approx \frac{\Delta\sigma_{av}}{\Delta\varepsilon} = \frac{EA_{el}}{A} \qquad (2.9.13)$$

Note that the ratio A_{el}/A can easily be calculated in the numerical procedure described above at any stress level σ_{av}^i by forming the ratio A_{el}^i/A at the end of Step 4.

2.10 COLUMN CURVES OF IDEALIZED STEEL I-SECTION

In Section 2.8, we discussed the column curves for aluminum columns. For aluminum columns, the nonlinear stress–strain behavior is due primarily to material nonlinearity. This nonlinear stress–strain relationship can be approximated closely by the Ramberg–Osgood equation [Eq.

(2.8.4)]. Upon differentiation of this equation, an expression for E_t can be obtained [Eq. (2.8.5)]. With the tangent modulus known, an aluminum-column curve can be constructed by the procedure outlined in that section.

For steel columns, the stress–strain behavior of a coupon is elastic–perfectly plastic as shown in Fig. 2.38. As a result, the tangent modulus can not be obtained directly from this curve. Instead, it should be obtained from the slope of a stub-column stress–strain curve (Fig. 2.38).

If a stub-column stress–strain curve (obtained either experimentally or numerically) is available, then the critical load of the steel column can be determined directly from the tangent modulus formula [Eq. (2.7.12)], with E_t determined from the slope of the stub-column stress–strain curve. However, if a stub-column stress–strain curve is not available, an analytical approach can be used to find P_{cr} for a given value of KL/r in the inelastic range (for KL/r in the elastic range, the critical load P_{cr} is the Euler load, P_e). In this approach, an idealized I-section is used. An idealized I-section is an I-section with negligible web thickness. In addition, an assumed residual stress distribution is also used. For illustration purposes, an idealized I-section with linear, varying residual stress is shown in Fig. 2.42. In the figure, σ_{rc} denotes the value of compressive residual stress at the toe of the flanges and σ_{rt} the value of tensile residual stress at the flange-web junctions. To maintain self-equilibrium in the absence of an externally applied force, the tensile and compression forces resulting from the tensile and compressive residual stresses must be in self-balance. If the flanges of the idealized I-section are identical and are of uniform thickness, we must have

$$\sigma_{rc} = \sigma_{rt} = \sigma_r \qquad (2.10.1)$$

Now, if an external force is applied concentrically to this column, portions of the cross section will yield, leaving only part of the cross section elastic. The yielded portion cannot carry any additional load for an idealized elastic–perfectly plastic, stress–strain relationship. As a result, only the elastic core (the elastic portion of the cross section) will be effective in resisting the applied load. In other words, only the elastic part of the cross section will provide the flexural rigidity to the column as it buckles. Based on this argument, the critical load of the column can be written as

$$P_{cr} = \frac{\pi^2 E I_e}{(KL)^2} = \frac{I_e}{I} P_e \qquad (2.10.2)$$

where

I_e = moment of inertia of the elastic core of the column cross section
P_e = Euler buckling load

2.10 Column Curves of Idealized Steel I-Section

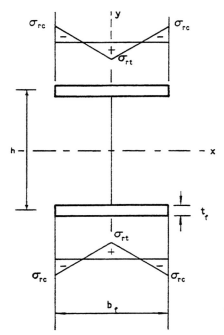

FIGURE 2.42 Assumed residual-stress distribution of an idealized I-section

The ratio I_e/I in Eq. (2.10.2) depends on (*i*) the axis of bending of the column, (*ii*) the shape of the cross section, and (*iii*) the distribution of residual stresses over the cross section.

For an idealized I-section with a linear variation of residual stresses (Fig. 2.42), the ratio I_e/I can be established as follows:

Strong Axis Bending

If the moment of inertia about the centroidal axes of the flanges are ignored, the ratio I_e/I about the strong ($x - x$) axis can be written as

$$\frac{I_e}{I} = \frac{2(b_{fe}t_f)h^2/4}{2(b_f t_f)h^2/4} = \frac{b_{fe}}{b_f} \tag{2.10.3}$$

The quantities b_{fe}, t_f, and h are as shown in Fig. 2.43.

Weak Axis Bending

The ratio I_e/I for the column bend about the weak ($y - y$) axis is

$$\frac{I_e}{I} = \frac{2t_f b_{fe}^3/12}{2t_f b_f^3/12} = \left(\frac{b_{fe}}{b_f}\right)^3 \tag{2.10.4}$$

Using Eqs. (2.10.3) and (2.10.4), we obtain from Eq. (2.10.2):

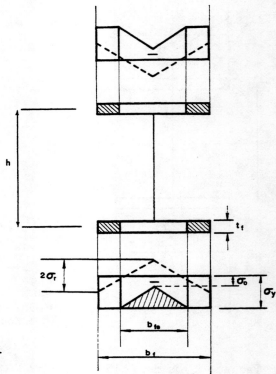

FIGURE 2.43 Partially plastified cross section

For strong axis bending

$$P_{cr} = \frac{b_{fe}}{b_f} P_e \qquad (2.10.5)$$

For weak axis bending

$$P_{cr} = \left(\frac{b_{fe}}{b_f}\right)^3 P_e \qquad (2.10.6)$$

It is obvious from Eqs. (2.10.5) and (2.10.6) that the critical loads are different for the same column bent about different axes. Since neither P_{cr} nor b_{fe} is known, we need another equation to relate P_{cr} and b_{fe}. This equation can be obtained by writing the expression for the axial force acting on the column that corresponds to a partially plastified cross section (Fig. 2.43).

$$P = 2[\sigma_y b_f t_f - \tfrac{1}{2}(\sigma_y - \sigma_0) b_{fe} t_f] \qquad (2.10.7)$$

In Eq. (2.10.7), σ_0 is the stress at the middle of the flange. From

2.10 Column Curves of Idealized Steel I-Section

similar triangles, it can be shown then

$$\frac{\sigma_y - \sigma_0}{b_{fe}/2} = \frac{2\sigma_r}{b_f/2} \tag{2.10.8}$$

or

$$\sigma_y - \sigma_0 = 2\sigma_r \frac{b_{fe}}{b_f} \tag{2.10.9}$$

Substituting Eq. (2.10.9) into Eq. (2.10.7) gives

$$P = 2\left[\sigma_y b_f t_f - \sigma_r \frac{b_{fe}^2 t_f}{b_f}\right] \tag{2.10.10}$$

or

$$P = A\left[\sigma_y - \sigma_r \left(\frac{b_{fe}}{b_f}\right)^2\right] \tag{2.10.11}$$

where $A = 2b_f t_f$ is the area of the cross section of the idealized I-section. Rearranging and realizing that $\sigma_{av} = P/A$ is the average stress over the cross section, we have

$$\frac{b_{fe}}{b_f} = \sqrt{\frac{\sigma_y - \sigma_{av}}{\sigma_r}} \tag{2.10.12}$$

On substituting Eq. (2.10.12) into Eqs. (2.10.5) and (2.10.6), we have for strong axis bending

$$P_{cr} = \sqrt{\frac{\sigma_y - \sigma_{av}}{\sigma_r}} P_e \tag{2.10.13}$$

or divide both sides by $A\sigma_y$

$$\frac{\sigma_{cr}}{\sigma_y} = \sqrt{\frac{\sigma_y - \sigma_{av}}{\sigma_r}} \bigg/ \lambda_c^2 \tag{2.10.14}$$

and, for weak axis bending,

$$P_{cr} = \left(\frac{\sigma_y - \sigma_{av}}{\sigma_r}\right)^{3/2} P_e \tag{2.10.15}$$

or

$$\frac{\sigma_{cr}}{\sigma_y} = \left(\frac{\sigma_y - \sigma_{av}}{\sigma_r}\right)^{3/2} \bigg/ \lambda_c^2 \tag{2.10.16}$$

in which $\lambda_c = \frac{1}{\pi}\sqrt{\frac{\sigma_y}{E}}(KL/r)$ is the *slenderness parameter*.

For hot-rolled, wide-flange sections $\sigma_r \approx 0.3\sigma_y$. Using this value for σ_r and specifying a value of $\sigma_{cr}(= \sigma_{av})$, a corresponding value for λ_c can be determined. By repeating this process, a column curve for σ_{cr}/σ_y versus λ_c can be plotted. Figure 2.44 shows such a plot. Note the difference in

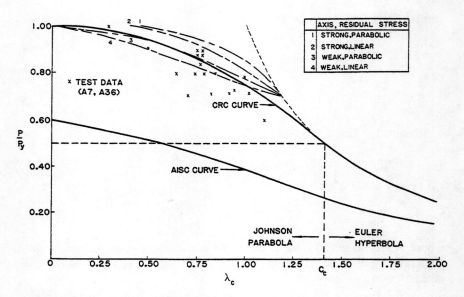

FIGURE 2.44 Column curves (theoretical, CRC, and AISC)

load-carrying capacity of the column for bending about different axes. In general, for hot-rolled, wide-flange sections, the load-carrying capacity of the column is larger for strong axis bending than that for weak axis bending. This is because the detrimental effect of compressive residual stress at the tips of the flanges is more pronounced for weak axis bending than that of strong axis bending.

Also shown in the figure are the column curves for I-sections with *parabolic* residual-stress distributions. Again, it is obvious from the figure that the distributions of residual stresses in a cross section have an influence on column strength. Extensive research at Lehigh[12–15] has shown that the distributions of residual stresses for hot-rolled I-sections usually fall between that of the linear and parabolic types. As a result, they represent upper and lower bounds to the strength of hot-rolled H-columns.

2.11 DESIGN CURVES FOR STEEL COLUMNS

2.11.1 Column Design Curves

Column Research Council Curve

On the basis of both column curves developed previously for the idealized I-shaped columns with linear and parabolic residual stress distributions as well as test results of a number of small and medium-size

2.11 Design Curves for Steel Columns

hot-rolled, wide-flange shapes of mild structural steel, the Column Research Council (CRC) recommended in the first and second editions of the Guide[16] a parabola of the form

$$\sigma_{cr} = \sigma_y - B\left(\frac{KL}{r}\right)^2 \qquad (2.11.1)$$

to represent the column strength in the inelastic range. The column strength in the elastic range, however, is represented by the Euler formula. The point of demarcation between inelastic and elastic behavior is chosen to be $\sigma_{cr} = 0.5\sigma_y$. The number 0.5 is chosen as a conservative measure of the maximum value of compressive residual stress present in hot-rolled, wide-flange shapes, which is about $0.3\sigma_y$. To obtain a smooth transition from the parabola to the Euler curve as well as to maintain a compromise between the strength of columns bent about the strong and weak axes, the constant B in Eq. (2.11.1) is chosen to be $\sigma_y^2/4\pi^2 E$. The slenderness ratio that corresponds to $\sigma_{cr} = 0.5\sigma_y$ is designated as C_c where

$$C_c = \sqrt{\frac{2\pi^2 E}{\sigma_y}} \qquad (2.11.2)$$

Thus, for columns with slenderness ratios less than or equal to C_c, the CRC curve assumes the shape of a parabola and for slenderness ratio exceeding C_c, the CRC curve takes the shape of a hyperbola, i.e.,

$$\sigma_{cr} = \begin{cases} \sigma_y\left[1 - \dfrac{(KL/r)^2}{2C_c^2}\right] & \dfrac{KL}{r} \leq C_c \\ \dfrac{\pi^2 E}{\left(\dfrac{KL}{r}\right)^2} & \dfrac{KL}{r} > C_c \end{cases} \qquad (2.11.3)$$

For comparison purposes, Eq. (2.11.3) is rewritten in its load form in terms of the nondimensional quantities P/P_y and λ_c, in which P_y is the yield load given by $P_y = A\sigma_y$ and λ_c is the slenderness parameter given by $\lambda_c = (KL/r)\sqrt{\sigma_y/\pi^2 E}$

$$\frac{P}{P_y} = \begin{cases} 1 - 0.25\lambda_c^2 & \lambda_c \leq \sqrt{2} \\ \lambda_c^{-2} & \lambda_c > \sqrt{2} \end{cases} \qquad (2.11.4)$$

The CRC curve is plotted in Fig. 2.45 in its nondimensional form [Eq. (2.11.4)].

AISC Allowable Stress Design Curve

The CRC curve divided by a variable factor of safety of

$$\frac{5}{3} + \frac{3}{8}\left(\frac{KL/r}{C_c}\right) - \frac{1}{8}\left(\frac{KL/r}{C_c}\right)^3 = \frac{5}{3} + \frac{3}{8}\left(\frac{\lambda_c}{\sqrt{2}}\right) - \frac{1}{8}\left(\frac{\lambda_c}{\sqrt{2}}\right)^3 \qquad (2.11.5)$$

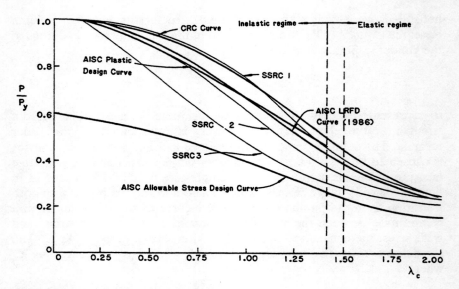

FIGURE 2.45 Column-design curves

in the inelastic range and a constant factor of safety of 23/12 in the elastic range gives the AISC Allowable Stress Design (ASD) curve. The factors of safety are employed to account for geometrical imperfections and load eccentricities that are unavoidable in real columns. The AISC-ASD curve is also plotted in Fig. 2.45. The ASD column curve is used in conjunction with the ASD format given by

$$\frac{R_n}{F.S.} \geq \sum_{i=1}^{m} Q_{ni} \qquad (2.11.6)$$

where

R_n = nominal resistance. (For column design, $R_n/F.S.$ is represented by the ASD column curve)
Q_n = service loads.

AISC Plastic Design Curve

The ASD curve multiplied by a factor of 1.7 forms the AISC Plastic Design (PD) curve (Fig. 2.45). In plastic design, only the inelastic regime of the curve is used because of the slenderness requirement. The design

2.11 Design Curves for Steel Columns

format for plastic design of columns is thus

$$\frac{1.7R_n}{F.S.} \geq \gamma \sum_{i=1}^{m} Q_{ni} \qquad (2.11.7)$$

where γ is the *load factor* used in the present AISC-PD Specification. The values for γ are $\gamma = 1.7$ for live and dead loads only, and $\gamma = 1.3$ for live and dead loads acting in conjunction with wind or earthquake loads.

Structural Stability Research Council Curves

Both the ASD curve and PD curve are originated from the CRC curve, which was developed on the basis of the bifurcation concept that assumes the column to be perfectly straight. Although the effect of residual stress is explicitly accounted for, the effect of geometrical imperfections is only accounted for implicitly by applying a variable factor of safety to the basic strength curve.

Realizing that perfectly straight columns are rarely encountered in real life, researchers[17,18] investigated theoretically and experimentally the strength and stability of initially crooked imperfect columns. It is evident from the results of these studies that the strengths of different types of steel columns, which are the result of different manufacturing and fabrication processes, different sizes and steel grades, and different axes of bending, may vary considerably, so that multiple-design curves may be desirable.

On the basis of a computer model developed for a geometrically imperfect column with an initial out-of-straightness at midheight equal to 0.001L, and with actual measured values of residual stresses, a set of three multiple-column strength curves from a total of 112 columns being investigated was developed.[17] Each of these curves is representative of the strength of a related category of columns. In the categories covered by these column curves we find hot-rolled and cold-straightened members, wide-flange and box shapes, as well as round bars and members composed of welded plates.

The Structural Stability Research Council (SSRC), in its third edition of the Guide,[19] presents these three column curves along with the former one (CRC curve).

The expressions for the three SSRC curves are the following:

Curve 1

$$\frac{P}{P_y} = \begin{cases} 1 \text{ (yield level)} & (0 \leq \lambda_c \leq 0.15) \\ 0.990 + 0.122\lambda_c - 0.367\lambda_c^2 & (0.15 \leq \lambda_c \leq 1.2) \\ 0.051 + 0.801\lambda_c^{-2} & (1.2 \leq \lambda_c \leq 1.8) \\ 0.008 + 0.942\lambda_c^{-2} & (1.8 \leq \lambda_c \leq 2.8) \\ \lambda_c^{-2} \text{ (Euler buckling)} & (\lambda_c \geq 2.8) \end{cases} \qquad (2.11.8a)$$

Curve 2

$$\frac{P}{P_y} = \begin{cases} 1 \text{ (yield level)} & (0 \leq \lambda_c \leq 0.15) \\ 1.035 - 0.202\lambda_c - 0.222\lambda_c^2 & (0.15 \leq \lambda_c \leq 1.0) \\ -0.111 + 0.636\lambda_c^{-1} + 0.087\lambda_c^{-2} & (1.0 \leq \lambda_c \leq 2.0) \\ 0.009 + 0.877\lambda_c^{-2} & (2.0 \leq \lambda_c \leq 3.6) \\ \lambda_c^{-2} \text{ (Euler buckling)} & (\lambda_c \geq 3.6) \end{cases} \quad (2.11.8b)$$

Curve 3

$$\frac{P}{P_y} = \begin{cases} 1 \text{ (yield level)} & (0 \leq \lambda_c \leq 0.15) \\ 1.093 - 0.622\lambda_c & (0.15 \leq \lambda_c \leq 0.8) \\ -0.128 + 0.707\lambda_c^{-1} - 0.102\lambda_c^{-2} & (0.8 \leq \lambda_c \leq 2.2) \\ 0.008 + 0.792\lambda_c^{-2} & (2.2 \leq \lambda_c \leq 5.0) \\ \lambda_c^{-2} \text{ (Euler buckling)} & (\lambda_c \geq 5.0) \end{cases} \quad (2.11.8c)$$

These equations were obtained by curve-fitting a parabola or hyperbola to the designated characteristic column curves that are the arithmetic mean curves of the three divided categories. These column curves are used in conjunction with the *Column Selection Table* shown in Table 2.3.

The curves in each category as developed based on the *stability analysis*. In the stability analysis, the complete load-deflection behavior of the column is traced from the start of loading to the ultimate state. Hence, a stability analysis is also known as a load-deflection analysis. The peak point of this load-deflection curve is the maximum load the column can carry. As mentioned previously, the stability analysis is quite different from that of the bifurcation analysis. In the bifurcation analysis, the load that corresponds to the state of bifurcation of equilibrium is calculated using an eigenvalue analysis. However, for columns that are initially crooked, lateral deflection begins as soon as the load is applied and so there is no distinct point of bifurcation. Stability analysis of columns will be discussed in Chapter 6. Usually, recourse must be had to numerical method for solutions.

For comparisons, the three SSRC curves are also plotted with the CRC, ASD, and PD curves in Fig. 2.45. It can be seen that these curves belly down in the intermediate slenderness range $(0.75 < \lambda_c < 1.25)$ because of the combined maximum detrimental effects of both residual stresses and initial crookedness on the column strength as predicted by the computer model. Tests of real columns have shown that the detrimental effects of residual stresses and initial crookedness are not always synergistic and so the SSRC curves with "belly down" in the intermediate slenderness range will be too conservative for most columns in building frames.

Table 2.3 SSRC Multiple Column Curves Selection Table (Numbers in parentheses may be subjected to later change)

Fabrication details				Bending axis	A7 A36	A 242	A572 50	A572 65	A 441	A 514	Steel grade (ASTM) A514 FL A441 WEB	Hybrid A514 FL A36 WEB	Hybrid A441 FL A36 WEB
Rolled wide-flange		Light		Major	2	1	(1)	(1)	(1)	1	—	—	—
				Minor	2	2	(2)	(2)	(2)	1	—	—	—
		Heavy		Major	3	(2)	(2)	(2)	(2)	1	—	—	—
				Minor	3	(2)	(2)	(2)	(2)	1	—	—	—
Welded built-up H	Light	Flame-cut		Major	2	(2)	2	(2)	(2)	1	1	1	2
				Minor	2	(2)	2	(2)	(2)	2	—	—	2
		Universal mill		Major	3	(2)	3	(2)	2	1	(1)	(1)	2
				Minor	3	(3)	(3)	(2)	(3)	2	(2)	(2)	2
	Heavy	Flame-cut		Major	2	(2)	2	(2)	2	(1)	(1)	(1)	(2)
				Minor	2	(2)	2	(2)	2	(2)	(2)	(2)	(2)
		Universal mill		Major	3	(2)	(2)	(2)	2	(1)	(2)	(2)	(2)
				Minor	3	(2)	(2)	(2)	2	(2)	(2)	(2)	(3)
Massive solid					—	(1)	(1)	(1)	(1)	(1)	—	—	—
Tubes		Extruded rolled			—	(1)	(1)	(1)	(1)	(1)	—	—	—
		Welded			—	(2)	(1)	(1)	(1)	(1)	—	—	—
Extruded rolled				Major	(1)	(1)	(1)	(1)	(1)	(1)	—	—	—
				Minor	(1)	(1)	(1)	(1)	(1)	(1)	—	—	—
Box		Flame-cut		Major	2	(2)	(2)	(1)	(2)	1	—	—	—
				Minor	2	(2)	(2)	(1)	(2)	1	—	—	—
		Universal mill		Major	2	(2)	(2)	(1)	(2)	(1)	—	—	—
				Minor	2	(2)	(2)	(1)	(2)	(1)	—	—	—
Stress-relieved shapes (all types)				Major	1	1	1	1	1	1	1	1	1
				Minor	1	1	1	1	1	1	1	1	1

Cold-straightened (gagged or roller-straightened) are designed according to the column curve immediately above the curve for the corresponding unstraightened shape. This is not valid for shapes already assigned to curve 1.

AISC Load and Resistance Factor Design Curve

As a result, the AISC Load and Resistance Factor Design (LRFD) Specification[20] adopts the following curve

$$\frac{P}{P_y} = \begin{cases} \exp[-0.419\lambda_c^2] & \lambda_c \leq 1.5 \\ 0.877\lambda_c^{-2} & \lambda_c > 1.5 \end{cases} \quad (2.11.9)$$

to represent column strength. Note that only one curve is recommended for the whole range of possible column strengths. In the development of this curve, the following assumptions were made:

1. The column has small end restraints corresponding to an end-restraint parameter $G = 10$ (see Chapter 4) or an effective length factor $K = 0.96$.
2. The column has an initial crookedness sinusoidal in shape and has an amplitude of $(1/1500)L$ at midheight.
3. The axial force is applied at the centroid of the column end cross sections.

This LRFD curve is plotted in Fig. 2.45 together with all other curves described above. Note that the LRFD column curve as represented by Eq. (2.11.9) is comparable to SSRC Curve 2, especially in the range $0 \leq \lambda \leq 1.0$. The LRFD format is

$$\phi R_n \geq \sum_{i=1}^{m} \gamma_i Q_{ni} \quad (2.11.10)$$

where

R_n = nominal resistance
Q_n = nominal load effects
ϕ = resistance factor (see Table 1.2)
γ = load factor (see Table 1.1)

Note that the LRFD format has the features of both the ASD and PD formats in that factors of safety are applied to both the load and resistance terms to account for the variabilities and uncertainties in predicting these values. Furthermore, these load and resistance factors (ϕ, γ) are evaluated based on *first-order probabilistic approach*. Since different types of loads have different degrees of uncertainties, different load factors are used for different types of loads (e.g., 1.6 for live load, 1.2 for dead load, etc.); therefore, the LRFD format represents a more rational design approach.

The expressions for various column curves described above together with the three state-of-the-art design formats (ASD, PD, and LRFD) are summarized in Tables 2.4 and 2.5.

2.11 Design Curves for Steel Columns

Table 2.4 Summary of Column Curves

Column curves	Column equations	
CRC curve	$\dfrac{P}{P_y} = 1 - \dfrac{\lambda_c^2}{4}$	$\lambda_c \leq \sqrt{2}$
	$\dfrac{P}{P_y} = \dfrac{1}{\lambda_c^2}$	$\lambda_c > \sqrt{2}$
AISC-ASD curve	$\dfrac{P}{P_y} = \dfrac{1 - \dfrac{\lambda_c^2}{4}}{\dfrac{5}{3} + \dfrac{3}{8}\left(\dfrac{\lambda_c}{\sqrt{2}}\right) - \dfrac{1}{8}\left(\dfrac{\lambda_c}{\sqrt{2}}\right)^3}$	$\lambda_c \leq \sqrt{2}$
	$\dfrac{P}{P_y} = \dfrac{12}{23}\dfrac{1}{\lambda_c^2}$	$\lambda_c > \sqrt{2}$
AISC-PD curve	$\dfrac{P}{P_y} = \dfrac{1.7\left(1 - \dfrac{\lambda_c^2}{4}\right)}{\dfrac{5}{3} + \dfrac{3}{8}\left(\dfrac{\lambda_c}{\sqrt{2}}\right) - \dfrac{1}{8}\left(\dfrac{\lambda_c}{\sqrt{2}}\right)^3} \leq 1.0$	$\lambda_c \leq \sqrt{2}$
AISC-LRFD curve	$\dfrac{P}{P_y} = \exp(-0.419\lambda_c^2)$	$\lambda_c \leq 1.5$
	$\dfrac{P}{P_y} = \dfrac{0.877}{\lambda_c^2}$	$\lambda_c > 1.5$

2.11.2 Single Equation for Multiple-Column Curves

Although multiple-column curves give a more realistic representation of column strengths, the use of these curves in design is rather cumbersome. For example, for the SSRC *multiple-column curves*, each curve is represented by nine or ten coefficients. It is therefore desirable to have a single equation that can be used to represent all these curves. In the present section, we will discuss two mathematical equations that can be

Table 2.5 Summary of Design Formats

ASD	$\dfrac{R_n}{F.S.} \geq \sum_{i=1}^{m} Q_{ni}$
PD	$R_n \geq \gamma \sum_{i=1}^{m} Q_n$
LRFD	$\phi R_n \geq \sum_{i=1}^{m} \gamma_i Q_n$

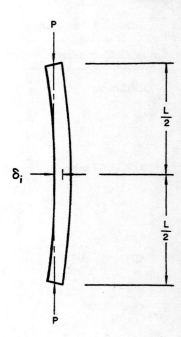

FIGURE 2.46 Physical model of imperfect column

used to represent these multiple-column curves. Both are developed based on the postulation that an initially crooked column with the initial crookedness at the midheight of the column equal to δ_i (Fig. 2.46) will fail under the combined action of axial force and (secondary) bending moment (arising from the $P - \delta$ effect) according to the criterion

$$\frac{P}{P_u} + \frac{M}{M_u} = 1 \qquad (2.11.11)$$

in which

P = applied axial force
M = bending moment arising from the $P - \delta$ effect [Eqs. (2.6.20) and (2.6.22)]

$$M = \frac{P\delta_i}{1 - P/P_e} \qquad (2.11.12)$$

P_u = ultimate axial capacity of the member in the absence of M
M_u = ultimate moment capacity of the member in the absence of P

Rondal–Maquoi Mathematical Form[21]

The Rondal–Maquoi mathematical expression can be developed by assuming that the ultimate strength of a column is reached when yielding

2.11 Design Curves for Steel Columns

occurs at the most severely stressed fiber, i.e., when

$$\frac{P}{P_y} + \frac{M}{M_y} = 1 \qquad (2.11.13)$$

Using Eq. (2.11.12), we have

$$\frac{P}{P_y} + \frac{P\delta_i}{\left(1 - \dfrac{P}{P_e}\right)M_y} = 1 \qquad (2.11.14)$$

or

$$\frac{P}{P_y} + \frac{P\delta_i}{\left(1 - \dfrac{P}{P_y}\dfrac{P_y}{P_e}\right)M_y} = 1 \qquad (2.11.15)$$

Substituting $M_y = S\sigma_y$ and $P_y/P_e = \lambda_c^2$ into Eq. (2.11.15) and defining $\eta = \delta_i A/S$ gives

$$\frac{P}{P_y} + \frac{\dfrac{P}{P_y}}{1 - \dfrac{P}{P_y}\lambda_c^2}\eta = 1 \qquad (2.11.16)$$

Solving for P/P_y yields

$$\frac{P}{P_y} = \frac{(1 + \eta + \lambda_c^2) - \sqrt{(1 + \eta + \lambda_c^2)^2 - 4\lambda_c^2}}{2\lambda_c^2} \qquad (2.11.17)$$

Equation (2.11.17) is the Rondal–Maquoi equation. By setting

$$\eta = \alpha(\lambda - 0.15) \qquad (2.11.18)$$

where

$$\alpha = \begin{cases} 0.103 & \text{(SSRC Curve 1)} \\ 0.293 & \text{(SSRC Curve 2)} \\ 0.622 & \text{(SSRC Curve 3)} \end{cases} \qquad (2.11.19)$$

the three SSRC curves can all be closely approximated (Fig. 2.47).

Lui–Chen Mathematical Form[22]

The Lui–Chen equation can be developed by setting

$$P_u = P_t \qquad (2.11.20)$$
$$M_u = M_m \qquad (2.11.21)$$

where

P_t = tangent modulus load of the column
M_m = flow moment (Fig. 2.48).

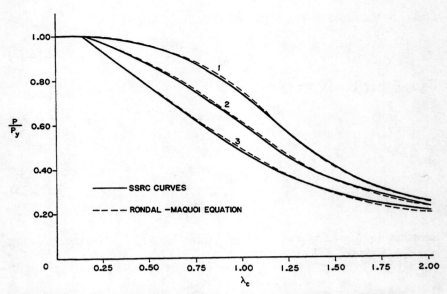

FIGURE 2.47 SSRC multiple-column curves and Rondal–Maquoi equation

FIGURE 2.48 Schematic representation of average flow moment M_m

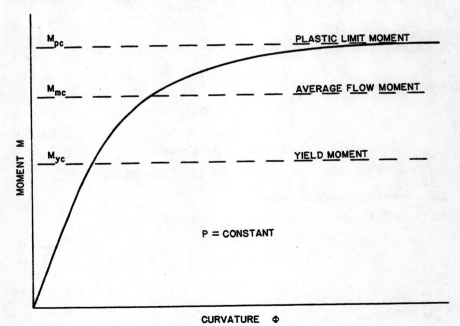

2.11 Design Curves for Steel Columns

Thus, the criterion of failure for the column is

$$\frac{P}{P_t} + \frac{M}{M_m} = 1 \tag{2.11.22}$$

Substituting Eq. (2.11.12) into Eq. (2.11.22) yields

$$\frac{P}{P_t} + \frac{P\delta_i}{\left(1 - \dfrac{P}{P_e}\right) M_m} = 1 \tag{2.11.23}$$

The flow moment can be expressed as the product of the *plastic section modulus* Z and an *average flow stress* σ_0

$$M_m = Z\sigma_0 \tag{2.11.24}$$

The flow stress is such that $\sigma_y/f \leq \sigma_0 \leq \sigma_y$, in which f is the *shape factor* of the cross section. The shape factor is defined as the ratio of the plastic section modulus Z to the elastic section modulus S.

In view of Eq. (2.11.24), Eq. (2.11.23) can be written as

$$\frac{P}{P_t} + \frac{P\delta_i}{\left(1 - \dfrac{P}{P_e}\right) Z\sigma_0} = 1 \tag{2.11.25}$$

or

$$\left(\frac{P}{P_y}\right)\left(\frac{P_y}{P_e}\right)\left(\frac{P_e}{P_t}\right) + \frac{P\delta_i}{\left(1 - \dfrac{P}{P_y}\dfrac{P_y}{P_e}\right)\left(\dfrac{Z}{S}\right) S \left(\dfrac{\sigma_0}{\sigma_y}\right) \sigma_y} = 1 \tag{2.11.26}$$

where S = elastic section modulus.

Defining

$$\bar{E} = \frac{E}{E_t} = \frac{P_e}{P_t} \tag{2.11.27}$$

$$f = \frac{Z}{S} \tag{2.11.28}$$

$$\bar{\sigma}_y = \frac{\sigma_0}{\sigma_y} \tag{2.11.29}$$

and realizing that

$$\frac{P_y}{P_e} = \lambda_c^2 \tag{2.11.30}$$

where $\lambda_c = \dfrac{1}{\pi}\sqrt{\dfrac{\sigma_y}{E}}\left(\dfrac{KL}{r}\right)$, Eq. (2.11.26) can be written as

$$\left(\frac{P}{P_y}\right)\lambda_c^2 \bar{E} + \frac{P\delta_i}{\left(1-\dfrac{P}{P_y}\lambda_c^2\right)f\tilde{\sigma}_y S \sigma_y} = 1 \qquad (2.11.31)$$

or

$$\left(\frac{P}{P_y}\right)\lambda_c^2 \bar{E} + \frac{\dfrac{P}{P_y}}{1-\dfrac{P}{P_y}\lambda_c^2}\bar{\eta} = 1 \qquad (2.11.32)$$

in which

$$\bar{\eta} = \frac{\delta_i A}{f\tilde{\sigma}_y S} \qquad (2.11.33)$$

is the *imperfection parameter*.

Solving Eq. (2.11.32) for P/P_y gives

$$\frac{P}{P_y} = \frac{\bar{\eta} + (1+\bar{E})\lambda_c^2 - \sqrt{[\bar{\eta} + (1+\bar{E})\lambda_c^2]^2 - 4\bar{E}\lambda_c^4}}{2\bar{E}\lambda_c^4} \qquad (2.11.34)$$

Equation (2.11.34) is the Lui–Chen equation for all column strength curves.

The maximum load a column can carry is a function of $\bar{\eta}$, $\bar{E}$, and λ_c. Any column curve can be generated using this equation provided that the parameters $\bar{\eta}$, $\bar{E}$, and λ_c are known.

Now, expressing the initial crookedness δ_i of a column as a fraction of the column length L

$$\delta_i = \gamma L \qquad (2.11.35)$$

and realizing that

$$S = \frac{I}{c} = \frac{Ar^2}{c} \qquad (2.11.36)$$

where

c = distance from neutral axis to extreme fiber of the cross section
r = radius of gyration

Equation (2.11.33) can be written as

$$\bar{\eta} = \frac{c\gamma L}{f\tilde{\sigma}_y r^2} \qquad (2.11.37)$$

2.11 Design Curves for Steel Columns

or, in terms of λ_c (with $K = 1$ for pinned-ended column)

$$\bar{\eta} = \pi\sqrt{\frac{E}{\sigma_y}}(\gamma)\left(\frac{1}{f}\right)\left(\frac{c}{r}\right)\left(\frac{\lambda_c}{\bar{\sigma}_y}\right) \qquad (2.11.38)$$

Note that this imperfection parameter reflects the effects of yield stress (σ_y), geometric imperfection (γ), axis of bending (f), cross-sectional shape (c, r), and slenderness ratio (λ_c) on the load-carrying capacity of columns.

The average flow stress σ_0 depends on the degree of plastification of the cross section and is a function of the load level and column types. Since for a column the degree of plastification depends on the load level and the load level is a function of the slenderness ratio of the column, this flow stress can be thought of as a function of the slenderness parameter λ_c. If λ_c is very large, P/P_y is very small, the problem resembles a beam problem and the *plastic limit moment* M_{pc} will govern the ultimate state, so σ_0 will approach σ_y. On the other hand, if λ_c is very small, P/P_y will approach unity, the problem resembles an axially loaded short-column problem and the yield moment M_{yc} will govern the ultimate state, so σ_0 will approach σ_y/f where f is the shape factor.

On the basis of this argument, the following expression for the flow stress is proposed:

$$\sigma_0 = \left(\frac{1}{\beta\lambda_c^2 + f}\right)\sigma_y \qquad (2.11.39)$$

The constant β can be determined from experiments or from calibration against existing column curves. For small and medium-size hot-rolled, wide-flange shapes, the value of β for strong axis bending can be taken as -0.378 and for weak axis bending as -0.308. By substituting Eq. (2.11.39) into Eq. (2.11.38), we can write

$$\bar{\eta} = \left[\pi\sqrt{\frac{E}{\sigma_y}}(\gamma)\left(\frac{1}{f}\right)\left(\frac{c}{r}\right)\right](\beta\lambda_c^3 + f\lambda_c) \qquad (2.11.40)$$

For a given column, the terms inside the brackets of Eq. (2.11.40) are known, so it can be written in the general form as

$$\bar{\eta} = \bar{a}\lambda_c^3 + \bar{b}\lambda_c \qquad (2.11.41)$$

where

$$\bar{a} = \pi\sqrt{\frac{E}{\sigma_y}}\frac{\gamma c \beta}{fr} \qquad (2.11.42)$$

$$\bar{b} = \pi\sqrt{\frac{E}{\sigma_y}}\frac{\gamma c}{r} \qquad (2.11.43)$$

The modulus ratio $\bar{E}$ can be evaluated if the tangent modulus E_t is

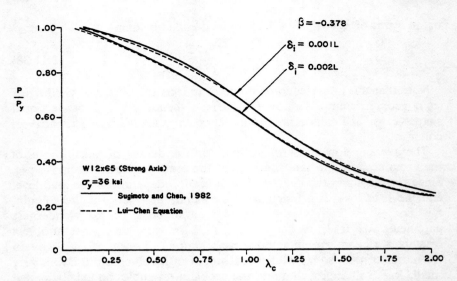

FIGURE 2.49 Comparison of Lui–Chen equation with computer model by Sugimoto and Chen

known either from an experimental or a theoretical approach. Recall that the tangent modulus is the slope of the nonlinear stress–strain curve. This nonlinearity is due to material for aluminum columns but for steel columns, it is due to residual stresses existing in the steel cross sections. Thus, this modulus ratio will reflect material nonlinearity and imperfec-

FIGURE 2.50 Comparison of Lui–Chen equation with computer model by Sugimoto and Chen

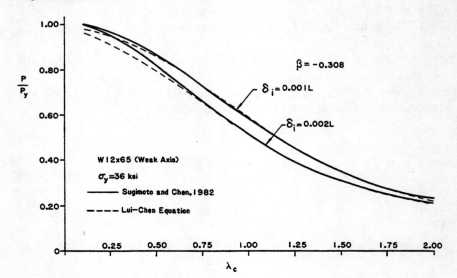

Table 2.6 Values of $\bar{a}$ and $\bar{b}$ for SSRC Multiple-Column Curves

SSRC curve	$\bar{a}$	$\bar{b}$
1	0.002	−0.001
2	−0.036	0.159
3	−0.092	0.453

tions. For small and medium-size hot-rolled, wide-flange shapes, this modulus ratio can be conveniently taken as the ratio of the Euler curve to the CRC curve, i.e.,

$$\bar{E} = \frac{E}{E_t} = \frac{P_e}{P_{CRC}} = \begin{cases} \dfrac{4}{(4-\lambda_c^2)\lambda_c^2} & (\lambda_c \leq \sqrt{2}) \\ 1 & (\lambda_c > \sqrt{2}) \end{cases} \quad (2.11.44)$$

Figure 2.49 shows two column curves, generated numerically by Sugimoto and Chen[23] for a W12 × 65 section, bent about the section's strong axis with initial imperfection δ_i at midheight equals to 0.001L and 0.002L. Also shown in the figure are the two curves generated using the Lui–Chen equation. It can be seen that good agreement is generally observed between the numerically generated curves and the curves predicted using the mathematical equation. A similar comparison for the same column bent about its weak axis is shown in Fig. 2.50. Again, good agreement is generally observed.

Equation (2.11.34) can also be used to approximate the SSRC multiple-column curves. By using the values of $\bar{a}$ and $\bar{b}$ shown in Table 2.6, the SSRC column curves can be closely approximated (Fig. 2.51).

2.12 SUMMARY

For a perfectly straight column that buckles in the elastic range, the differential equation of equilibrium can be written as

$$EIy'' + Py = \pm V_0 x \pm M_0 \quad (2.12.1)$$

where

E = elastic modulus
I = moment of inertia of the cross section
P = axial force
V_0 = end shear
M_0 = end moment

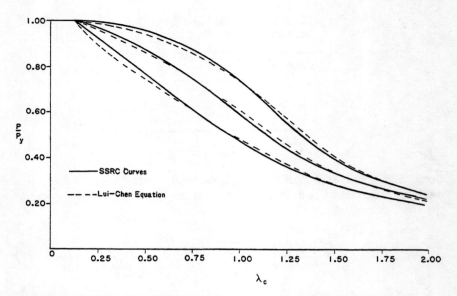

FIGURE 2.51 Comparison of SSRC curves with Lui–Chen equation

The general solution to Eq. (2.12.1) is

$$y = A \sin kx + B \cos kx \pm \frac{V_0}{P} x \pm \frac{M_0}{P} \qquad (2.12.2)$$

where $k = \sqrt{P/EI}$ and A and B are unknown coefficients.

Since there are more unknowns in Eq. (2.12.2) than geometrical boundary conditions available in the problem, there is no unique solution for this equation. This class of problem is known as the *eigenvalue problem*. In an eigenvalue problem, nontrivial solutions for the dependent variable exist only for certain values called eigenvalues. The nontrivial solutions that correspond to these eigenvalues are called eigenvectors. Because there are more unknowns than boundary conditions, only the shape and not the amplitude of the eigenvector can be determined. For the case of a column, the eigenvectors are the buckling modes of the column and the eigenvalues are the values of k. The lowest value of k gives the critical load of the column.

At the critical load, the column ceases to be stable in its initial straight position. A small lateral disturbance that occurs at the critical load will cause a lateral deflection that will not disappear as the disturbance is removed.

Equation (2.12.1) is a second-order linear differential equation with constant coefficients. This equation depends on the end conditions of the column, since V_0 and M_0 are different for different end conditions. A

2.12 Summary

more convenient form that is independent of end conditions can be obtained by differentiating Eq. (2.12.1) twice to give

$$EIy^{IV} + Py'' = 0 \qquad (2.12.3)$$

whose general solution is

$$y = A \sin kx + B \cos kx + Cx + D \qquad (2.12.4)$$

Alternatively, Eq. (2.12.3) can be obtained by considering equilibrium of an infinitesimal segment of a column. Note that there are five unknowns (k, A, B, C, and D) in Eq. (2.12.4) and there are only four boundary conditions (geometrical and natural). Again, this is an eigenvalue problem. The lowest eigenvalue to Eq. (2.12.4) will thus give the critical load of the column.

For end-restrained columns, it is convenient to modify the unbraced length of the column to an equivalent length of a pinned-ended column so that the column curves prepared for pinned-ended columns can be used directly for the restrained case. This can be achieved by multiplying the actual length of the end-restrained column by an *effective length factor* defined as

$$K = \sqrt{P_e/P_{cr}} \qquad (2.12.5)$$

The key phenomenon, that is associated with a column instability, is known as the $P - \delta$ *effect*. This effect arises as the axial force P is acting through the displacement δ of the member relative to its chord. The result of this effect is an increase in lateral deflection and moment in the column. This $P - \delta$ effect can be studied conveniently by analyzing an eccentrically loaded or initially crooked column. From an elastic analysis, it can be shown that the moment in these columns can be obtained by simply multiplying the first-order moment by the *amplification factor*

$$A_F = \begin{cases} \sec\left(\frac{\pi}{2}\sqrt{P/P_e}\right) & \text{(for eccentrically loaded columns)} \\ \dfrac{1}{1 - (P/P_e)} & \text{(for initially crooked columns)} \end{cases} \qquad (2.12.6)$$

For perfectly straight columns that buckle in the inelastic range, the critical load can be obtained by simply replacing the elastic modulus E by an effective modulus E_{eff} where

$$E_{\text{eff}} = \begin{cases} E_t & \text{(according to the tangent modulus theory)} \\ E_r & \text{(according to the reduced modulus theory)} \end{cases}$$

The use of this effective modulus can, approximately, take into account the *material inelasticity*. In the *tangent modulus theory*, no strain reversal is allowed, whereas in the *reduced modulus theory*, a complete strain reversal is allowed, and where a complete strain reversal is assumed to occur on the convex side of the column as it buckles. Although the reduced modulus theory is theoretically correct, the validity of the tangent modulus theory for predicting the buckling strength of real columns is demonstrated and explained by the *Shanley's inelastic column theory*.

Since the tangent modulus load represents a lower bound to the buckling strength of real columns, and since it is easier to evaluate than the reduced modulus load, the tangent modulus theory is used extensively to develop column strength curves for the purpose of design. For instance, the CRC column curve was developed on the basis of the tangent modulus concept. By introducing a safety factor to the CRC curve to account for imperfections of the columns and load eccentricities, one can develop the AISC-ASD curve, which is contained in the present ASD Specification.[3]

The tangent modulus concept is based on the eigenvalue or bifurcation analysis. In using the eigenvalue analysis one must assume that the column is geometrically perfect. Columns in reality are never perfect. As a result, an alternate and more elaborate approach that explicitly takes into account the effect of geometrical imperfections in the columns may be desirable. This approach is known as the stability or load-deflection analysis. In contrast to the eigenvalue analysis, in which only the load that corresponds to the point of bifurcation can be obtained, the load-deflection analysis permits us to trace the complete load deflection response of the column from the start of loading to failure. Because of the complexity in calculation inherent in the load-deflection analysis, recourse to numerical techniques is inevitable. The SSRC multiple-column curves have been developed on the basis of an extensive load-deflection analysis. The SSRC curve 2 represents the column strength of medium-sized hot-rolled, wide-flange shapes frequently used in building construction. This curve forms the basis of the new AISC-LRFD column curve, which is contained in the present LRFD Specification.[20]

PROBLEMS

2.1 Find the ratio of the critical loads that corresponds to the first two buckling modes of the two columns shown in Fig. P2.1. Sketch the deflected shapes of the columns.

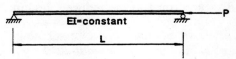

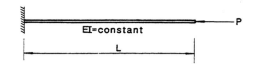

FIGURE P2.1

2.2 Find the buckling load of the rectangular section pinned at both ends (Fig. P2.2).

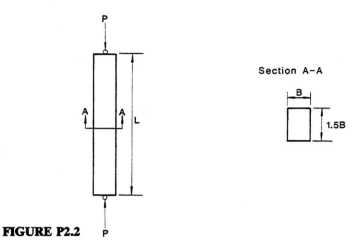

FIGURE P2.2

2.3 Find the buckling load of the fixed–free stepped column shown below (Fig. P2.3).

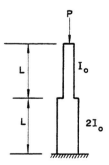

FIGURE P2.3

2.4 Find the buckling loads of the columns shown in Fig. P2.4a–c.

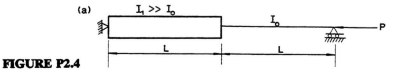

FIGURE P2.4

(b)

```
|//// ——————————L——————————△——0.5L——→ P
                         EI=constant
```

(c)

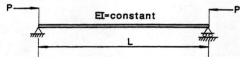

FIGURE P2.4   EI=constant

2.5 For the eccentrically loaded column shown in Fig. P2.5, find an approximate expression for the amplification factor A_F in the form $A_F = a/(1 - P/P_e)$ where $a = f(P/P_e)$. Compare it with the exact expression $A_F = \sec[(\pi/2)\sqrt{P/P_e}]$.

```
P——  ———————EI=constant———————  ——P
     △————————————L————————————△
```

FIGURE P2.5

2.6 Discuss the assumptions used and the limitations of
   a. the Secant formula
   b. the Perry–Robertson formula

2.7 Plot the tangent modulus column curve for an aluminum alloy column with $n = 8.0$, $\sigma_{0.2} = 22.78$ ksi (157 MPa), $E = 10,181$ ksi (1.6×10^6 MPa).

2.8 Plot the reduced modulus column curve for the aluminum column in the above problem with the following cross section (Fig. P2.8).
   a. Find the ratio P_r/P_t at $\lambda_0 = 0.4$ and $\lambda_0 = 1.2$ where $\lambda_0 = 1/\pi\sqrt{(\sigma_{0.2}/E)}\,(L/r)$.
   b. Approximate the stress–strain behavior of the aluminum alloy by two straight lines and re-evaluate the ratio P_r/P_t at $\lambda_0 = 0.4$ and $\lambda_0 = 1.2$. What conclusions can you draw upon comparison with the values obtained in part (a)?

2.9 State the basic assumptions made in the development of the
   a. CRC column curve
   b. SSRC multiple-column curves

2.10 Describe the inter-relationship and design format of
   a. AISC-ASD column curve
   b. AISC-PD column curve
   c. AISC-LRFD column curve
   How is the concept of "safety factor" incorporated in these design formats?

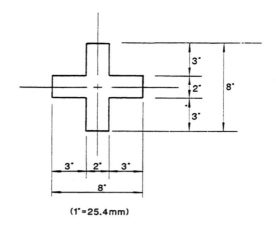

FIGURE P2.8

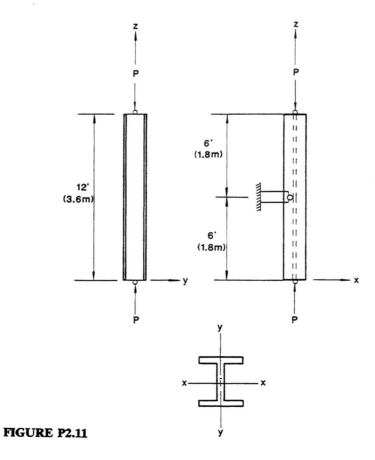

FIGURE P2.11

2.11 Design a W-section pinned at both ends and braced at midheight in the weak direction (Fig. P2.11) with an occupancy live load (L_n) of 60 kips (267 kN), a roof live load (L_r) of 40 kips (178 kN), and a dead load (D_n) of 60 kips (267 kN) based on
  a. the ASD format
  b. the PD format
  c. the LRFD format

2.12 Derive the reduced modulus of elasticity E_r for an idealized I-section shown in Fig. P2.12, in which it is assumed that one-half of the cross-section area is concentrated in each flange and the area of the web is disregarded.

FIGURE P2.12

2.13 Find the buckling load of the column shown in Fig. P2.13.

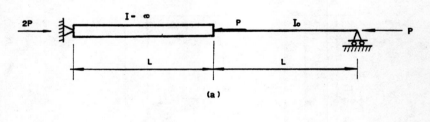

FIGURE P2.13

2.14 Find the critical load P for the structure shown in Fig. P2.14.

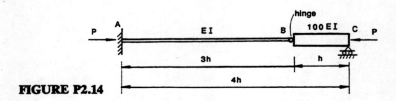

FIGURE P2.14

REFERENCES

1. Euler, L. De curvis elasticis. Lausanne and Geneva, 1744, pp. 267–268. (The Euler formula was derived in a later paper: Sur le Force de Colonnes, Memoires de l'Academie Royale des Sciences et Belles Lettres, Vol. 13, Berlin, 1759.)
2. Wang, C. T. Applied Elasticity. McGraw-Hill, New York, 1953.
3. Specification for the Design, Fabrication and Erection of Structural Steel for Buildings. AISC, Chicago, Illinois, November 1978.
4. Timoshenko, S. P., and Gere, J. M. Theory of Elastic Stability. Second edition. Engineering Societies Monographs, McGraw-Hill, New York, 1961.
5. Ayrton, W. E., and Perry, J. On struts. The Engineer. 62: 464 1886.
6. Robertson, A. The strength of struts, ICE Selected Engineering Paper, 28, 1925.
7. Engesser, F. Zeitschrift für Architektur und Ingenieurwesen. 35:455, 1889. Also, Schweizerische Bauzeitung, 26:24, 1895.
8. Shanley, F. R. Inelastic column theory. Journal of the Aeronautical Sciences. Vol. 14 (May): 261–264, 1947.
9. Considere, A. Resistance des Pièces Comprimées. Congrès International de Procedes de Construction, Paris. 3:371, 1891.
10. Ramberg, W., and Osgood, W. R. Description of stress–strain curves by three parameters. National Advisory Committee on Aeronautics, Technical Note No. 902, 1943.
11. Batterman, R. H., and Johnston, B. G. Behavior and maximum strength of metal columns. Journal of the Structural Division. ASCE, Vol. 93, No. ST2, April 1967, pp. 205–231.
12. Huber, A. W. Residual Stresses in Wide Flange Beams and Columns. Fritz Engineering Laboratory Report No. 220A.25. Lehigh University, Bethlehem, PA, July 1956.
13. Ketter, R. L. The influence of residual stresses on the strength of structural members. Welding Research Council Bulletin No. 44, November 1958.
14. Beedle, L. S., and Tall, L. Basic column strength. Proceedings of the ASCE, Vol. 86, No. ST7, July 1960, p. 139–173.
15. Tebedge, N., and Tall, L. Residual Stresses in Structural Steel Shapes—A Summary of Measured Values. Fritz Engineering Laboratory Report No. 337.34. Lehigh University, Bethlehem, PA, 1973.
16. Johnston, B. G., ed. Guide to Design Criteria for Metal Compression Members. Column Research Council, 1960. Second edition. John Wiley & Sons, New York, 1973.
17. Bjorhovde, R. Deterministic and Probabilistic Approaches to the Strength of Steel Columns. Ph.D. dissertation, Department of Civil Engineering, Lehigh University, Bethlehem, PA, 1972.
18. Sfintesco, D., ed. European Convention of Constructional Steelworks Manual on the Stability of Steel Structures. Second edition. ECCS, Paris, 1976.
19. Johnston, B. G., ed. Structural Stability Research Council Guide to Stability Design Criteria for Metal Structures. Third edition. John Wiley & Sons, New York, 1976.

20. Load and Resistance Factor Design Specification for Structural Steel Buildings. AISC, Chicago, November, 1986.
21. Rondal, J., and Maquoi, R. Single equation for SSRC column strength curves. Journal of the Structural Division, ASCE, 105(ST1): 247–250, 1979.
22. Lui, E. M., and Chen, W. F. Simplified approach to the analysis and design of columns with imperfections. Engineering Journal. AISC, Vol. 21, second quarter, 1984, pp. 99–117.
23. Sugimoto, H., and Chen, W. F. Small end restraint effects on strength of H-columns. Journal of the Structural Division. ASCE. 108(ST3): 661–681, 1982.

Further Reading

Bleich, F. Buckling Strength of Metal Structures Engineering Societies Monographs. McGraw-Hill, New York, 1952.

Chen, W. F., and Atsuta, T. Theory of Beam-Columns. Vol. 1: In-Plane Behavior and Design, 1976; and Vol. 2: Space Behavior and Design. McGraw-Hill, New York, 1977.

Chajes, A. Principles of Structural Stability Theory. Prentice-Hall, Englewood Cliffs, NJ, 1974.

Hoff, N. J. The Analysis of Structures. John Wiley & Sons, New York, 1956.

Chapter 3

BEAM-COLUMNS

3.1 INTRODUCTION

A *beam-column* is a structural member that is subjected to both bending and compression. In reality, all members in a frame are beam-columns. However, if the axial force effect in a member is negligible compared to the bending effect, it is more convenient to analyze and design that member as a *beam*. On the other hand, if the bending effect in a member is secondary compared to the axial force effect, it is more convenient to treat such a member as a column and analyze and design it accordingly. Thus, beams and columns are regarded as special cases of beam-columns.

Because for a beam-column both the bending and axial effects are significant, the analysis of this type of member involves the features of both the *deflection problem* as a beam and the *stability problem* as a column. As a beam problem, the bending moments induced in the member by the application of end moments, or by in-span transverse loadings, or by a combination of both, will cause lateral deflections. These bending moments and lateral deflections are called *primary bending moments and deflections*. As a column problem, the axial force at certain critical values will cause instability of the member. In the case of beam-columns, the axial force will act through the lateral deflection caused by the bending effect to produce additional lateral deflection and moment in the member. To distinguish between the deflections (and moments) induced by the bending and axial-force effects, it is customary to refer to the deflection (and moment) caused by the primary-bending effect as *primary* deflection (and moment), and to refer to the additional deflection (and moment) caused by the axial-force effect as *secondary deflection (and moment)*. Note that the words primary and secondary are used solely for the purpose of convenience and not for the purpose of representing the relative importance of the two effects. In fact, the secondary deflection (and moment) of a beam-column caused by the axial-force effect is sometimes more significant than that caused by the primary bending effect.

Although the analysis of a beam-column is more complicated than that of a beam or a column, closed-form solutions of most beam-columns are available so long as they stay within the realm of purely *elastic* behavior in which the moment can be related to the curvature by a *linear* relationship. (See, for example, *Theory of Elastic Stability* by Timoshenko and Gere.[1]) If yielding or inelasticity occurs in the member, the moment-curvature relationship becomes *nonlinear*. In such cases the use of formal mathematics for the solution of the governing differential equations become intractable and recourse must be had to numerical methods to obtain solutions. In some cases, however, closed-form solutions are still possible if one makes drastically simplified assumptions regarding the stress–strain behavior of the material, cross-sectional geometry, and deflection shape of the member. For more general cases, however, only numerical solutions with recourse to computer routines are possible. (See, for example, the two-volume work by Chen and Atsuta.[2,3])

In this chapter, we will show in detail the elastic solutions of a simply supported beam-column under three types of loadings—that of (1) uniformly distributed, (2) concentrated, and (3) end moments—in order to demonstrate the solution procedures and general behavior of a typical beam-column problem. Afterward, we will develop the general governing differential equation of a beam-column under general loading conditions. This, in turn, will be followed by the solution of an elastic-plastic beam-column under equal and opposite end moments. We will then conclude the chapter with a discussion of design equations, which will be based on the approaches used by the AISC for beam-columns.

Since the behavior of a beam-column is different depending on whether there is a relative translation between the member ends, our discussions of beam-columns will address separately the *nonsway* versus the *sway* case. In addition, we will distinguish between cases in which the beam-column is treated as an *individual* member or as a member *in a frame*. To take into account the effects of other members on the beam-column under consideration in a frame, the concept of *effective length factor* will be used again. The AISC effective length alignment charts for nonsway and sway cases will be discussed on the basis of certain simple assumptions.

3.2 BEAM-COLUMN WITH UNIFORMLY DISTRIBUTED LATERAL LOAD

3.2.1 The Closed-Form Solution (Fig. 3.1)

To begin our discussion of the elastic behavior of a beam-column, let us consider a simply supported beam-column subjected to an axial force P and uniformly distributed lateral load of intensity w as shown in Fig. 3.1a. A free-body diagram of a segment of the beam-column of length x from the left support is shown in Fig. 3.1b. The external moment acting on the cut section, is

$$M_{ext} = Py - \frac{w}{2}x^2 + \frac{wL}{2}x \qquad (3.2.1)$$

If elastic behavior is assumed and if the material obeys Hooke's Law, the internal moment M_{int} is related to the bending curvature y'' by the linear relationship

$$M_{int} = -EIy'' \qquad (3.2.2)$$

where the negative sign indicates that the curvature or the rate of change of slope $y'' = dy'/dx$ is decreasing with increasing x as shown in Fig. 3.1b.

For equilibrium, the external moment must be balanced by the internal moment. Therefore, by equating Eqs. (3.2.1) and (3.2.2), we have, upon rearranging,

$$EIy'' + Py = \frac{w}{2}x^2 - \frac{wL}{2}x \qquad (3.2.3)$$

FIGURE 3.1 Beam-column with uniformly distributed lateral load

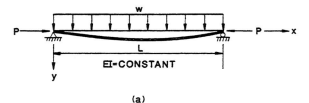

or

$$y'' + k^2 y = \frac{w}{2EI} x^2 - \frac{wL}{2EI} x \qquad (3.2.4)$$

where

$$k^2 = \frac{P}{EI} \qquad (3.2.5)$$

The general solution to Eq. (3.2.4) consists of a complementary solution y_c and a particular solution y_p, i.e.,

$$y = y_c + y_p \qquad (3.2.6)$$

The complementary solution that satisfies the homogeneous differential equation

$$y'' + k^2 y = 0 \qquad (3.2.7)$$

has the general form

$$y_c = A \sin kx + B \cos kx \qquad (3.2.8)$$

The particular solution that satisfies Eq. (3.2.4) can be obtained by either the *method of undetermined coefficients* or by *the method of variation of parameters*. We shall use the method of undetermined coefficients for this example.

In the method of undetermined coefficients, since the right-hand side of Eq. (3.2.4) is a polynomial, we assume the particular solution to be a polynomial with the highest order equal to that of the polynomial in the right-hand side of Eq. (3.2.4)

$$y_p = C_1 x^2 + C_2 x + C_3 \qquad (3.2.9)$$

in which C_1, C_2, and C_3 are the undetermined coefficients.

Taking derivatives of Eq. (3.2.9), we obtain

$$y'_p = 2C_1 x + C_2 \qquad (3.2.10)$$

$$y''_p = 2C_1 \qquad (3.2.11)$$

and so

$$y''_p + k^2 y_p = 2C_1 + k^2(C_1 x^2 + C_2 x + C_3) \qquad (3.2.12)$$

Rearranging, we have

$$y''_p + k^2 y_p = (C_1 k^2) x^2 + (C_2 k^2) x + (2C_1 + C_3 k^2) \qquad (3.2.13)$$

Upon comparison with Eq. (3.2.4), we can write

$$C_1 k^2 = \frac{w}{2EI} \qquad (3.2.14)$$

3.2 Beam-Column with Uniform Distribution Lateral Load

$$C_2 k^2 = -\frac{wL}{2EI} \qquad (3.2.15)$$

$$2C_1 + C_3 k^2 = 0 \qquad (3.2.16)$$

from which

$$C_1 = \frac{w}{2EIk^2} \qquad (3.2.17)$$

$$C_2 = -\frac{wL}{2EIk^2} \qquad (3.2.18)$$

$$C_3 = -\frac{2C_1}{k^2} = -\frac{w}{EIk^4} \qquad (3.2.19)$$

Hence, the particular solution is

$$y_p = \frac{w}{2EIk^2} x^2 - \frac{wL}{2EIk^2} x - \frac{w}{EIk^4} \qquad (3.2.20)$$

Substituting the complementary solution Eq. (3.2.8) and the particular solution Eq. (3.2.20) into Eq. (3.2.6) gives the general solution to Eq. (3.2.4) as

$$y = A \sin kx + B \cos kx + \frac{w}{2EIk^2} x^2 - \frac{wL}{2EIk^2} x - \frac{w}{EIk^4} \qquad (3.2.21)$$

from which

$$y' = Ak \cos kx - Bk \sin kx + \frac{w}{EIk^2} x - \frac{wL}{2EIk^2} \qquad (3.2.22)$$

The constants A and B can be obtained by considering the boundary conditions

$$y(0) = 0, \qquad y'\left(\frac{L}{2}\right) = 0 \qquad (3.2.23)$$

Using the first boundary condition, we find

$$B = \frac{w}{EIk^4} \qquad (3.2.24)$$

and using the second boundary condition, we have

$$A = \frac{w}{EIk^4} \tan \frac{kL}{2} \qquad (3.2.25)$$

Thus, Eq. (3.2.21) can be written as

$$y = \frac{w}{EIk^4}\left[\tan\frac{kL}{2}\sin kx + \cos kx - 1\right] - \frac{w}{2EIk^2}x(L-x) \quad (3.2.26)$$

Introducing the notation

$$u = \frac{kL}{2}. \quad (3.2.27)$$

Eq. (3.2.26) can be written as

$$y = \frac{wL^4}{16EIu^4}\left[\tan u \sin\frac{2ux}{L} + \cos\frac{2ux}{L} - 1\right]$$

$$- \frac{wL^2}{8EIu^2}x(L-x) \quad (3.2.28)$$

from which the moment distribution along the length of the member is

$$M = -EIy'' = \frac{wL^2}{4u^2}\left[\tan u \sin\frac{2ux}{L} + \cos\frac{2ux}{L} - 1\right] \quad (3.2.29)$$

3.2.2 The Calculation of y_{max}

The maximum deflection of the member occurs at midspan and is expressed by

$$y_{max} = y\left(\frac{L}{2}\right) = \frac{wL^4}{16EIu^4}\left[\frac{1-\cos u}{\cos u}\right] - \frac{wL^4}{32EIu^2}$$

$$= \frac{5wL^4}{384EI}\left[\frac{12(2\sec u - u^2 - 2)}{5u^4}\right]$$

$$= y_0\left[\frac{12(2\sec u - u^2 - 2)}{5u^4}\right] \quad (3.2.30)$$

where $y_0 = 5wL^4/384EI$ is the maximum lateral deflection that would exist if the uniform lateral load w were acting alone (i.e., if P were absent). The effect of the axial force on the maximum deflection is manifested in the term in the square bracket in Eq. (3.2.30). As seen from Fig. 3.2, if $u = 0$ (i.e., $P = 0$) the term in the brackets reduces to unity, and as u increases (i.e., P increases) the value of this term increases. Finally, at $u = \pi/2$ (i.e., $P = \pi^2 EI/L^2$, the Euler buckling load) the value of the term approaches infinity. In other words, as P approaches the Euler load, the lateral deflection of the member increases without bound or, to put it in still another way, the bending stiffness of the member vanishes as P approaches the Euler load. Thus, for an elastic

3.2 Beam-Column with Uniform Distribution Lateral Load

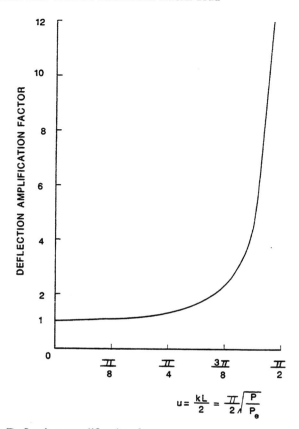

FIGURE 3.2 Deflection amplification factor

system, the critical load can be obtained by reference to the state at which the tangent stiffness of the system vanishes. This concept will be used in the next chapter to determine the critical loads of elastic frameworks. Let us now return to the beam-column problem. It can clearly be seen in Fig. 3.2 that the effect of axial force on the lateral deflection of the member depends on the magnitude of the axial force. The term in the brackets in Eq. (3.2.30) can be regarded as an *amplification factor*, which amplifies the deflection of the member when an axial force is acting in conjunction with the lateral force. The larger the value of the axial force, the greater will be the amplification.

Another useful observation can be made from Eq. (3.2.30). For a constant u (i.e., if P remains unchanged), the lateral deflection y is directly proportional to the applied lateral load w. In other words, the deflections (and bending moments) are *linear* functions with respect to the lateral applied loads. Thus, the total lateral deflection resulting from different load combinations can be obtained simply as the sum of the deflections resulting from each individual load application with the same axial load. This is known as the *principle of superposition*. The principle of superposition has been used extensively for the special case of beam problems for which $P = 0$. Here, it shows that this principle holds also for beam-column problems for which $P \neq 0$ *provided that (1) the axial force in the general case of the member remains constant, and (2) the same axial force is applied to each of the component cases*. We shall use this principle later in the chapter to evaluate the fixed-end moments of beam-columns (Section 3.5).

Although the lateral deflection is directly proportional to the applied lateral load for a constant axial force, this deflection varies nonlinearly with the axial force. This is true even if the lateral load remains unchanged. Also, the proportionality between lateral deflection and lateral load will be destroyed if the axial force varies during the application of the lateral load.

For the purpose of design application, it is more convenient to simplify the expression of Eq. (3.2.30). Expanding sec u in a power series,

$$\sec u = 1 + \frac{1}{2}u^2 + \frac{5}{24}u^4 + \frac{61}{720}u^6 + \frac{277}{8064}u^8 + \cdots \tag{3.2.31}$$

and substituting this series into Eq. (3.2.30), we obtain

$$y_{\max} = y_0[1 + 0.4067u^2 + 0.1649u^4 + \cdots] \tag{3.2.32}$$

Since

$$u = \frac{kL}{2} = \frac{L}{2}\sqrt{\frac{P}{EI}} = \frac{\pi}{2}\sqrt{\frac{P}{P_e}} \tag{3.2.33}$$

Equation (3.2.32) can be written as

$$y_{\max} = y_0\left[1 + 1.003\left(\frac{P}{P_e}\right) + 1.004\left(\frac{P}{P_e}\right)^2 + \cdots\right] \tag{3.2.34}$$

or, approximately

$$y_{\max} \approx y_0\left[1 + \left(\frac{P}{P_e}\right) + \left(\frac{P}{P_e}\right)^2 + \cdots\right] = y_0\left[\frac{1}{1 - \left(\frac{P}{P_e}\right)}\right] \tag{3.2.35}$$

in which the term in the square brackets is the *design amplification factor* for the lateral deflection. Table 3.1 shows a comparison of the theoretical amplification factor [the term in brackets in Eq. (3.2.30)] and the design amplification factor [the term in brackets in Eq. (3.2.35)]. As can be seen, for small values of P/P_e, which is generally the case for the

3.2 Beam-Column with Uniform Distribution Lateral Load

Table 3.1 Theoretical and Design Deflection Amplification Factors for a Uniformly Loaded Beam-Column

$u = \dfrac{kL}{2} = \dfrac{\pi}{2}\sqrt{\dfrac{P}{P_e}}$	Theoretical Eq. (3.2.30)	Design Eq. (3.2.35)
0	1.000	1.000
0.20	1.016	1.016
0.40	1.070	1.069
0.60	1.173	1.171
0.80	1.354	1.350
1.00	1.690	1.681
1.20	2.400	2.402
1.40	4.822	4.863
$\pi/2$	∞	∞

axial-load conditions in most beam-columns in real structures, the two expressions give very comparable results.

3.2.3 The Calculation of M_{max}

In addition to knowing the maximum deflection, it is also important for an engineer or designer to know the maximum moment in the beam-column. For a uniformly loaded beam-column, the maximum moment occurs at midspan. Therefore, from the moment expression Eq. (3.2.29), the maximum moment is

$$M_{max} = M\left(\frac{L}{2}\right) = \frac{wL^2}{4u^2}[\sec u - 1]$$

$$= \frac{wL^2}{8}\left[\frac{2(\sec u - 1)}{u^2}\right]$$

$$= M_0\left[\frac{2(\sec u - 1)}{u^2}\right] \quad (3.2.36)$$

where $M_0 = wL^2/8$ is the maximum moment that would exist if the lateral load w were acting alone (i.e., if P were absent). The term in the square brackets thus represents the *moment amplification factor,* which magnifies the primary moment in the member due to the presence of an axial force.

Note that another way the maximum moment can be obtained is by realizing that M_{max} consists of two components: the primary moment M_0 caused by the lateral load w and the secondary moment caused by the axial load P acting through the maximum lateral deflection y_{max}, i.e.,

$$M_{max} = M_0 + Py_{max} \quad (3.2.37)$$

Using the expression for y_{max} [Eq. (3.2.30)] in Eq. (3.2.37), it can easily be shown that Eq. (3.2.36) is obtainable.

For the purpose of design application, we shall again simplify Eq. (3.2.36) by using the power series expansion for $\sec u$ [Eq. (3.2.31)] in Eq. (3.2.36).

$$M_{max} = M_0[1 + 0.4167u^2 + 0.1694u^4 + 0.06870u^6 + \cdots] \quad (3.2.38)$$

Using the expression for u in Eq. (3.2.33), it can be shown that

$$M_{max} = M_0\left[1 + 1.028\left(\frac{P}{P_e}\right) + 1.031\left(\frac{P}{P_e}\right)^2 \right.$$
$$\left. + 1.032\left(\frac{P}{P_e}\right)^3 + \cdots \right] \quad (3.2.39)$$

or

$$M_{max} = M_0\left\{1 + \left[1.028\left(\frac{P}{P_e}\right)\right]\left[1 + 1.003\left(\frac{P}{P_e}\right) \right.\right.$$
$$\left.\left. + 1.004\left(\frac{P}{P_e}\right)^2 + \cdots \right]\right\} \quad (3.2.40)$$

or, approximately

$$M_{max} \approx M_0\left\{1 + \left[1.028\left(\frac{P}{P_e}\right)\right]\left[1 + \left(\frac{P}{P_e}\right) + \left(\frac{P}{P_e}\right)^2 + \cdots \right]\right\}$$

$$= M_0\left\{1 + \left[1.028\left(\frac{P}{P_e}\right)\right]\left[\frac{1}{1-\left(\frac{P}{P_e}\right)}\right]\right\}$$

$$= M_0\left[\frac{1 + 0.028(P/P_e)}{1 - (P/P_e)}\right]$$

$$\approx M_0\left[\frac{1}{1 - P/P_e}\right] \quad (3.2.41)$$

where the term in the square brackets is the *design moment amplification factor*. Table 3.2 shows a comparison of the theoretical moment amplification factor [the term in brackets in Eq. (3.2.36)] with the design moment amplification factor [the term in brackets in Eq. (3.2.41)]. It can be seen that the two expressions give very comparable results.

3.3 BEAM-COLUMN WITH A CONCENTRATED LATERAL LOAD

3.3.1 The Closed-Form Solution (Fig. 3.3)

Figure 3.3a shows a simply supported beam-column acted on by a concentrated lateral load Q at a distance a from the left end and an axial

3.3 Beam-Column with a Concentrated Lateral Load

Table 3.2 Theoretical and Design Moment Amplification Factors for a Uniformly Loaded Beam-Column

$u = \dfrac{kL}{2} = \dfrac{\pi}{2}\sqrt{\dfrac{P}{P_e}}$	Theoretical Eq. (3.2.36)	Design Eq. (3.2.41)
0	1.000	1.000
0.20	1.017	1.016
0.40	1.071	1.069
0.60	1.176	1.171
0.80	1.360	1.350
1.00	1.702	1.681
1.20	2.444	2.402
1.40	4.983	4.863
$\pi/2$	∞	∞

force P. Referring to the free-body diagram in Fig. 3.3b, the differential equations for this beam-column can be written as

$$-EIy'' = \frac{Q(L-a)}{L}x + Py \quad \text{for} \quad 0 \leq x \leq a \qquad (3.3.1a)$$

$$-EIy'' = Qa\left(\frac{L-x}{L}\right) + Py \quad \text{for} \quad a \leq x \leq L \qquad (3.3.1b)$$

FIGURE 3.3 Beam-column with a concentrated lateral load

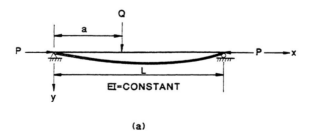

(a)

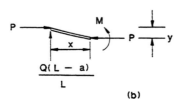

(b)

Rearranging and using the expression $k^2 = P/EI$, we obtain

$$y'' + k^2 y = \frac{-Q(L-a)}{LEI} x \quad \text{for} \quad 0 \leq x \leq a \quad (3.3.2a)$$

$$y'' + k^2 y = -\frac{Qa(L-x)}{LEI} \quad \text{for} \quad a \leq x \leq L \quad (3.3.2b)$$

The general solutions are

$$y = A \sin kx + B \cos kx - \frac{Q(L-a)}{LEIk^2} x \quad \text{for} \quad 0 \leq x \leq a \quad (3.3.3a)$$

$$y = C \sin kx + D \cos kx - \frac{Qa(L-x)}{LEIk^2} \quad \text{for} \quad a \leq x \leq L \quad (3.3.3b)$$

from which

$$y' = Ak \cos kx - Bk \sin kx - \frac{Q(L-a)}{LEIk^2} \quad \text{for} \quad 0 \leq x \leq a \quad (3.3.4a)$$

$$y' = Ck \cos kx - Dk \sin kx + \frac{Qa}{LEIk^2} \quad \text{for} \quad a \leq x \leq L \quad (3.3.4b)$$

Using the boundary conditions that there are no lateral displacements at the supports

$$y(0) \quad \text{in Eq. (3.3.3a)} = 0 \quad (3.3.5)$$

$$y(L) \quad \text{in Eq. (3.3.3b)} = 0 \quad (3.3.6)$$

and the continuity conditions that displacement y and slope y' must be continuous at the point of load application Q

$$y(a) \text{ in Eq. (3.3.3a)} = y(a) \text{ in Eq. (3.3.3b)} \quad (3.3.7)$$

$$y'(a) \text{ in Eq. (3.3.4a)} = y'(a) \text{ in Eq. (3.3.4b)} \quad (3.3.8)$$

the four constants A, B, C, and D can be determined as

$$A = \frac{Q \sin k(L-a)}{EIk^3 \sin kL} \quad (3.3.9)$$

$$B = 0 \quad (3.3.10)$$

$$C = \frac{-Q \sin ka}{EIk^3 \tan kL} \quad (3.3.11)$$

$$D = \frac{Q \sin ka}{EIk^3} \quad (3.3.12)$$

Substituting these constants into the deflection functions Eqs. (3.3.3a)

3.3 Beam-Column with a Concentrated Lateral Load

and (3.3.3b) gives

$$y = \frac{Q}{EIk^3} \frac{\sin k(L-a)}{\sin kL} \sin kx$$
$$- \frac{Q(L-a)}{LEIk^2} x \quad \text{for} \quad 0 \le x \le a \quad (3.3.13a)$$

$$y = -\frac{Q \sin ka}{EIk^3 \tan kL} \sin kx + \frac{Q \sin ka}{EIk^3} \cos kx$$
$$- \frac{Qa(L-x)}{LEIk^2} \quad \text{for} \quad a \le x \le L \quad (3.3.13b)$$

from which

$$y' = \frac{Q}{EIk^2} \frac{\sin k(L-a)}{\sin kL} \cos kx - \frac{Q(L-a)}{LEIk^2} \quad \text{for} \quad 0 \le x \le a \quad (3.3.14a)$$

$$y' = \frac{-Q \sin ka}{EIk^2 \tan kL} \cos kx - \frac{Q \sin ka}{EIk^2} \sin kx$$
$$+ \frac{Qa}{LEIk^2} \quad \text{for} \quad a \le x \le L \quad (3.3.14b)$$

and

$$y'' = -\frac{Q}{EIk} \frac{\sin k(L-a)}{\sin kL} \sin kx \quad \text{for} \quad 0 \le x \le a \quad (3.3.15a)$$

$$y'' = \frac{Q \sin ka}{EIk \tan kL} \sin kx - \frac{Q \sin ka}{EIk} \cos kx \quad \text{for} \quad a \le x \le L \quad (3.3.15b)$$

3.3.2 The Calculation of M_{max} and y_{max}

Consider now the special case in which the concentrated lateral load Q acts at midspan. By setting $a = L/2$ and $x = L/2$ in Eq. (3.3.13a or b) and Eq. (3.3.15a or b), we obtain the maximum deflection and maximum moment, respectively, as

$$y_{max} = \frac{QL^3}{48EI} \left[\frac{3(\tan u - u)}{u^3} \right] = y_0 \left[\frac{3(\tan u - u)}{u^3} \right] \quad (3.3.16)$$

$$M_{max} = \frac{QL}{4} \left[\frac{\tan u}{u} \right] = M_0 \left[\frac{\tan u}{u} \right] \quad (3.3.17)$$

in which $u = kL/2$, y_0 and M_0 are, respectively, the maximum deflection and moment that would exist if the axial force P were absent. The terms in brackets in Eqs. (3.3.16) and (3.3.17) are thus the *theoretical deflection and moment amplification factors.*

To simplify the expressions for the maximum deflection equation (3.3.16) and the maximum moment equation (3.3.17), we use the power series expansion for $\tan u$

$$\tan u = u + \frac{1}{3}u^3 + \frac{2}{15}u^5 + \frac{17}{315}u^7 + \cdots \quad (3.3.18)$$

Upon substituting Eq. (3.3.18) into Eqs. (3.3.16) and (3.3.17) and simplifying, it can be shown that these equations can be written approximately as

$$y_{\max} \approx y_0 \left[\frac{1}{1-\left(\frac{P}{P_e}\right)} \right] \quad (3.3.19)$$

$$M_{\max} \approx M_0 \left[\frac{1-0.18(P/P_e)}{1-(P/P_e)} \right] \approx M_0 \left[\frac{1-0.2(P/P_e)}{1-(P/P_e)} \right] \quad (3.3.20)$$

in which the terms in brackets in the above equations are the *design deflection and moment amplification factors,* respectively.

Tables 3.3 and 3.4 show a numerical comparison of the theoretical and design deflection and moment amplification factors, respectively. Good correlation between the theoretical and design amplification factors are observed.

At this point, the reader should recognize the similarity in form of the deflection amplification factors in Eqs. (3.2.35) and (3.3.19) and the

Table 3.3 Theoretical and Design Deflection Amplification Factor for a Beam-Column with a Concentrated Lateral Load at Midspan

$u = \dfrac{kL}{2} = \dfrac{\pi}{2}\sqrt{\dfrac{P}{P_e}}$	Theoretical Eq. (3.3.16)	Design Eq. (3.3.19)
0	1.000	1.000
0.20	1.016	1.016
0.40	1.068	1.069
0.60	1.169	1.171
0.80	1.346	1.350
1.00	1.672	1.681
1.20	2.382	2.402
1.40	4.808	4.863
$\pi/2$	∞	∞

3.4 Beam-Columns Subjected to End Moments

Table 3.4 Theoretical and Design Moment Amplification Factors for a Beam-Column with a Concentrated Lateral Load at Midspan

$u = \dfrac{kL}{2} = \dfrac{\pi}{2}\sqrt{\dfrac{P}{P_e}}$	Theoretical Eq. (3.3.17)	Design Eq. (3.3.20)
0	1.000	1.000
0.20	1.014	1.013
0.40	1.057	1.055
0.60	1.140	1.137
0.80	1.287	1.280
1.00	1.557	1.545
1.20	2.143	2.122
1.40	4.141	4.090
$\pi/2$	∞	∞

similarity in form of the moment amplification factors in Eqs. (3.2.41) and (3.3.20) for the simply supported beam-column under uniformly distributed and midspan concentrated lateral loads. We shall take advantage of these similarities in developing design formulas for beam-columns. This will be discussed later.

3.4 BEAM-COLUMNS SUBJECTED TO END MOMENTS

3.4.1 The Closed-Form Solution (Fig. 3.4)

So far, we have considered only the cases in which the primary bending moments in the beam-columns are caused by in-span lateral loads. In this section, we shall consider the case in which the primary bending moment is caused by end moments in the beam-column. Shown in Fig. 3.4a is a beam-column acted on by end couples M_A and M_B at the left and right ends of the member, respectively, and acted on by an axial force P. Using the free-body diagram of a segment of beam-column of length x from the left end (Fig. 3.4b), the external moment acting on the cut section is

$$M_{\text{ext}} = M_A + Py - \frac{M_A + M_B}{L}x \qquad (3.4.1)$$

Equating this to the internal moment of $-EIy''$ and rearranging, we have

$$EIy'' + Py = \frac{M_A + M_B}{L}x - M_A \qquad (3.4.2)$$

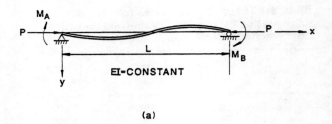

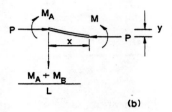

FIGURE 3.4 Beam-column with end couples (double-curvature bending)

or, using $k^2 = P/EI$, we can write

$$y'' + k^2 y = \frac{M_A + M_B}{LEI} x - \frac{M_A}{EI} \qquad (3.4.3)$$

The general solution is

$$y = A \sin kx + B \cos kx + \frac{M_A + M_B}{LEIk^2} x - \frac{M_A}{EIk^2} \qquad (3.4.4)$$

The constants A and B can be evaluated by using the boundary conditions

$$y(0) = 0, \qquad y(L) = 0 \qquad (3.4.5)$$

From the first boundary condition, we obtain

$$B = \frac{M_A}{EIk^2} \qquad (3.4.6)$$

and from the second boundary condition, we obtain

$$A = -\frac{1}{EIk^2 \sin kL} (M_A \cos kL + M_B) \qquad (3.4.7)$$

3.4 Beam-Columns Subjected to End Moments

Therefore, Eq. (3.4.4) can now be written as

$$y = -\frac{(M_A \cos kL + M_B)}{EIk^2 \sin kL} \sin kx + \frac{M_A}{EIk^2} \cos kx$$
$$+ \frac{M_A + M_B}{LEIk^2} x - \frac{M_A}{EIk^2} \qquad (3.4.8)$$

from which

$$y' = -\frac{(M_A \cos kL + M_B)}{EIk \sin kL} \cos kx - \frac{M_A}{EIk} \sin kx + \frac{M_A + M_B}{LEIk^2} \qquad (3.4.9)$$

and

$$y'' = \frac{(M_A \cos kL + M_B)}{EI \sin kL} \sin kx - \frac{M_A}{EI} \cos kx \qquad (3.4.10)$$

and

$$y''' = \frac{k(M_A \cos kL + M_B)}{EI \sin kL} \cos kx + \frac{kM_A}{EI} \sin kx \qquad (3.4.11)$$

To determine the location of the maximum moment, we set the shear force $(-EIy''')$, or Eq. (3.4.11), equal to zero. In doing so, we obtain the location $\bar{x}$

$$\tan k\bar{x} = \frac{-(M_A \cos kL + M_B)}{M_A \sin kL} \qquad (3.4.12)$$

From Fig. 3.5, it can be seen that

$$\sin k\bar{x} = \frac{(M_A \cos kL + M_B)}{\sqrt{M_A^2 + 2M_A M_B \cos kL + M_B^2}} \qquad (3.4.13)$$

$$\cos k\bar{x} = \frac{-M_A \sin kL}{\sqrt{M_A^2 + 2M_A M_B \cos kL + M_B^2}} \qquad (3.4.14)$$

FIGURE 3.5 Trigonometric relationship

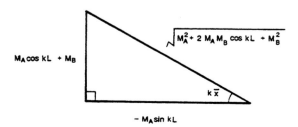

(Note: For $0 \leq k\bar{x} \leq kL = \pi\sqrt{P/P_e} \leq \pi$, we have $\sin k\bar{x} \geq 0$ and $\cos k\bar{x} \leq 0$)

The maximum moment is obtained by using the above expressions for $\sin k\bar{x}$ and $\cos k\bar{x}$ in the expression $M = -EIy''$ where y'' is given in Eq. (3.4.10). Thus

$$M_{max} = \frac{-(M_A \cos kL + M_B)^2}{\sin kL \sqrt{M_A^2 + 2M_A M_B \cos kL + M_B^2}}$$
$$- \frac{M_A^2 \sin kL}{\sqrt{M_A^2 + 2M_A M_B \cos kL + M_B^2}}$$
$$= -\frac{\sqrt{M_A^2 + 2M_A M_B \cos kL + M_B^2}}{\sin kL}$$
$$= -M_B \left[\sqrt{\frac{(M_A/M_B)^2 + 2(M_A/M_B)\cos kL + 1}{\sin^2 kL}} \right] \quad (3.4.15)$$

The minus sign that appears in Eq. (3.4.15) simply indicates that M_{max} causes tension on the top fiber of the cross section.

If M_B is the larger of the two end moments, then the terms in the brackets in Eq. (3.4.15) represent the moment amplification factor for the beam-column subjected to end moments M_A, M_B and an axial force P. Note that this amplification factor depends not only on the magnitude of the axial force, but also on the magnitude of the ratio of the end moments.

For members bent in *double curvature*, sometimes the maximum moment occurs at the end and is therefore equal to M_B, as shown in Fig. 3.6. If this is the case, the amplification factor in Eq. (3.4.15) becomes meaningless because this theoretical maximum moment occurs outside the length of the beam-column. To check whether Eq. (3.4.15) is applicable then for a given value of M_A, M_B, and P, one should also evaluate $\bar{x}$ from Eq. (3.4.12). If the calculated value of $\bar{x}$ does not fall within the range $0 \leq \bar{x} \leq L$, Eq. (3.4.15) is not applicable and the maximum moment occurs at the end and is equal to the larger of the two end moments (see Problem 3.6).

It should also be mentioned that in the development of Eq. (3.4.15) only member overall stability is considered. Failure due to lateral torsional buckling or buckling due to unwinding from double to single curvature is not considered. The phenomenon of lateral torsional buckling is the subject of discussion in Chapter 5. The phenomenon of buckling due to unwinding from double to single curvature is beyond the scope of this book, but is discussed in detail by Ketter.[4] Generally speaking, this type of buckling will occur if the ratio M_A/M_B lies in the range 0.5 to 1.0. As a result, Eq. (3.4.15) is not applicable if M_A/M_B is between 0.5 and 1.0.

3.4 Beam-Columns Subjected to End Moments

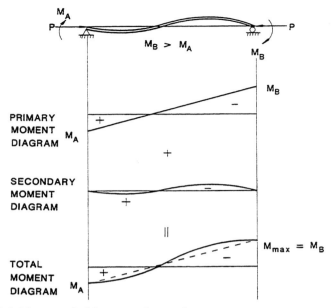

FIGURE 3.6 M_{max} equals M_B at member end

Although Eq. (3.4.15) has been developed for a member bent in double curvature, the same expression can also be used for a member bent in single curvature (Fig. 3.7) simply by replacing M_B by $-M_B$. Thus, for members bent in single curvature, the expression for maximum moment is

$$M_{max} = M_B \left[\sqrt{\frac{(M_A/M_B)^2 - 2(M_A/M_B)\cos kL + 1}{\sin^2 kL}} \right] \quad (3.4.16)$$

Here, just as in the case of members bent in double curvature, the maximum moment for a certain combination of M_A, M_B, and P, of a beam-column bent in single curvature may occur at the member end rather than within the members (Fig. 3.8). Thus, to check the validity of Eq. (3.4.16), one should evaluate $\bar{x}$ from the equation

$$\tan k\bar{x} = \frac{-(M_A \cos kL - M_B)}{M_A \sin kL} \quad (3.4.17)$$

If the calculated value of $\bar{x}$ falls outside the range $0 \le \bar{x} \le L$, the maximum moment occurs at the member end. Note that Eq. (3.4.17) is the same as Eq. (3.4.12), except that M_B has been replaced by $-M_B$.

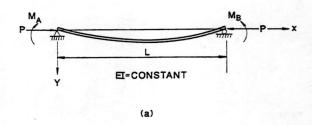

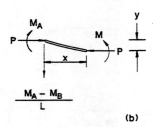

(b)

FIGURE 3.7 Beam-column with end couples (single-curvature bending)

FIGURE 3.8 M_{max} equals M_B at member end

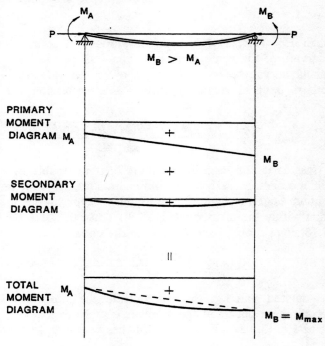

3.4 Beam-Columns Subjected to End Moments

Equations (3.4.15) and (3.4.16) can be written in a combined form as

$$M_{max} = |M_B| \left[\sqrt{\frac{(M_A/M_B)^2 + 2(M_A/M_B)\cos kL + 1}{\sin^2 kL}} \right] \quad (3.4.18)$$

where M_A/M_B is *positive* if the member is bent on *double curvature* and is *negative* if it is bent in *single curvature*. The absolute value for M_B is used in the coefficient of Eq. (3.4.18) because we are interested only in the magnitude, not the direction of M_{max}.

A special case for a beam column bent in single curvature is the case in which the end moments are equal and opposite, i.e., if $M_A = -M_B = M$, as shown in Fig. 3.9. For this case the maximum moment is given by substituting $M_A/M_B = -1$ in Eq. (3.4.18)

$$M_{max} = M \sqrt{\frac{2(1 - \cos kL)}{\sin^2 kL}} \quad (3.4.19)$$

and its location is always at midspan as depicted in the figure.

FIGURE 3.9 Beam-column with equal and opposite end couples

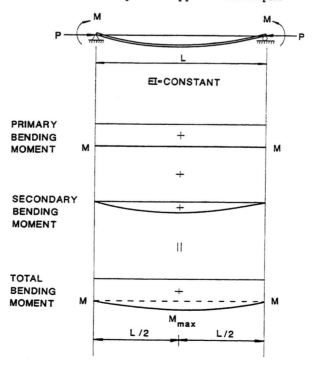

3.4.2 Concept of Equivalent Moment

Equation (3.4.18) is the expression for the maximum moment for the general case of a beam-column subjected to unequal end moments. The maximum moment may occur at a member's end, and be equal to the larger of the two end moments, or it may occur somewhere within the member whose magnitude is given by Eq. (3.4.18) and whose location is determined by Eq. (3.4.12) or Eq. (3.4.17). For the purpose of design, one needs to know whether the maximum moment occurs at the end or away from the ends, and also the location of the maximum moment if it should occur away from the ends. To eliminate these calculations, the *concept of equivalent moment* is introduced in design practice.

The concept of equivalent moment is shown schematically in Fig. 3.10. The end moments M_A and M_B that act on the member are replaced by a pair of equal and opposite equivalent moment M_{eq}. The magnitude of the equivalent moment is such that the maximum moment produced by it will be equal to that produced by the actual end moments M_A and M_B. Mathematically, one can obtain the equivalent moment by setting $M = M_{eq}$ in Eq. (3.4.19) and equate this to the M_{max} in Eq. (3.4.18).

$$|M_B|\left[\sqrt{\frac{(M_A/M_B)^2 + 2(M_A/M_B)\cos kL + 1}{\sin^2 kL}}\right]$$

$$= M_{eq}\left[\sqrt{\frac{2(1-\cos kL)}{\sin^2 kL}}\right] \quad (3.4.20)$$

from which we solve for

$$M_{eq} = \left[\sqrt{\frac{(M_A/M_B)^2 + 2(M_A/M_B)\cos kL + 1}{2(1-\cos kL)}}\right]|M_B|$$

$$= C_m |M_B| \quad (3.4.21)$$

in which C_m is the *equivalent moment factor*.

FIGURE 3.10 Schematic representation of the concept of equivalent moment

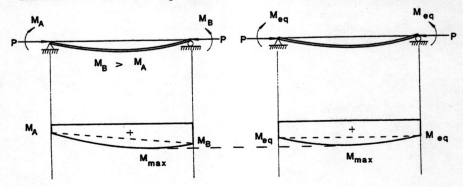

3.4 Beam-Columns Subjected to End Moments

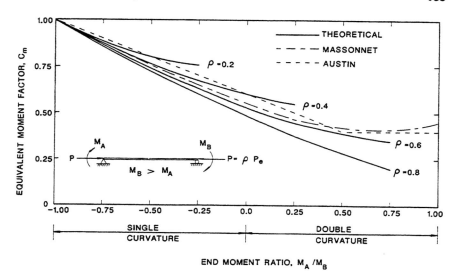

FIGURE 3.11 Comparison of various expressions for C_m.

As can be seen from Eq. (3.4.21), the equivalent moment factor C_m is a function of the moment ratio M_A/M_B and the axial force P. Simplified expressions for C_m that eliminate its dependency on the axial force have been proposed by Massonnet[5] and Austin.[6] The Massonnet expression is

$$C_m = \sqrt{0.3(M_A/M_B)^2 - 0.4(M_A/M_B) + 0.3} \qquad (3.4.22)$$

and the Austin expression is

$$C_m = 0.6 - 0.4(M_A/M_B) \geq 0.4 \qquad (3.4.23)$$

The various expressions for C_m are plotted in Fig. 3.11. As can be seen, the simplified expressions give a rather good approximation to the theoretical one. Because of its simplicity, the Austin expression was adopted in the AISC/ASD,[7] and LRFD[8] Specifications for the design of steel structures. Note that in Fig. 3.11 the curves for the theoretical C_m, each for a given value of $\rho = P/P_e$, terminates when the larger of the two end moments (i.e., M_B) represents the maximum moment of the member.

3.4.3 The Calculation of M_{max}

For a beam-column subjected to end moments only, the following steps are thus taken to evaluate M_{max} for design.

1. Evaluate C_m from Eq. (3.4.23).

2. Calculate the equivalent moment by multiplying the C_m factor by the larger of the two end moments (Eq. 3.4.21).
3. Calculate the maximum moment from Eq. (3.4.19) with $M = M_{eq}$.

Equation (3.4.19) can be further simplified as follows. Using the trigonometric identities

$$1 - \cos kL = 2 \sin^2 \frac{kL}{2} \quad (3.4.24)$$

and

$$\sin^2 kL = 4 \sin^2 \frac{kL}{2} \cos^2 \frac{kL}{2} \quad (3.4.25)$$

Eq. (3.4.19) can be written as

$$M_{max} = M_{eq} \sec \frac{kL}{2} \approx M_{eq}\left(\frac{1}{1 - \frac{P}{P_e}}\right) \quad (3.4.26)$$

and since $M_{eq} = C_m M_B$, the maximum moment can now be computed by the simple formula

$$M_{max} = \left(\frac{C_m}{1 - \frac{P}{P_e}}\right) M_B = A_F M_B \quad (3.4.27)$$

In summary, to obtain the maximum moment for a *nonsway* beam-column subjected to end moments only, one need only multiply the larger of the two end moments by a factor $A_F = C_m/[1 - [(P/P_e)]$. For Eq. (3.4.27) to have physical meaning, the moment magnification factor must be greater than or equal to unity, otherwise, the larger end moment M_B will be taken as M_{max}. This fact is observed in the LRFD Specification,[8] but not in the ASD Specification.[7]

3.5 SUPERPOSITION OF SOLUTIONS

3.5.1 Simply Supported Beam-Column (Fig. 3.12)

As mentioned previously, the *principle of superposition* holds for beam-columns so long as the axial force remains constant and the same axial force is applied to each component of the solution. To demonstrate the use of this principle, we will analyze the beam-column shown in Fig. 3.12a. This simply supported beam-column is subjected to a uniformly distributed lateral load of w and a concentrated lateral load of Q at midspan. A constant axial force P is also applied to the member. The

3.5 Superposition of Solutions

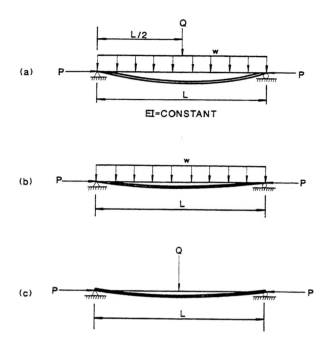

FIGURE 3.12 Beam-column loaded by uniformly distributed and concentrated lateral loads

deflected shape of this beam-column can be obtained by superposing the deflected shapes of Fig. 3.12b and c, i.e.,

$$y = \text{Eq. (3.2.26)} + \text{Eq. (3.3.13a)}\big|_{a=L/2}$$

$$= \frac{w}{EIk^4}\left[\tan\frac{kL}{2}\sin kx + \cos kx - 1\right]$$

$$-\frac{w}{2EIk^2}x(L-x) + \frac{Q}{EIk^3}\sin\frac{kL}{2}\left(\frac{1}{\sin kL}\right)\sin kx - \frac{Q}{2EIk^2}x$$

$$= \frac{1}{EIk^4}\left[\left(w\tan\frac{kL}{2} + \frac{Qk}{2}\sec\frac{kL}{2}\right)\sin kx \right.$$

$$\left. + w\cos kx + \frac{1}{2}wk^2x^2 - \left(\frac{wLk^2}{2} + \frac{Q}{2}k^2\right)x - w\right] \quad (3.5.1)$$

To obtain the maximum moment, we set the shear force (or $-EIy'''$) equal to zero (where a prime denotes derivative with respect to x) to obtain the location $\bar{x}$ and then backsubstitute the value $\bar{x}$ that we obtained this way into the expression $M = -EIy''$. However, for this example there is a simpler approach. Knowing that the maximum moment occurs at midspan for both the uniformly loaded case (Fig. 3.12b) and the concentrated loaded case (Fig. 3.12c), we can just add the maximum moment for these two cases together to obtain the maximum moment of the combined loading case (Fig. 3.12a). Thus

$$M_{max} = \text{Eq. (3.2.36)} + \text{Eq. (3.3.17)}$$
$$= \frac{wL^2}{8}\left[\frac{2(\sec u - 1)}{u^2}\right] + \frac{QL}{4}\left[\frac{\tan u}{u}\right] \quad (3.5.2)$$

3.5.2 Fixed-Ended Beam-Columns

Another application of the principle of superposition is to determine the fixed-end moments of a beam-column.

Uniform Load Case (Fig. 3.13)

We will now consider a beam-column fixed at both ends and loaded by a uniformly distributed lateral load w and an axial force P (Fig. 3.13a). We

FIGURE 3.13 Fixed-end moments of a uniformly loaded beam-column

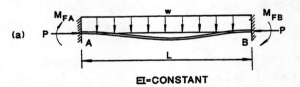

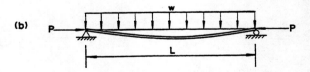

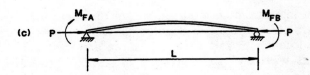

3.5 Superposition of Solutions

will determine the fixed-end moments (M_{FA}, M_{FB}) of this member using the principle of superposition. The beam-column in Fig. 3.13a can be decomposed into that of Fig. 3.13b and Fig. 3.13c. To satisfy the continuity condition of zero slopes at the built-in ends, the algebraic sum of the rotation at the ends produced by the uniform load w (Fig. 3.13b) and that produced by the end moments (Fig. 3.13c) must be zero. Because of symmetry, we need only consider half of the member. By taking the derivative of Eq. (3.2.28) and evaluating the resulting equation at $x = 0$, we obtain the A end rotation produced by the uniform load w as

$$y'_w(0) = \frac{wL^3}{24EI}\left[\frac{3(\tan u - u)}{u^3}\right] \quad (3.5.3)$$

and from Eq. (3.4.9), by setting $M_A = M_{FA}$, $M_B = M_{FB} = -M_{FA}$ and $x = 0$, we obtain the A end rotation produced by the end moments M_{FA} and M_{FB} as

$$y'_M(0) = \frac{M_{FA}L}{EI}\left[\frac{1 - \cos kL}{kL \sin kL}\right] \quad (3.5.4)$$

Using the trigonometric identities of

$$1 - \cos kL = 2\sin^2 kL/2,$$

we can write Eq. (3.5.4) with

$$u = \frac{kL}{2}$$

as

$$y'_M(0) = \frac{M_{FA}L}{2EI}\left[\frac{\tan u}{u}\right] \quad (3.5.5)$$

Since the continuity condition at the fixed-end requires that

$$y'_w(0) + y'_M(0) = 0 \quad (3.5.6)$$

the substitution of Eqs. (3.5.3) and (3.5.5) into Eq. (3.5.6) gives the fixed-end moment

$$M_{FA} = -\frac{wL^2}{12}\left[\frac{3(\tan u - u)}{u^2 \tan u}\right] \quad (3.5.7)$$

The minus sign in Eq. (3.5.7) indicates that the direction of M_{FA} is opposite to that shown in Fig. 3.13. Note that the term $wL^2/12$ is the fixed-end moment for a member subjected to uniformly distributed load only. Thus, the terms in the brackets represent the effect of axial force on the fixed-end moment of the member.

Concentrated Load Case (Fig. 3.14)

The same procedure can be used to evaluate the fixed-end moment of a beam-column subjected to a concentrated lateral load Q acting at midspan of the member (Fig. 3.14a).

Again, because of symmetry, we only need consider half of the member. From Eq. (3.3.14a), by setting $a = L/2$, we obtain the A end rotation due to the lateral load Q as

$$y'_Q(0) = \frac{Q}{EIk^2} \frac{\sin \frac{kL}{2}}{\sin kL} - \frac{Q}{2EIk^2}$$

$$= \frac{QL^2}{8EIu^2}[\sec u - 1] \qquad (3.5.8)$$

where $u = kL/2$.

The end rotation at the A end due to the end moments M_{FA}

FIGURE 3.14 Fixed-end moments of a beam-column with a concentrated transverse load at midspan

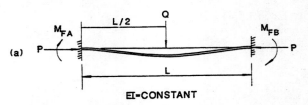

EI=CONSTANT

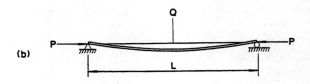

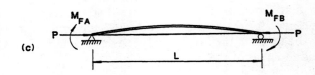

and $M_{FB} = -M_{FA}$ is obtained from Eq. (3.5.5)

$$y'_M(0) = \frac{M_{FA}L}{2EI}\left[\frac{\tan u}{u}\right] \quad (3.5.5)$$

Since the slope at the fixed-end must be zero, we must satisfy the condition

$$y'_Q(0) + y'_M(0) = 0 \quad (3.5.9)$$

Upon substituting Eqs. (3.5.8) and (3.5.5) into (3.5.9), we can determine the fixed-end moment as

$$M_{FA} = \frac{-QL}{8}\left[\frac{2(\sec u - 1)}{u \tan u}\right] = \frac{-QL}{8}\left[\frac{2(1 - \cos u)}{u \sin u}\right] \quad (3.5.10)$$

The minus sign indicates that the direction of M_{FA} is opposite to that shown in Fig. 3.14. The term $QL/8$ that appeared in Eq. (3.5.10) represents the fixed-end moment of the member when P is absent. The terms in the brackets thus represent the effect of axial force on the fixed-end moment of the member.

3.6 BASIC DIFFERENTIAL EQUATIONS

Up to this point, we have derived the differential equation of the beam-column by considering moment equilibrium of a segment of the member cut at a distance x from the left support for a given beam-column problem. The resulting linear differential equation has been second-order. In this section, we will develop the basic differential equations of a beam-column subjected to general lateral loadings.

As seen in Fig. 3.15a, an initially straight beam-column is subjected to an axial force P and a lateral load $w(x)$ along its entire span. To examine the stability of the member, we have to consider the equilibrium state in the deflected configuration. Figure 3.15b shows an infinitesimal element of the deflected member of projected length dx. The longitudinal force P and transverse force V are shown with their directions parallel and normal to the *undeflected* axis of the member, respectively. Summing forces horizontally, we have

$$P - \left(P + \frac{dP}{dx}dx\right) = 0$$

or

$$\frac{dP}{dx} = 0 \quad (3.6.1)$$

Summing forces vertically, we have

$$V - \left(V + \frac{dV}{dx}dx\right) - w(x)\,dx = 0$$

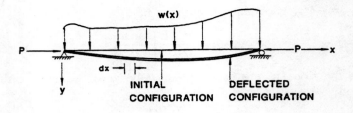

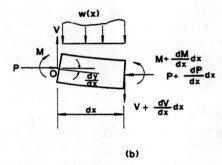

FIGURE 3.15 General differential equation of a beam-column

or

$$\frac{dV}{dx} = -w(x) \tag{3.6.2}$$

Summing moments about point 0, we have

$$-M + \left(M + \frac{dM}{dx}dx\right) - \left(V + \frac{dV}{dx}dx\right)dx$$
$$-\left(P + \frac{dP}{dx}dx\right)\frac{dy}{dx}dx - w(x)\,dx\,\frac{dx}{2} = 0$$

or

$$\frac{dM}{dx}dx - V\,dx - \frac{dV}{dx}(dx)^2 - P\frac{dy}{dx}dx$$
$$-\frac{dP}{dx}\frac{dy}{dx}(dx)^2 - w(x)\frac{(dx)^2}{2} = 0$$

3.6 Basic Differential Equations

or, neglecting the higher order terms involving $(dx)^2$, we obtain

$$\frac{dM}{dx} - V - P\frac{dy}{dx} = 0 \quad \text{or} \quad \frac{dM}{dx} = V + P\frac{dy}{dx} \tag{3.6.3}$$

Equation (3.6.1) states that there is no change in longitudinal force in the segment. That is to say, the force P remains constant in the element. Equation (3.6.2) states that the rate of change in transverse force across the segment is equal numerically to the magnitude of the applied transverse load $w(x)$, which is assumed to be constant along the infinitesimal length dx. Equation (3.6.3) states that the rate of change in moment across the segment is equal numerically to the transverse force V plus the longitudinal force P times the change in slope dy/dx of the infinitesimal element.

For small deflection, we can write $M = -EI\, d^2y/dx^2$. Upon substitution into Eq. (3.6.3) and rearranging, we obtain

$$EI\frac{d^3y}{dx^3} + P\frac{dy}{dx} = -V \tag{3.6.4}$$

or using $k^2 = P/EI$

$$\frac{d^3y}{dx^3} + k^2\frac{dy}{dx} = -\frac{V}{EI} \tag{3.6.5}$$

If we differentiate Eq. (3.6.4) with respect to x and substitute Eq. (3.6.2) into the resulting equation, we have

$$EI\frac{d^4y}{dx^4} + P\frac{d^2y}{dx^2} = w(x) \tag{3.6.6}$$

or with $k^2 = P/EI$

$$\frac{d^4y}{dx^4} + k^2\frac{d^2y}{dx^2} = \frac{w(x)}{EI} \tag{3.6.7}$$

Equation (3.6.5) is the basic differential equation of a beam-column relating the lateral deflection y, the axial thrust P, and the transverse force V. The general solution to this equation is

$$y = A \sin kx + B \cos kx + C + f(x) \tag{3.6.8}$$

where $f(x)$ is the particular solution of the differential equation. Equation (3.6.7) is the basic differential equation of a beam-column relating the lateral deflection y, the axial thrust P, and the transverse loading $w(x)$. The general solution to this equation is

$$y = A \sin kx + B \cos kx + Cx + D + f(x) \tag{3.6.9}$$

where $f(x)$ is the particular solution of the differential equation. The following examples will be used to demonstrate the use of Eqs. (3.6.5) and (3.6.7).

3.6.1 Fixed-Fixed Beam-Column with Concentrated Load at Midspan

The beam-column under investigation is shown in Fig. 3.16. Because of symmetry, only half of the member is considered in the analysis. Since the transverse force acting on any cut section is constant and is equal to $Q/2$, we have, from Eq. (3.6.5),

$$\frac{d^3y}{dx^3} + k^2 \frac{dy}{dx} = -\frac{Q}{2EI} \qquad (3.6.10)$$

From Eq. (3.6.8), the general solution is

$$y = A \sin kx + B \cos kx + C - \frac{Q}{2EIk^2} x \qquad (3.6.11)$$

The constants A, B, and C are determined from the boundary conditions

$$y(0) = 0, \qquad y'(0) = 0, \qquad y'\left(\frac{L}{2}\right) = 0 \qquad (3.6.12)$$

FIGURE 3.16 M_{max} of a beam-column with a concentrated transverse load at midspan

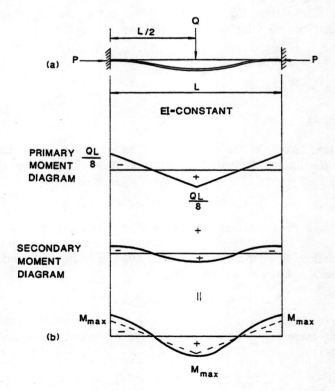

3.6 Basic Differential Equations

Using these boundary conditions, we find

$$A = \frac{Q}{2EIk^3} \tag{3.6.13}$$

$$B = \frac{Q}{2EIk^3 \sin\frac{kL}{2}} \left(\cos\frac{kL}{2} - 1\right) \tag{3.6.14}$$

$$C = \frac{Q}{2EIk^3 \sin\frac{kL}{2}} \left(1 - \cos\frac{kL}{2}\right) \tag{3.6.15}$$

With these constants, the deflection function (3.6.11) is fully determined

$$y = \frac{Q}{2EIk^3}\left[\sin kx + \left(\cos\frac{kL}{2} - 1\right)\frac{\cos kx}{\sin\frac{kL}{2}} + \frac{\left(1 - \cos\frac{kL}{2}\right)}{\sin\frac{kL}{2}} - kx\right] \tag{3.6.16}$$

The deflection at midspan is

$$y\left(\frac{L}{2}\right) = \frac{QL^3}{192EI}\left[\frac{12(2 - 2\cos u - u \sin u)}{u^3 \sin u}\right] \tag{3.6.17}$$

where $u = kL/2$

The term $QL^3/192EI$, which appeared in Eq. (3.6.17), represents the midspan deflection when the axial force is not present. Thus, the terms in the brackets represent the effect of axial force on the primary deflection of the member.

The maximum moment for this beam-column occurs at midspan and at the ends (see Fig. 3.16b) and is equal to

$$M_{max} = -EIy''(0) = \frac{-QL}{8}\left[\frac{2(1-\cos u)}{u \sin u}\right] \tag{3.6.18}$$

Note the correspondence of Eq. (3.6.18) with Eq. (3.5.10).

For the purpose of design, it is more convenient to approximate Eq. (3.6.18) in the format of Eqs. (3.2.41) and (3.3.20), i.e.,

$$M_{max} = M_0\left[\frac{1 + \Psi(P/P_{ek})}{1 - (P/P_{ek})}\right] \tag{3.6.19}$$

in which

M_0 = maximum primary moment (maximum moment when the axial force is absent)

P_{ek} = critical load of the beam-column considering the end conditions when the lateral force is absent

Ψ = constant

Table 3.5 Theoretical and Design Moment Amplification Factors for a Fixed-Ended Beam-Column Loaded by a Concentrated Lateral Load at Midspan

$u = \dfrac{kL}{2} = \dfrac{\pi}{2K}\sqrt{\dfrac{P}{P_{ek}}}$	Theoretical Eq. (3.6.18)	Design Eq. (3.6.19)
0	1.000	1.000
0.20	1.003	1.003
0.40	1.014	1.013
0.60	1.031	1.030
0.80	1.057	1.055
1.00	1.093	1.090
1.20	1.140	1.137
1.40	1.203	1.198
π	∞	∞

For the present case, $M_0 = QL/8$, $P_{ek} = \pi^2 EI/(KL)^2$ in which $K = 0.5$. The value of Ψ can be determined by equating Eq. (3.6.18) to (3.6.19), and solving for Ψ

$$\Psi = \frac{1}{(P/P_{ek})}\left\{\left|\frac{2(\cos u - 1)}{u \sin u}\right|[1-(P/P_{ek})] - 1\right\} \quad (3.6.20)$$

in which

$$u = \frac{kL}{2} = \frac{L}{2}\sqrt{\frac{P}{EI}} = \frac{\pi}{2K}\sqrt{\frac{P}{P_{ek}}}$$

It can be shown that the value of Ψ does not vary too much for various values of P/P_{ek}. As shown in Table 3.5, for $\Psi = -0.2$, Eq. (3.6.19) gives a good approximation to Eq. (3.6.18).

3.6.2 Fixed-Fixed Beam-Column with Uniformly Distributed Loads

The differential equation for this case is from Eq. (3.6.7)

$$\frac{d^4y}{dx^4} + k^2\frac{d^2y}{dx^2} = \frac{w}{EI} \quad (3.6.21)$$

From Eq. (3.6.9), the general solution is

$$y = A \sin kx + B \cos kx + Cx + D + \frac{w}{2EIk^2}x^2 \quad (3.6.22)$$

The constants A, B, C, and D are determined from the boundary

3.6 Basic Differential Equations

conditions

$$y(0) = 0, \quad y'(0) = 0, \quad y(L) = 0, \quad y'(L) = 0 \quad (3.6.23)$$

or, alternatively, by making use of the condition of symmetry, the four boundary conditions are

$$y(0) = 0, \quad y'(0) = 0, \quad y'\left(\frac{L}{2}\right) = 0, \quad y'''\left(\frac{L}{2}\right) = 0 \quad (3.6.24)$$

Using either set of these four conditions, we find

$$A = \frac{wL}{2EIk^3} \quad (3.6.25)$$

$$B = \frac{wL}{2EIk^3 \tan \frac{kL}{2}} \quad (3.6.26)$$

$$C = -\frac{wL}{2EIk^2} \quad (3.6.27)$$

$$D = -\frac{wL}{2EIk^3 \tan \frac{kL}{2}} \quad (3.6.28)$$

With these constants, Eq. (3.6.22) can be written as

$$y = \frac{wL}{2EIk^3}\left[\sin kx + \frac{\cos kx}{\tan \frac{kL}{2}} - kx - \frac{1}{\tan \frac{kL}{2}} + \frac{kx^2}{L}\right] \quad (3.6.29)$$

The deflection at midspan is

$$y\left(\frac{L}{2}\right) = \frac{wL^4}{384EI}\left[\frac{12(2 - 2\cos u - u \sin u)}{u^3 \sin u}\right] \quad (3.6.30)$$

Again, the term $wL^4/384EI$ represents the lateral deflection when the axial force P is absent, and the terms in square brackets represent the effect of axial thrust on the lateral deflection.

The maximum moment for this beam-column occurs at the fixed-end and is equal to

$$M_{\max} = -EIy''(0) = \frac{-wL^2}{12}\left[\frac{3(\tan u - u)}{u^2 \tan u}\right] \quad (3.6.31)$$

The reader should note the correspondence of Eq. (3.6.31) with Eq. (3.5.7). Again, for the purpose of design, it is more convenient to approximate Eq. (3.6.31) in the format of Eq. (3.6.19). With reference to

Table 3.6 Theoretical and Design Moment Amplification Factors for a Fixed-Ended Beam-Column Loaded by a Uniformly Distributed Lateral Load

$u = \dfrac{kL}{2} = \dfrac{\pi}{2K}\sqrt{\dfrac{P}{P_{ek}}}$	Theoretical Eq. (3.6.31)	Design Eq. (3.6.19)
0	0	0
0.20	1.003	1.002
0.40	1.011	1.010
0.60	1.025	1.023
0.80	1.045	1.042
1.00	1.074	1.068
1.20	1.111	1.102
1.40	1.161	1.149
π	∞	∞

Eq. (3.6.19), M_0 is now $wL^2/12$, $P_{ek} = \pi^2 EI/(KL)^2$ where $K = 0.5$. The value of Ψ is

$$\Psi = \frac{1}{(P/P_{ek})}\left\{\left|\frac{3(\tan u - u)}{u^2 \tan u}\right|[1 - (P/P_{ek})] - 1\right\} \qquad (3.6.32)$$

Again, the value of Ψ does not vary too much for various values of P/P_{ek}. By choosing a Ψ-value equal to -0.4, we will observe a good correlation between the theoretical M_{max} as expressed in Eq. (3.6.31) and the approximate M_{max} as expressed in Eq. (3.6.19) (Table 3.6).

3.7 SLOPE-DEFLECTION EQUATIONS

In this section, we will develop the *slope-deflection equations* for a beam-column. Consider the beam-column shown in Fig. 3.17; we now want to establish a relationship between the end moments (M_A, M_B) and the end rotations (θ_A, θ_B).

From Eq. (3.4.8), the deflection function for this beam-column has the

FIGURE 3.17 Beam-column subjected to end moments (without relative joint translation)

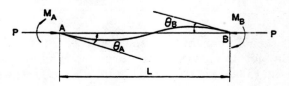

3.7 Slope-Deflection Equations

form

$$y = -\frac{(M_A \cos kL + M_B)}{EIk^2 \sin kL} \sin kx + \frac{M_A}{EIk^2} \cos kx$$
$$+ \frac{M_A + M_B}{LEIk^2} x - \frac{M_A}{EIk^2} \quad (3.7.1)$$

Rearranging, we have

$$y = -\frac{1}{EIk^2}\left[\frac{\cos kL}{\sin kL}\sin kx - \cos kx - \frac{x}{L} + 1\right]M_A$$
$$-\frac{1}{EIk^2}\left[\frac{1}{\sin kL}\sin kx - \frac{x}{L}\right]M_B \quad (3.7.2)$$

from which

$$y' = -\frac{1}{EIk}\left[\frac{\cos kL}{\sin kL}\cos kx + \sin kx - \frac{1}{kL}\right]M_A$$
$$-\frac{1}{EIk}\left[\frac{\cos kx}{\sin kL} - \frac{1}{kL}\right]M_B \quad (3.7.3)$$

Using Eq. (3.7.3), the end rotations θ_A and θ_B can be obtained as

$$\theta_A = y'(0) = -\frac{1}{EIk}\left[\frac{\cos kL}{\sin kL} - \frac{1}{kL}\right]M_A$$
$$-\frac{1}{EIk}\left[\frac{1}{\sin kL} - \frac{1}{kL}\right]M_B$$
$$= \frac{L}{EI}\left[\frac{\sin kL - kL \cos kL}{(kL)^2 \sin kL}\right]M_A$$
$$+ \frac{L}{EI}\left[\frac{\sin kL - kL}{(kL)^2 \sin kL}\right]M_B \quad (3.7.4)$$

$$\theta_B = y'(L) = -\frac{1}{EIk}\left[\frac{1}{\sin kL} - \frac{1}{kL}\right]M_A$$
$$-\frac{1}{EIk}\left[\frac{\cos kL}{\sin kL} - \frac{1}{kL}\right]M_B$$
$$= \frac{L}{EI}\left[\frac{\sin kL - kL}{(kL)^2 \sin kL}\right]M_A$$
$$+ \frac{L}{EI}\left[\frac{\sin kL - kL \cos kL}{(kL)^2 \sin kL}\right]M_B \quad (3.7.5)$$

Equations (3.7.4) and (3.7.5) can be written in matrix form as

$$\begin{bmatrix} \theta_A \\ \theta_B \end{bmatrix} = \begin{bmatrix} f_{11} & f_{12} \\ f_{21} & f_{22} \end{bmatrix} \begin{bmatrix} M_A \\ M_B \end{bmatrix} \quad (3.7.6)$$

where

$$f_{11} = f_{22} = \frac{L}{EI}\left[\frac{\sin kL - kL \cos kL}{(kL)^2 \sin kL}\right] \quad (3.7.7)$$

$$f_{12} = f_{21} = \frac{L}{EI}\left[\frac{\sin kL - kL}{(kL)^2 \sin kL}\right] \quad (3.7.8)$$

From Eq. (3.7.6), we can write

$$\begin{bmatrix} M_A \\ M_B \end{bmatrix} = \begin{bmatrix} f_{11} & f_{12} \\ f_{21} & f_{22} \end{bmatrix}^{-1} \begin{bmatrix} \theta_A \\ \theta_B \end{bmatrix} \quad (3.7.9)$$

or

$$\begin{bmatrix} M_A \\ M_B \end{bmatrix} = \begin{bmatrix} c_{11} & c_{12} \\ c_{21} & c_{22} \end{bmatrix} \begin{bmatrix} \theta_A \\ \theta_B \end{bmatrix} \quad (3.7.10)$$

where

$$c_{11} = c_{22} = \frac{EI}{L}\left[\frac{kL \sin kL - (kL)^2 \cos kL}{2 - 2\cos kL - kL \sin kL}\right] \quad (3.7.11)$$

$$c_{12} = c_{21} = \frac{EI}{L}\left[\frac{(kL)^2 - kL \sin kL}{2 - 2\cos kL - kL \sin kL}\right] \quad (3.7.12)$$

Equation (3.7.10) can be written in its expanded form as

$$M_A = \frac{EI}{L}(s_{ii}\theta_A + s_{ij}\theta_B) \quad (3.7.13)$$

$$M_B = \frac{EI}{L}(s_{ji}\theta_A + s_{jj}\theta_B) \quad (3.7.14)$$

where

$$s_{ii} = s_{jj} = \frac{c_{11}L}{EI} = \frac{c_{22}L}{EI} \quad (3.7.15)$$

$$s_{ij} = s_{ji} = \frac{c_{12}L}{EI} = \frac{c_{21}L}{EI} \quad (3.7.16)$$

are referred to as the *stability functions*.

Equations (3.7.13) and (3.7.14) are the slope-deflection equations for a

Table 3.7 Stability Functions ($kL = \pi\sqrt{P/P_e}$)

		compression		tension	
kL	P/P_e	s_{ii}	s_{ij}	s_{ii}	s_{ij}
0.	0.	4.0000	2.0000	4.0000	2.0000
0.0500	0.0003	3.9997	2.0001	4.0003	1.9999
0.1000	0.0010	3.9987	2.0003	4.0013	1.9997
0.1500	0.0023	3.9970	2.0008	4.0030	1.9993
0.2000	0.0041	3.9947	2.0013	4.0053	1.9987
0.2500	0.0063	3.9917	2.0021	4.0083	1.9979
0.3000	0.0091	3.9876	2.0028	4.0120	1.9970
0.3500	0.0124	3.9833	2.0039	4.0157	1.9956
0.4000	0.0162	3.9786	2.0054	4.0211	1.9946
0.4500	0.0205	3.9729	2.0068	4.0268	1.9932
0.5000	0.0253	3.9665	2.0084	4.0332	1.9917
0.5500	0.0306	3.9595	2.0102	4.0401	1.9900
0.6000	0.0365	3.9517	2.0121	4.0477	1.9881
0.6500	0.0428	3.9433	2.0143	4.0560	1.9861
0.7000	0.0496	3.9342	2.0166	4.0649	1.9839
0.7500	0.0570	3.9244	2.0191	4.0744	1.9816
0.8000	0.0648	3.9139	2.0218	4.0846	1.9791
0.8500	0.0732	3.9027	2.0246	4.0954	1.9764
0.9000	0.0821	3.8908	2.0277	4.1069	1.9737
0.9500	0.0914	3.8782	2.0309	4.1189	1.9707
1.0000	0.1013	3.8649	2.0344	4.1316	1.9677
1.0500	0.1117	3.8508	2.0380	4.1449	1.9645
1.1000	0.1226	3.8360	2.0419	4.1588	1.9611
1.1500	0.1340	3.8205	2.0460	4.1734	1.9577
1.2000	0.1459	3.8043	2.0502	4.1885	1.9541
1.2500	0.1583	3.7873	2.0547	4.2042	1.9503
1.3000	0.1712	3.7695	2.0594	4.2205	1.9465
1.3500	0.1847	3.7510	2.0644	4.2374	1.9425
1.4000	0.1986	3.7317	2.0695	4.2549	1.9384
1.4500	0.2130	3.7116	2.0749	4.2729	1.9342
1.5000	0.2280	3.6907	2.0806	4.2916	1.9299
1.5500	0.2434	3.6690	2.0865	4.3107	1.9255
1.6000	0.2594	3.6466	2.0926	4.3305	1.9210
1.6500	0.2758	3.6233	2.0990	4.3508	1.9163
1.7000	0.2928	3.5991	2.1057	4.3716	1.9116
1.7500	0.3103	3.5741	2.1127	4.3929	1.9068
1.8000	0.3283	3.5483	2.1199	4.4148	1.9019
1.8500	0.3468	3.5216	2.1275	4.4373	1.8969
1.9000	0.3658	3.4940	2.1353	4.4602	1.8919
1.9500	0.3853	3.4655	2.1434	4.4836	1.8867
2.0000	0.4053	3.4361	2.1519	4.5076	1.8815
2.0500	0.4258	3.4058	2.1607	4.5320	1.8762
2.1000	0.4468	3.3745	2.1699	4.5569	1.8708
2.1500	0.4684	3.3422	2.1794	4.5823	1.8654
2.2000	0.4904	3.3090	2.1893	4.6082	1.8599
2.2500	0.5129	3.2748	2.1996	4.6345	1.8544
2.3000	0.5360	3.2395	2.2102	4.6613	1.8488

(continued)

Table 3.7 Stability Functions ($kL = \pi\sqrt{P/P_e}$) (continued)

kL	P/P_e	compression s_{ii}	s_{ij}	tension s_{ii}	s_{ij}
2.3500	0.5595	3.2032	2.2213	4.6886	1.8431
2.4000	0.5836	3.1659	2.2328	4.7163	1.8374
2.4500	0.6082	3.1274	2.2447	4.7444	1.8317
2.5000	0.6333	3.0878	2.2572	4.7730	1.8259
2.5500	0.6588	3.0471	2.2701	4.8020	1.8201
2.6000	0.6849	3.0052	2.2834	4.8314	1.8142
2.6500	0.7115	2.9622	2.2974	4.8612	1.8083
2.7000	0.7386	2.9179	2.3118	4.8915	1.8024
2.7500	0.7662	2.8723	2.3268	4.9221	1.7965
2.8000	0.7944	2.8254	2.3425	4.9531	1.7905
2.8500	0.8230	2.7772	2.3587	4.9845	1.7845
2.9000	0.8521	2.7276	2.3756	5.0162	1.7785
2.9500	0.8817	2.6766	2.3932	5.0484	1.7725
3.0000	0.9119	2.6242	2.4115	5.0809	1.7665
3.0500	0.9425	2.5703	2.4305	5.1137	1.7605
3.1000	0.9737	2.5148	2.4503	5.1469	1.7544
3.1500	1.0054	2.4577	2.4709	5.1805	1.7484
3.2000	1.0375	2.3990	2.4924	5.2143	1.7424
3.2500	1.0702	2.3385	2.5148	5.2485	1.7363
3.3000	1.1034	2.2763	2.5382	5.2831	1.7303
3.3500	1.1371	2.2122	2.5626	5.3179	1.7243
3.4000	1.1713	2.1463	2.5880	5.3530	1.7183
3.4500	1.2060	2.0783	2.6146	5.3885	1.7123
3.5000	1.2412	2.0083	2.6424	5.4242	1.7063
3.5500	1.2769	1.9362	2.6714	5.4603	1.7003
3.6000	1.3131	1.8618	2.7017	5.4966	1.6944
3.6500	1.3498	1.7851	2.7335	5.5332	1.6884
3.7000	1.3871	1.7060	2.7668	5.5701	1.6825
3.7500	1.4248	1.6243	2.8016	5.6073	1.6766
3.8000	1.4631	1.5400	2.8382	5.6447	1.6708
3.8500	1.5018	1.4528	2.8765	5.6823	1.6649
3.9000	1.5411	1.3627	2.9168	5.7203	1.6591
3.9500	1.5809	1.2696	2.9592	5.7584	1.6533
4.0000	1.6211	1.1731	3.0037	5.7968	1.6476
4.0500	1.6619	1.0733	3.0507	5.8355	1.6419
4.1000	1.7032	0.9698	3.1001	5.8744	1.6362
4.1500	1.7450	0.8624	3.1523	5.9135	1.6305
4.2000	1.7873	0.7510	3.2074	5.9528	1.6249
4.2500	1.8301	0.6353	3.2656	5.9923	1.6193
4.3000	1.8734	0.5149	3.3273	6.0321	1.6138
4.3500	1.9172	0.3897	3.3926	6.0720	1.6083
4.4000	1.9616	0.2592	3.4619	6.1122	1.6028
4.4500	2.0064	0.1231	3.5356	6.1526	1.5974
4.5000	2.0518	−0.0191	3.6140	6.1931	1.5920
4.5500	2.0976	−0.1678	3.6975	6.2339	1.5867
4.6000	2.1440	−0.3234	3.7866	6.2748	1.5814
4.6500	2.1908	−0.4867	3.8819	6.3159	1.5761

3.8 Modified Slope-Deflection Equations

Table 3.7 (continued)

		compression		tension	
kL	P/P_e	s_{ii}	s_{ij}	s_{ii}	s_{ij}
4.7000	2.2382	−0.6582	3.9839	6.3572	1.5709
4.7500	2.2861	−0.8387	4.0934	6.3987	1.5658
4.8000	2.3344	−1.0289	4.2112	6.4403	1.5606
4.8500	2.3833	−1.2299	4.3381	6.4821	1.5556
4.9000	2.4327	−1.4427	4.4751	6.5241	1.5505
4.9500	2.4826	−1.6685	4.6235	6.5662	1.5456
5.0000	2.5330	−1.9087	4.7845	6.6085	1.5406
5.0500	2.5839	−2.1651	4.9599	6.6509	1.5357
5.1000	2.6354	−2.4394	5.1514	6.6934	1.5309
5.1500	2.6873	−2.7341	5.3613	6.7362	1.5261
5.2000	2.7397	−3.0516	5.5921	6.7790	1.5213
5.2500	2.7927	−3.3953	5.8470	6.8220	1.5166
5.3000	2.8461	−3.7689	6.1297	6.8652	1.5120
5.3500	2.9001	−4.1770	6.4447	6.9084	1.5074
5.4000	2.9545	−4.6254	6.7977	6.9518	1.5028
5.4500	3.0095	−5.1210	7.1957	6.9953	1.4983
5.5000	3.0650	−5.6727	7.6472	7.0390	1.4938
5.5500	3.1209	−6.2916	8.1635	7.0827	1.4894
5.6000	3.1774	−6.9923	8.7589	7.1266	1.4851
5.6500	3.2344	−7.7937	9.4524	7.1706	1.4807
5.7000	3.2919	−8.7215	10.2693	7.2147	1.4765
5.7500	3.3499	−9.8106	11.2447	7.2590	1.4722
5.8000	3.4084	−11.1107	12.4279	7.3033	1.4680
5.8500	3.4675	−12.6943	13.8915	7.3477	1.4639
5.9000	3.5270	−14.6717	15.7455	7.3922	1.4598
5.9500	3.5870	−17.2192	18.1662	7.4369	1.4558
6.0000	3.6476	−20.6379	21.4544	7.4816	1.4518
6.0500	3.7086	−25.4868	26.1690	7.5264	1.4478
6.1000	3.7702	−32.9355	33.4794	7.5714	1.4439
6.1500	3.8322	−45.9092	46.3106	7.6164	1.4401
6.2000	3.8948	−74.3671	74.6217	7.6615	1.4363
6.2500	3.9579	−188.3001	188.4032	7.7067	1.4325

beam-column that is not subjected to transverse loadings and relative joint translation (or sidesway). Note that when P approaches zero, we see that $kL = (\sqrt{P/EI})L$ approaches zero, and by using the L'Hospital's rule, it can be shown that s_{ii} reduces to 4 and s_{ij} reduces to 2 (see Problem 3.3). Values for s_{ii} and s_{ij} for various values of kL are shown in Table 3.7 and plotted in Fig. 3.18.

3.8 MODIFIED SLOPE-DEFLECTION EQUATIONS

Equations (3.7.13) and (3.7.14) are valid provided that the following conditions are observed.

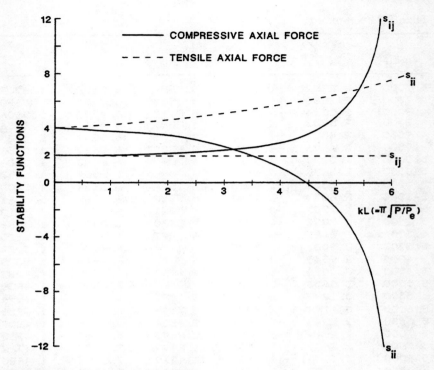

FIGURE 3.18 Plot of stability functions

1. The beam is prismatic.
2. There is no relative joint displacement between the two ends of the member, i.e., the member does not sway.
3. The member is continuous, i.e., there is no internal hinge or discontinuity in the member.
4. There is no in-span transverse loadings on the member.
5. The axial force in the member is compressive.

If these conditions are not satisfied, modifications to the slope-deflection equations are necessary. Some of these modifications to special cases of beam-columns are described below.

3.8.1 Member with Sway

If there is a relative joint translation between the member ends, designated as Δ in Fig. 3.19, the slope-deflection equations are modified

3.8 Modified Slope-Deflection Equations

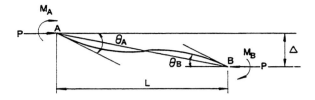

FIGURE 3.19 Beam-column subjected to end moments (with relative joint translation)

as

$$M_A = \frac{EI}{L}\left[s_{ii}\left(\theta_A - \frac{\Delta}{L}\right) + s_{ij}\left(\theta_B - \frac{\Delta}{L}\right)\right]$$
$$= \frac{EI}{L}\left[s_{ii}\theta_A + s_{ij}\theta_B - (s_{ii} + s_{ij})\frac{\Delta}{L}\right] \quad (3.8.1)$$

$$M_B = \frac{EI}{L}\left[s_{ij}\left(\theta_A - \frac{\Delta}{L}\right) + s_{ii}\left(\theta_B - \frac{\Delta}{L}\right)\right]$$
$$= \frac{EI}{L}\left[s_{ij}\theta_A + s_{ii}\theta_B - (s_{ii} + s_{ij})\frac{\Delta}{L}\right] \quad (3.8.2)$$

3.8.2 Member with a Hinge at One End

If a hinge is present at one end of the member—as in Fig. 3.20a, in which the B end is hinged—the moment there is zero, i.e.,

$$M_B = \frac{EI}{L}(s_{ij}\theta_A + s_{ii}\theta_B) = 0 \quad (3.8.3)$$

from which

$$\theta_B = -\frac{s_{ij}}{s_{ii}}\theta_A \quad (3.8.4)$$

Upon substituting Eq. (3.8.4) into Eq. (3.7.13), we have

$$M_A = \frac{EI}{L}\left(s_{ii} - \frac{s_{ij}^2}{s_{ii}}\right)\theta_A \quad (3.8.5)$$

Note that θ_B has been condensed out of Eq. (3.7.13) in Eq. (3.8.5). Thus, by using Eq. (3.8.5), the degrees of freedom used for the analysis can be reduced if the member is hinged at one end.

If the member is hinged at the A rather than the B end, as shown in

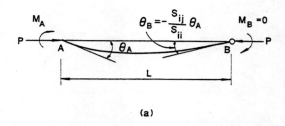

(a)

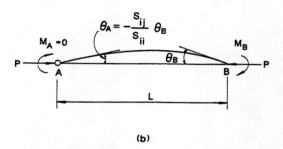

(b)

FIGURE 3.20 Beam-column subjected to end moments (with one end hinged)

Fig. 3.20b, Eq. (3.8.5) is still valid, provided that the subscript A is changed to B.

3.8.3 Member with Elastically Restrained Ends

A member may not be connected rigidly to other members at its ends, but may be connected instead to such members by a linear elastic spring, as in Fig. 3.21, with spring constants R_{kA} and R_{kB} at the A and B ends, respectively. The additional end rotations introduced as the result of the linear spring are M_A/R_{kA} and M_B/R_{kB}. If we denote the total end

FIGURE 3.21 Beam-column subjected to end moments (with elastically restrained ends)

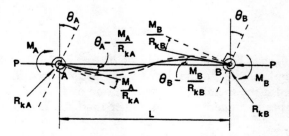

3.8 Modified Slope-Deflection Equations

rotations at joints A and B by θ_A and θ_B, respectively, as in the preceding cases, then the member end rotations, with respect to its chord, will be $(\theta_A - M_A/R_{kA})$ and $(\theta_B - M_B/R_{kB})$. As a result, the modified slope-deflection equations are modified to

$$M_A = \frac{EI}{L}\left[s_{ii}\left(\theta_A - \frac{M_A}{R_{kA}}\right) + s_{ij}\left(\theta_B - \frac{M_B}{R_{kB}}\right)\right] \quad (3.8.6)$$

$$M_B = \frac{EI}{L}\left[s_{ij}\left(\theta_A - \frac{M_A}{R_{kA}}\right) + s_{jj}\left(\theta_B - \frac{M_B}{R_{kB}}\right)\right] \quad (3.8.7)$$

Solving Eqs. (3.8.6) and (3.8.7) simultaneously for M_A and M_B gives

$$M_A = \frac{EI}{LR^*}\left[\left(s_{ii} + \frac{EIs_{ii}^2}{LR_{kB}} - \frac{EIs_{ij}^2}{LR_{kB}}\right)\theta_A + s_{ij}\theta_B\right] \quad (3.8.8)$$

$$M_B = \frac{EI}{LR^*}\left[s_{ij}\theta_A + \left(s_{ii} + \frac{EIs_{ii}^2}{LR_{kA}} - \frac{EIs_{ij}^2}{LR_{kA}}\right)\theta_B\right] \quad (3.8.9)$$

where

$$R^* = \left(1 + \frac{EIs_{ii}}{LR_{kA}}\right)\left(1 + \frac{EIs_{ii}}{LR_{kB}}\right) - \left(\frac{EI}{L}\right)^2 \frac{s_{ij}^2}{R_{kA}R_{kB}} \quad (3.8.10)$$

In writing Eqs. (3.8.8) to (3.8.10), the equality $s_{jj} = s_{ii}$ has been used. Note that as R_{kA} and R_{kB} approach infinity, Eqs. (3.8.8) and (3.8.9) reduce to Eqs. (3.7.13) and (3.7.14), respectively.

3.8.4 Member with Transverse Loadings

For members subjected to transverse loadings, the slope-deflection equations (3.7.13) and (3.7.14) must be modified by adding an extra term for the fixed-end moment of the member

$$M_A = \frac{EI}{L}(s_{ii}\theta_A + s_{ij}\theta_B) + M_{FA} \quad (3.8.11)$$

$$M_B = \frac{EI}{L}(s_{ij}\theta_A + s_{jj}\theta_B) + M_{FB} \quad (3.8.12)$$

The fixed-end moments M_{FA} and M_{FB} can be obtained by a procedure outlined in Section 3.5. Table 3.8 gives the expressions for the fixed-end moments of three commonly encountered cases of transverse loadings. The solutions for the first two cases have been derived in Section 3.5. The last case is left as an exercise for the reader (see Problem 3.10).

Table 3.8 Expressions for Fixed-End Moments $\left(u = \frac{L}{2}\sqrt{\frac{P}{EI}}\right)$

Case	Fixed-End Moments
Uniformly distributed load w over span L, EI-constant, axial force P at ends, moments M_{FA}, M_{FB}	$M_{FA} = -M_{FB} = -\dfrac{wL^2}{12}\left[\dfrac{3(\tan u - u)}{u^2 \tan u}\right]$
Concentrated load Q at midspan ($L/2$), EI-constant	$M_{FA} = -M_{FB} = -\dfrac{QL}{8}\left[\dfrac{2(1-\cos u)}{u \sin u}\right]$
Concentrated load Q at distance a from left, b from right, EI-constant	$M_{FA} = -\dfrac{QL}{d}\left[\dfrac{2ub}{L}\cos 2u - 2u\cos\dfrac{2ub}{L} - \sin 2u \right.$ $\left. + \sin\dfrac{2ua}{L} + \sin\dfrac{2ub}{L} + \dfrac{2ua}{L}\right]$ $M_{FB} = \dfrac{QL}{d}\left[\dfrac{2ua}{L}\cos 2u - 2u\cos\dfrac{2ua}{L} - \sin 2u \right.$ $\left. + \sin\dfrac{2ub}{L} + \sin\dfrac{2ua}{L} + \dfrac{2ub}{L}\right]$ $d = 2u(2 - 2\cos 2u - 2u \sin 2u)$

3.8.5 Member with Tensile Axial Force

For members subjected to an axial force that is tensile rather than compressive, Eqs. (3.7.13) and (3.7.14) are still valid provided that the stability functions defined in Eqs. (3.7.15) and (3.7.16) are redefined as

$$s_{ii} = s_{jj} = \frac{(kL)^2 \cosh kL - kL \sinh kL}{2 - 2\cosh kL + kL \sinh kL} \qquad (3.8.13)$$

$$s_{ij} = s_{ji} = \frac{kL \sinh kL - (kL)^2}{2 - 2\cosh kL + kL \sinh kL} \qquad (3.8.14)$$

3.8.6 Member Bent in Single Curvature with $\theta_B = -\theta_A$

For the member shown in Fig. 3.22a, the slope-deflection equations reduce to

$$M_A = \frac{EI}{L}(s_{ii} - s_{ij})\theta_A \qquad (3.8.15)$$

$$M_B = -M_A \qquad (3.8.16)$$

3.9 Inelastic Beam-Columns

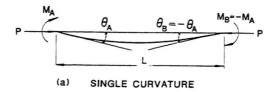

(a) SINGLE CURVATURE

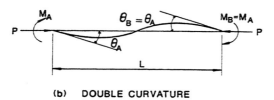

(b) DOUBLE CURVATURE

FIGURE 3.22 Beam-column subjected to end moments (single-curvature and double-curvature bending)

3.8.7 Member Bent in Double Curvature with $\theta_B = \theta_A$

For the member shown in Fig. 3.22b, the slope-deflection equations reduce to

$$M_A = \frac{EI}{L}(s_{ii} + s_{ij})\theta_A \quad (3.8.17)$$

$$M_B = M_A \quad (3.8.18)$$

3.9 INELASTIC BEAM-COLUMNS

Our discussion so far has been limited to the case in which the member remains fully elastic. In other words, no yielding of material has taken place in any part of the member. The assumption of a fully elastic behavior is justified to some extent for the member under service loading conditions. However, for failure behavior, we must include *inelasticity* in the analysis.

The inclusion of inelasticity in an analysis makes the problem much more complex because the governing differential equations become highly nonlinear. In many instances, closed-form solutions are intractable and recourse must be had to numerical techniques to obtain solutions.[2-4,9] In the following, we shall show that if certain simplifying assumptions are made, approximate solutions to some specific cases of inelastic beam-columns can be obtained analytically.[10,11]

In this section, the behavior and failure load of an eccentrically loaded beam-column of rectangular cross section will be investigated. The derivation follows closely to that given in reference 11.

The three basic assumptions used in the following derivation are the following:

1. The deflected shape of the member follows a half-sine wave (Fig. 3.23a).
2. The equilibrium condition is established only at midspan of the member.
3. The stress–strain relationship is assumed to be elastic–perfectly plastic (Fig. 3.23b).

In the process of analysis, we will need to use the nonlinear relationship between the internal moment (M) and the curvature

FIGURE 3.23 Elastic-plastic analysis of an eccentrically loaded rectangular member

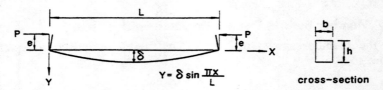

(a) ECCENTRICALLY LOADED RECTANGULAR MEMBER

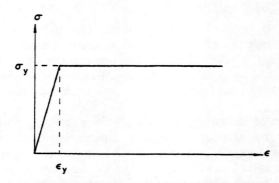

(b) ELASTIC–PERFECTLY PLASTIC STRESS–STRAIN RELATIONSHIP

3.9 Inelastic Beam-Columns

($\Phi = -y''$) with the presence of an axial force (P). It is therefore necessary to develop this relationship before proceeding to the analysis.

3.9.1 $M-\Phi-P$ Relationship

Figure 3.24, a–c, shows a series of strain and stress diagrams that correspond to three stages of loading sequences: the elastic (no yielding), the primary plastic (yielding in compression zone only), and the secondary plastic (yielding in both compression and tension zones),

FIGURE 3.24 Strain and stress diagrams

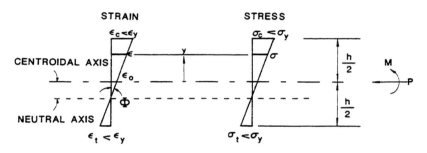

(a) CASE 1

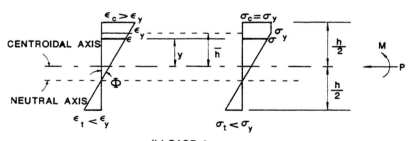

(b) CASE 2

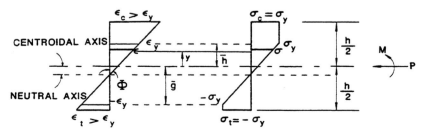

(c) CASE 3

respectively. For convenience, they are designated in the following as Case 1, Case 2, and Case 3, respectively. The M–Φ–P relationship for each case will be developed separately as follows:

Case 1: Elastic (Fig. 3.24a)

KINEMATICS

From the kinematic assumption that plane sections remain plane after bending, we write

$$\varepsilon = \varepsilon_0 + \Phi y \quad \text{for} \quad -\frac{h}{2} \leq y \leq \frac{h}{2} \tag{3.9.1}$$

where ε_0 is the axial strain at the centroid of the cross section.

STRESS–STRAIN RELATION

Since the entire cross section is elastic, the stress is related to the strain by

$$\sigma = E\varepsilon \quad \text{for} \quad -\frac{h}{2} \leq y \leq \frac{h}{2} \tag{3.9.2}$$

EQUILIBRIUM

The axial force P and the internal moment M are related to the stress σ by

$$P = \int_A \sigma \, dA = \int_{-h/2}^{h/2} E\varepsilon b \, dy$$

$$= \int_{-h/2}^{h/2} E(\varepsilon_0 + \Phi y) b \, dy \tag{3.9.3}$$

$$M = \int_A \sigma y \, dA = \int_{-h/2}^{h/2} E\varepsilon y b \, dy$$

$$= \int_{-h/2}^{h/2} E(\varepsilon_0 + \Phi y) y b \, dy \tag{3.9.4}$$

In Eqs. (3.9.3) and (3.9.4), b is the width and h is the height of the cross section (Fig. 3.23a).

By performing the necessary integrations, we obtain

$$P = EA\varepsilon_0 \tag{3.9.5}$$

and

$$M = EI\Phi \tag{3.9.6}$$

Equation (3.9.5) indicates that the axial force P is equal to the axial stiffness EA times the axial strain ε_0 at the centroid of the cross section.

Equation (3.9.6) is the familiar elastic beam moment-curvature relationship of $M = -EIy''$.

Introducing the notations

$$P_y = A\sigma_y = bh\sigma_y \tag{3.9.7}$$

$$M_y = S\sigma_y = \frac{bh^2}{6}\sigma_y \tag{3.9.8}$$

$$\Phi_y = \frac{2\varepsilon_y}{h} = \frac{2\sigma_y}{Eh} \tag{3.9.9}$$

$$p = \frac{P}{P_y} \tag{3.9.10}$$

$$m = \frac{M}{M_y} \tag{3.9.11}$$

$$\phi = \frac{\Phi}{\Phi_y} \tag{3.9.12}$$

Equations (3.9.5) and (3.9.6) can be written in a nondimensional form as

$$p = E\varepsilon_0/\sigma_y \tag{3.9.13}$$

and

$$m = \phi \tag{3.9.14}$$

The range of applicability of Eq. (3.9.14) is for $\phi \leq (1-p)$.

Case 2: Primary Plastic (Fig. 3.24b)

KINEMATICS

$$\varepsilon = \varepsilon_0 + \Phi y \quad \text{for} \quad -\frac{h}{2} \leq y \leq \frac{h}{2} \tag{3.9.1}$$

Since at $y = \bar{h}$, $\varepsilon = \varepsilon_y$, we have

$$\varepsilon_y = \varepsilon_0 + \Phi\bar{h} \tag{3.9.15}$$

from which we obtain

$$\varepsilon_0 = \varepsilon_y - \Phi\bar{h} \tag{3.9.16}$$

and substituting this into Eq. (3.9.1) leads to

$$\varepsilon = \varepsilon_y - (\bar{h} - y)\Phi \quad \text{for} \quad -\frac{h}{2} \leq y \leq \frac{h}{2} \tag{3.9.17}$$

Stress–Strain Relation

$$\sigma = E\varepsilon \quad \text{for} \quad -\frac{h}{2} \le y \le \bar{h} \qquad (3.9.18)$$

$$\sigma = \sigma_y \quad \text{for} \quad \bar{h} \le y \le \frac{h}{2} \qquad (3.9.19)$$

Equilibrium

$$\begin{aligned}P &= \int_A \sigma\, dA \\ &= \int_{-h/2}^{\bar{h}} E[\varepsilon_y - (\bar{h} - y)\Phi]b\, dy + \int_{\bar{h}}^{h/2} \sigma_y b\, dy \end{aligned} \qquad (3.9.20)$$

$$\begin{aligned}M &= \int_A \sigma y\, dA \\ &= \int_{-h/2}^{\bar{h}} E[\varepsilon_y - (\bar{h} - y)\Phi]yb\, dy + \int_{\bar{h}}^{h/2} \sigma_y by\, dy \end{aligned} \qquad (3.9.21)$$

By performing the necessary integrations and eliminating $\bar{h}$ from the resulting expressions (3.9.20) and (3.9.21), we obtain

$$m = 3(1-p) - \frac{2(1-p)^{3/2}}{\sqrt{\phi}} \qquad (3.9.22)$$

where m, ϕ, and p are as defined in Eqs. (3.9.11), (3.9.12), and (3.9.10), respectively.

The range of applicability of Eq. (3.9.22) is for $(1-p) \le \phi \le 1/(1-p)$.

Case 3: Secondary Plastic (Fig. 3.24c)

Kinematics

The strain at any point y from the centroidal axis is expressed by Eq. (3.9.17) as

$$\varepsilon = \varepsilon_y - (\bar{h} - y)\Phi \quad \text{for} \quad -\frac{h}{2} \le y \le \frac{h}{2} \qquad (3.9.17)$$

In addition, to facilitate the integration, we need to establish a relationship between the elastic-plastic boundaries $\bar{g}$ and $\bar{h}$. This can be achieved by using similar triangles in the strain diagram:

$$\bar{g} = \frac{2\varepsilon_y}{\Phi} - \bar{h} \qquad (3.9.23)$$

3.9 Inelastic Beam-Columns

STRESS–STRAIN RELATION

$$\sigma = -\sigma_y \quad \text{for} \quad -\frac{h}{2} \leq y \leq -\bar{g} \tag{3.9.24}$$

$$\sigma = E\varepsilon \quad \text{for} \quad -\bar{g} \leq y \leq \bar{h} \tag{3.9.25}$$

$$\sigma = \sigma_y \quad \text{for} \quad \bar{h} \leq y \leq \frac{h}{2} \tag{3.9.26}$$

EQUILIBRIUM

$$P = \int_A \sigma\, da$$

$$= \int_{-h/2}^{-\bar{g}} (-\sigma_y) b\, dy + \int_{-\bar{g}}^{\bar{h}} E[\varepsilon_y - (\bar{h} - y)\Phi] b\, dy$$

$$+ \int_{\bar{h}}^{h/2} \sigma_y b\, dy \tag{3.9.27}$$

$$M = \int_A \sigma y\, dA$$

$$= \int_{-h/2}^{-\bar{g}} (-\sigma_y) y b\, dy + \int_{-\bar{g}}^{\bar{h}} E[\varepsilon_y - (\bar{h} - y)\Phi] y b\, dy$$

$$+ \int_{\bar{h}}^{h/2} \sigma_y y b\, dy \tag{3.9.28}$$

By performing the necessary integrations, with $\bar{g}$ given in Eq. (3.9.23), and eliminating $\bar{h}$ from the resulting expressions (3.9.27) and (3.9.28), we obtain

$$m = \tfrac{3}{2}(1 - p^2) - \frac{1}{2\phi^2} \tag{3.9.29}$$

The range of applicability of Eq. (3.9.29) is for $\phi \geq 1/(1-p)$.

To sum up, the nondimensional moment-curvature-thrust $(m - \phi - p)$ relationships for a rectangular section can be written as the following:

Case 1:

$$m = \phi, \quad \text{for} \quad \phi \leq (1-p) \tag{3.9.30a}$$

Case 2:

$$m = 3(1-p) - \frac{2(1-p)^{3/2}}{\sqrt{\phi}}, \quad \text{for} \quad (1-p) \leq \phi \leq 1/(1-p) \tag{3.9.30b}$$

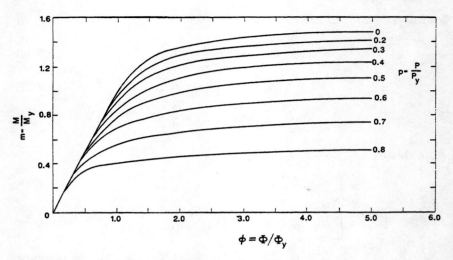

FIGURE 3.25 Moment-curvature-thrust relationships

Case 3:

$$m = \tfrac{3}{2}(1-p^2) - \frac{1}{2\phi^2}, \quad \text{for} \quad \phi \geq 1/(1-p) \qquad (3.9.30c)$$

Note that if the member is fully elastic (Case 1), the axial force has no effect on the moment-curvature relationship. However, as soon as yielding commences (Case 2 and Case 3), the moment-curvature relationships will be affected by the axial force. This explains why the analysis of an *inelastic* beam-column is much more complex than an elastic beam-column: because the moment-curvature relationships become nonlinear as yielding starts to occur in the member. Figure 3.25 shows some plots of the moment-curvature $(m - \phi)$ relationship for various values of p.

It should be remembered that the moment-curvature-thrust relationships developed above are valid only for rectangular cross sections with idealized elastic-plastic, stress–strain behavior. For general cross sections with general stress–strain behavior, it is often necessary to evaluate the integrals for the axial force P and internal bending moment M numerically. Numerical and approximate moment-curvature-thrust expressions for various cross-sectional shapes have been developed and are discussed in detail in reference 2.

3.9.2 Approximate Solution

With the moment-curvature-thrust relationships developed, we now proceed with the investigation of the inelastic behavior of an eccentrically

3.9 Inelastic Beam-Columns

loaded beam-column (Fig. 3.23a) by using the analysis method presented in reference 11. By assuming that the deflected shape of the member resembles a half-sine wave, we write

$$y = \delta \sin \frac{\pi x}{L} \qquad (3.9.31)$$

from which we obtain

$$y' = \delta \frac{\pi}{L} \cos \frac{\pi x}{L} \qquad (3.9.32)$$

and

$$y'' = -\delta \left(\frac{\pi}{L}\right)^2 \sin \frac{\pi x}{L} \qquad (3.9.33)$$

The curvature at midspan is

$$\Phi_m = -y''\left(\frac{L}{2}\right) = \delta \left(\frac{\pi}{L}\right)^2 \qquad (3.9.34)$$

where δ is the deflection at midspan.

The equilibrium condition at midspan (Fig. 3.26) is

$$M_0 + P\delta = M_m \qquad (3.9.35)$$

where

$$M_0 = Pe \qquad (3.9.36)$$

$$e = \text{load eccentricity} \qquad (3.9.37)$$

$$M_m = \text{moment at midspan} \qquad (3.9.38)$$

Denoting

$$m_0 = \frac{M_0}{M_y} \qquad (3.9.39)$$

$$m_m = \frac{M_m}{M_y} \qquad (3.9.40)$$

FIGURE 3.26 Free-body diagram of the eccentrically loaded member cut at midspan

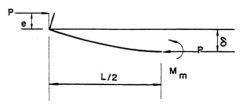

Equation (3.9.35) can be written as

$$m_0 + \frac{P\delta}{M_y} = m_m \qquad (3.9.41)$$

Since, from Eq. (3.9.34)

$$\frac{P\delta}{M_y} = \frac{P\left[\Phi_m\left(\frac{L}{\pi}\right)^2\right]}{EI\Phi_y} = \left(\frac{P}{P_e}\right)\left(\frac{\Phi_m}{\Phi_y}\right) \qquad (3.9.42)$$

We can write Eq. (3.9.41) as

$$m_0 + \frac{p}{p_e}\phi_m = m_m \qquad (3.9.43)$$

If we substitute the moment-curvature-thrust $(m - \phi - p)$ relationships [Eq. (3.9.30)] with $m = m_m$, $\phi = \phi_m$ (normalized moment and curvature at midspan) into Eq. (3.9.43), and rearrange, we have the following:

Case 1: $\phi_m \leq 1 - p$

$$m_0 = \phi_m\left(1 - \frac{p}{p_e}\right) \qquad (3.9.44a)$$

Case 2: $(1 - p) \leq \phi_m \leq 1/(1 - p)$

$$m_0 = 3(1 - p) - \frac{2(1-p)^{3/2}}{\sqrt{\phi_m}} - \frac{p}{p_e}\phi_m \qquad (3.9.44b)$$

Case 3: $\phi_m \geq 1/(1 - p)$

$$m_0 = \frac{3}{2}(1 - p^2) - \frac{1}{2\phi_m^2} - \frac{p}{P_e}\phi_m \qquad (3.9.44c)$$

Figure 3.27 shows a plot of the $m_0 - \phi_m$ curve for a simply-supported, eccentrically loaded member with the following dimensions and properties:

$$b = 1 \text{ in}, \qquad h = 2\sqrt{3} \text{ in}, \qquad L = 120 \text{ in}$$
$$e = 1.15 \text{ in}, \qquad \sigma_y = 34 \text{ ksi}, \qquad E = 30{,}000 \text{ ksi}$$

Notice that at approximately $m_0 = 0.47$ yielding starts at the compression side of the cross section. As a result, a noticeable decrease in stiffness of the member is observed. The degradation of stiffness continues until at approximately the peak moment $m_0 = 0.53$, the member is no longer able to resist an increase in load. Therefore, the

3.9 Inelastic Beam-Columns

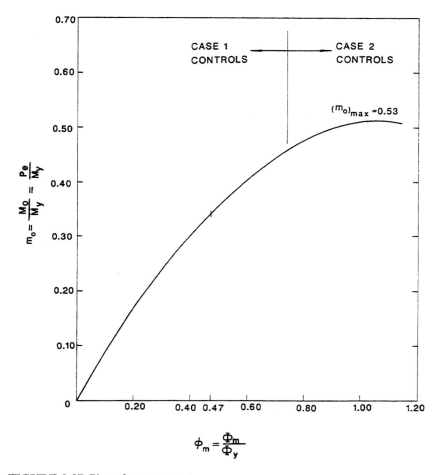

FIGURE 3.27 Plot of m_0 versus ϕ_m

maximum load the member can resist is

$$P = \frac{M_0}{e} = \frac{m_0 M_y}{e} = \frac{m_0\left(\dfrac{bh^2\sigma_y}{6}\right)}{e}$$

$$= \frac{(0.53)(1)(2\sqrt{3})^2(34)}{6(1.15)} = 31.3 \text{ kips}$$

For this case, the maximum load occurs in the primary plastic range [i.e., $(1-p) \leq \phi_m \leq 1/(1-p)$]. However, this is not always the case. Depending on the slenderness ratio and the magnitude of eccentricity, the failure load may occur in the secondary plastic range [i.e., $\phi_m > 1/(1-p)$]. In this example, the maximum end moment $(m_0)_{max}$ has been obtained graphically as the peak point of the m_0 versus ϕ_m curve. The same maximum moment can be obtained more conveniently by realizing that the peak point of the curve corresponds to the condition

$$\frac{dm_0}{d\phi_m} = 0 \qquad (3.9.45)$$

Thus, by setting the derivative of Eq. (3.9.44b, c) with respect to ϕ_m equal to zero, the value of ϕ_m that corresponds to $(m_0)_{max}$ for each relevant case can be calculated. Backsubstituting the value of ϕ_m so obtained into the corresponding equation will give the value of $(m_0)_{max}$. In doing so, a relation between $(m_0)_{max}$ and p can be established as the following:

If Case 2 controls, i.e., if $(1-p)^3 \leq \dfrac{p}{p_e} \leq 1$

$$(m_0)_{max} = 3(1-p)\left[1 - \left(\frac{p}{p_e}\right)^{1/3}\right] \qquad (3.9.46a)$$

If Case 3 controls, i.e., if $0 \leq \dfrac{p}{p_e} \leq (1-p)^3$

$$(m_0)_{max} = \frac{3}{2}\left[1 - p^2 - \left(\frac{p}{p_e}\right)^{2/3}\right] \qquad (3.9.46b)$$

Thus, by using Eq. (3.9.46a) or Eq. (3.9.46b), depending on the range of applicability of the equations, ultimate strength interaction curves for m_0 and p (i.e., curves giving the value of $(M_0)_{max}$ for a given value of P) can be developed. Figure 3.28 shows three such ultimate strength interaction curves for slenderness ratios $L/r = 20$, 60, and 120.

Although a rectangular section with idealized stress–strain relationship has been used here to obtain the interaction diagrams shown in the figure, the same procedure can be extended to obtain ultimate strength interaction diagrams for I-shaped sections with or without residual stress. However, the determination of ultimate loads for these sections involves considerably more effort and, in many cases, resort to numerical solution techniques is inevitable. Various numerical techniques to obtain ultimate strength interaction diagrams are summarized and presented in detail in references 2 and 3. Two such numerical techniques (Newmark's method, step-by-step numerical integration) will be presented in Chapter 6.

From the above example, one can see that although several simplifying assumptions concerning the member and material behavior have been used, the determination of the maximum load that a member can resist is

3.10 Design Interaction Equations

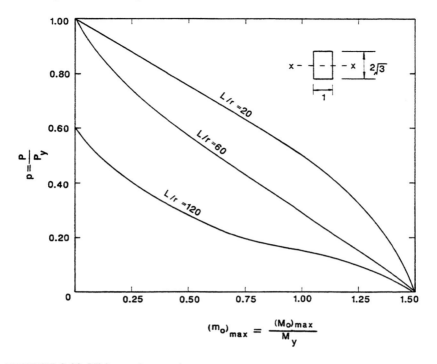

FIGURE 3.28 Ultimate interaction curves

still cumbersome. From a practical standpoint, it is more convenient if the maximum load can be determined approximately in a direct manner by simpler formulas. To this end, *design interaction formulas* provide just such a quick and easy way for estimating the maximum load-carrying capacity of a beam-column. Design interaction equations are the equations that relate the ratio of axial stress (force) in the member to the ultimate axial capacity of the member and the ratio of bending stress (moment) in the member to the ultimate bending stress (moment) of the member. These design interaction equations provide a convenient and direct means for designers to estimate the adequacy of members subjected to combined stresses (forces). Various forms of design interaction equations used for design purposes will be discussed in the following section.

3.10 DESIGN INTERACTION EQUATIONS

The interaction curves shown in Fig. 3.28 for rectangular cross sections based on Eq. (3.9.46a) or Eq. (3.9.46b) can be approximated by a simple linear interaction equation of the form

$$\frac{P}{P_u} + \frac{M_{max}}{M_u} = 1 \qquad (3.10.1)$$

where

P = axial force in the member
P_u = ultimate axial capacity of the member in the absence of primary bending moment
M_{max} = maximum moment in the member
M_u = ultimate moment capacity of the member in the absence of axial force

Depending on the design philosophy, the ultimate axial capacity of the member P_u can be represented by the CRC curve and SSRC curves or the LRFD curve (Chapter 2). The quantity M_{max}, however, is the amplified moment in the member and can be represented by

$$M_{max} = A_F M_0 \qquad (3.10.2)$$

where A_F is the *moment amplification factor* discussed earlier in the chapter.

For an eccentrically loaded member, which is equivalent to a member subjected to equal end moments and an axial force, the expression for A_F is (from Eq. 3.4.27)

$$A_F = \frac{C_m}{1 - \left(\frac{P}{P_{ek}}\right)} \qquad (3.10.3)$$

where C_m is given by Eq. (3.4.23).

Upon substituting Eq. (3.10.3) into Eq. (3.10.2) and then into Eq. (3.10.1), we obtain

$$\frac{P}{P_u} + \frac{C_m M_0}{M_u(1 - P/P_{ek})} = 1 \qquad (3.10.4)$$

Equation (3.10.4) is plotted as dotted lines in Fig. 3.29 by using P_u = LRFD curve (Eq. 2.11.9), $M_u = M_p$ the plastic moment capacity of the cross section, and $C_m = 1$ [Eq. (3.4.23) with $M_1/M_2 = -1$]. The theoretical ultimate strength interaction curves are given by Eq. (3.9.46a,b) and are plotted in Fig. 3.28. However, to obtain a direct comparison, these curves are replotted as solid lines in Fig. 3.29 using an abscissa of M_0/M_p rather than M_0/M_y. For rectangular cross sections, the plastic moment M_p and the yield moment M_y are related by $M_p = 1.5 M_y$ (Eq. 3.9.30c with $p = 0$ and $\phi = \infty$). As can be seen, except for low slenderness ratios the interaction equation (3.10.4) gives a good ap-

3.10 Design Interaction Equations

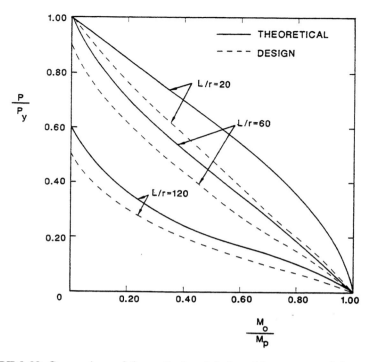

FIGURE 3.29 Comparison of theoretical and design ultimate strength interaction curves

proximation to the theoretical solution. For low slenderness ratios, Eq. (3.10.4) becomes too conservative. This is because the $P - \delta$ effect (i.e., the additional moment induced in the member as a result of the axial force acting through the lateral deflection) is not significant. Consequently, the presence of the term $1/[1 - (P/P_{ek})]$ which reflects the $P - \delta$ effect will render Eq. (3.10.1) too conservative. In fact, if the slenderness ratio of the member approaches zero, no instability will occur and a nonlinear interaction equation of the form

$$\left(\frac{P}{P_y}\right)^a + \frac{M_0}{bM_p} = 1 \qquad (3.10.5)$$

will give a much better approximation to the theoretical ultimate strength interaction curve. The constants a and b define the shape of the interaction curve. For rectangular sections, $a = 2$, $b = 1$, and Eq. (3.10.5) represents the exact ultimate strength interaction curve (Fig. 3.30). For I-section bent about the strong axis, $a = 1$, $b = 1.18$, and Eq. (3.10.5) is an approximate ultimate strength interaction equation (Fig. 3.31). For

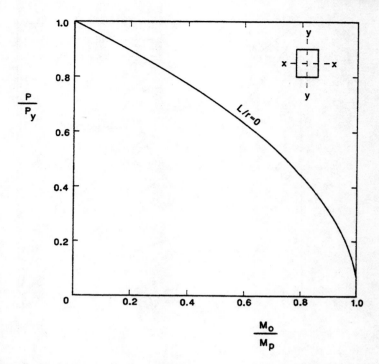

FIGURE 3.30 Ultimate strength interaction curve for rectangular cross sections with $L/r = 0$

I-section bent about the weak axis, $a = 2$, $b = 1.19$, and Eq. (3.10.5) represents an approximate interaction equation (Fig. 3.32).

It should be mentioned here that the use of $M_u = M_p$ in Eq. (3.10.4) implies that lateral torsional instability will not occur in the member. Lateral torsional instability is a phenomenon in which the member bends out of its plane of loading in addition to in-plane deflection as a result of insufficient lateral stiffness or bracing. The subject of lateral instability of beams will be discussed in Chapter 5. If lateral torsional instability occurs in the member, the ultimate moment capacity M_u of the member will be less than M_p as the member will fail by lateral torsional buckling before the plastic moment capacity of the member can be attained.

Furthermore, it should be noted that Eq. (3.10.4) can also be used for members whose primary bending moment results from transverse loading rather than from end moments. Nevertheless, in these cases where the primary moment results from transverse loading, the C_m value is defined as

$$C_m = 1 + \Psi P/P_{ek} \tag{3.10.6}$$

3.10 Design Interaction Equations

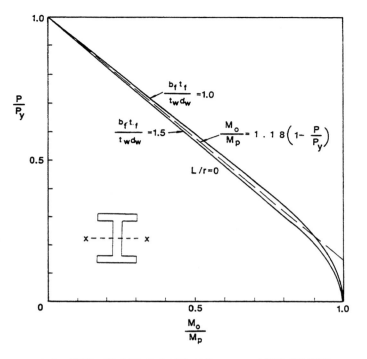

$b_f t_f$ = AREA OF ONE FLANGE AND $t_w d_w$ = AREA OF WEB

FIGURE 3.31 Ultimate strength interaction curve for I-sections bent about the strong axis with $L/r = 0$

This definition of C_m can be induced by comparing Eq. (3.6.19) with Eqs. (3.10.2) and (3.10.3). The values of Ψ and C_m for various transverse loading cases are shown in Table 3.9. Note that the Ψ values for Cases 1, 3, 4, and 6 have already been developed in detail in previous sections [see Eq. (3.2.41), Eq. (3.6.32), Eq. (3.3.20), and Eq. (3.6.20), respectively]. The determination of the Ψ values for Cases 2 and 5 is left as an exercise for the reader (see Problem 3.4). Note that for the two simply supported cases (Cases 1 and 4), the values of Ψ were determined by expanding the theoretical beam-column solutions for M_{max}, whereas for the other cases the values of Ψ were determined by comparing Eq. (3.6.19) with the corresponding theoretical solutions for M_{max}.

Equation (3.10.4) can also be extended to the case in which bending occurs in both axes. In such case, the interaction equation takes the form

$$\frac{P}{P_u} + \frac{C_{mx} M_{0x}}{M_{ux}(1 - P/P_{ex})} + \frac{C_{my} M_{0y}}{M_{uy}(1 - P/P_{ey})} = 1 \quad (3.10.7)$$

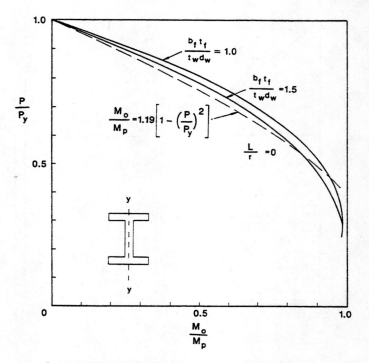

FIGURE 3.32 Ultimate strength interaction curve for I-sections bent about the weak axis with $L/r = 0$

in which the subscripts x and y refer to the action about the x and y axes of the member, respectively.

Equation (3.10.7) gives the general form of the biaxial bending interaction equation used in practical design when stability governs the limit state.[12] For member with low slenderness ratios or at support locations, yielding rather than instability may govern the limit state. In this case, an extension of Eq. (3.10.5) will be more appropriate. For example, for I-section, a conservative form by using $a = 1$ for both axes of bending can be written (see reference 12) as

$$\frac{P}{P_y} + \frac{M_{0x}}{1.18 M_{px}} + \frac{M_{0y}}{1.19 M_{py}} = 1 \qquad (3.10.8)$$

where M_{px} and M_{py} are plastic moment capacities of the I-section with respect to strong- and weak-axis bending, respectively.

In the following sections, interaction equations for various design formats will be summarized and discussed.

3.10 Design Interaction Equations

3.10.1 AISC/ASD Format

The AISC/ASD format is a stress-based design format. As a result, allowable stresses and service loads rather than ultimate strengths and maximum loads are used as the basis for design. Equations (3.10.7) and (3.10.8) can be converted to unit of stresses and a factor of safety applies to them to bring them into the service load range.

If stability controls, the interaction equation is

$$\frac{f_a}{F_a} + \frac{f_{bx}C_{mx}}{F_{bx}(1-f_a/F'_{ex})} + \frac{f_{by}C_{my}}{F_{by}(1-f_a/F'_{ey})} \leq 1.0 \quad (3.10.9)$$

where

$f_a = P/A_g$ = axial stress at service load

f_{bx}, f_{by} = flexural stresses at service load due to primary bending moment about the x and y axes, respectively

F_a = allowable compressive stress if the member is under axial compression only [= AISC/ASD column curve equation (2.11.4) divided by the area of the cross section]

F_{bx}, F_{by} = allowable flexural stresses about the x and y axes, respectively, if the member is loaded in bending only (see Chapter 5)

C_m = define as follows:
1. For members braced against joint translation and without transverse loading between supports, C_m is referred to as the *equivalent moment factor* and is defined in Eq. (3.4.23) as $C_m = 0.6 - 0.4\ M_A/M_B \geq 0.4$, where M_A/M_B is the ratio of the smaller to larger end moments. It is negative if the member is bent in single curvature and is positive if the member is bent in reverse curvature.
2. For members braced against joint translation with transverse loading between supports, C_m is referred to as the *moment reduction factor*. It is defined in Eq. (3.10.6) as $C_m = 1 + \Psi P/P_{ek}$ and it is an integral part of the moment magnifier [see Eqs. (3.2.41), (3.3.20), and (3.6.19)]. Table 3.9 gives the C_m values for various transverse loading cases. However, the AISC Specification suggests the use of $C_m = 0.85$ for members with restrained ends and $C_m = 1.0$ for members with unrestrained ends.
3. For members not braced against joint translation, C_m is considered to be 0.85. The number 0.85 is derived based on the model structure shown in Fig. 3.33a. For members in which sidesway is possible, a different type of secondary moment known as the $P-\Delta$ moment will be induced in the member. This $P-\Delta$ moment occurs as a

Table 3.9 Values for ψ and C_m

Case	ψ	C_m
1 (simply supported, uniform load)	0	1.0
2 (propped cantilever, uniform load)	−0.4	$1 - 0.4\, P/P_{ek}$
3 (fixed-fixed, uniform load)	−0.4	$1 - 0.4\, P/P_{ek}$
4 (simply supported, midspan point load)	−0.2	$1 - 0.2\, P/P_{ek}$
5 (propped cantilever, point load at L/2)	−0.3	$1 - 0.3\, P/P_{ek}$
6 (fixed-fixed, midspan point load)	−0.2	$1 - 0.2\, P/P_{ek}$

result of the axial force P acting through the sway Δ of the member. If we consider the free-body diagram of Fig. 3.33b, it roughly resembles a simply supported member loaded by a transverse load at midspan, so from Eq. (3.3.20), the value of C_m is $1 - 0.18 P/P_{ek}$. However, due to the errors involved in the approximation, the AISC Specification recommends the use of 0.85.

F'_{ex}, F'_{ey} = Critical elastic buckling stress about the x and y axis, respectively, divided by a factor of safety of 23/12 and evaluated using the effective length of the member. (The effective lengths of isolated members with idealized end conditions have been shown in Table 2.1. The effective lengths of members as parts of a frame will be discussed in the next chapter.)

If the yielding of material controls, the interaction equation is

$$\frac{f_a}{0.60 F_y} + \frac{f_{bx}}{F_{bx}} + \frac{f_{by}}{F_{by}} \leq 1.0 \qquad (3.10.10)$$

3.10 Design Interaction Equations

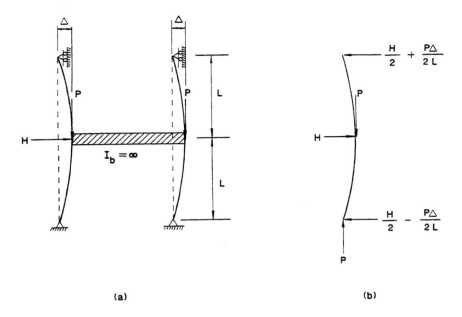

FIGURE 3.33 AISC/ASD model for sway frames

Note that the moment magnification factor $C_m/(1 - f_a/F'_{ek})$ is absent in Eq. (3.10.10) because this equation pertains to cases in which yielding rather than instability controls the design. This happens at support locations in braced frames and for members with low slenderness ratios in unbraced frames. The number 0.60 (=1/1.67) in the denominator of the first term that reflects the axial force effect is a safety factor applied to the CRC curve in order to obtain the AISC/ASD curve at $KL/r = 0$.

In actual design, both the stability [Eq. (3.10.9)] and yield [Eq. (3.10.10)] interaction equations should be checked. However, if the axial force in the member is small, say if $f_a/F_a \leq 0.15$, the AISC Specification allows the use of the following interaction equation, instead of Eqs. (3.10.9) and (3.10.10):

$$\frac{f_a}{F_a} + \frac{f_{bx}}{F_{bx}} + \frac{f_{by}}{F_{by}} \leq 1.0 \qquad (3.10.11)$$

Thus, in the design, the value f_a/F_a is first evaluated. If f_a/F_a is less than or equal to 0.15, Eq. (3.10.11) is used to check the adequacy of the section. If f_a/F_a is greater than 0.15, then both Eqs. (3.10.9) and (3.10.10) are used to check the adequacy of the section.

3.10.2 AISC/PD Format

The AISC/PD format provides two interaction equations for the design of beam-columns.

If yielding controls,

$$\frac{P}{P_y} + \frac{M}{1.18M_p} \leq 1.0 \quad (3.10.12)$$

where

$P_y = A_g F_y$ = yield load of the section where A_g is the area of the cross section

$M_p = Z F_y$ = full plastic moment capacity of the section where Z is plastic-section modulus

P and M are the factored axial force and moment (service loads time load factor). The load factor is 1.7 if only gravity loads are acting and is 1.3 if wind or earthquake loads is acting in conjunction with gravity loads. Note the correspondence of Eq. (3.10.12) with Eq. (3.10.5).

If stability controls,

$$\frac{P}{P_u} + \frac{C_m M_0}{M_m(1 - P/P_{ek})} \leq 1.0 \quad (3.10.13)$$

where

P_u = ultimate axial compressive strength of the axially loaded column taken as 1.7 times AISC/ASD column curve using the effective length of the column

M_m = maximum resisting moment in the absence of axial force, taken as M_p if the member is braced against lateral torsional buckling and taken as

$$M_m = \left[1.07 - \frac{(L/r_y)\sqrt{F_y}}{3160}\right] = M_p \leq M_d \quad (3.10.14)$$

if the member fails by lateral torsional buckling.

In Eq. (3.10.14) the units are inches and ksi.

$P_{ek} = \pi^2 EI/(KL)^2$

C_m = same as in AISC/ASD format

Again, in actual design, both the strength interaction equation [Eq. (3.10.12)] and the stability interaction equation [Eq. (3.10.13)] are used to check the adequacy of the section.

3.10.3 AISC/LRFD Format

The AISC/LRFD format based on the exact inelastic solutions of 82 beam-columns,[13] recommends the following interaction equations for sway and nonsway beam-columns.

For $P/\phi_c P_u \geq 0.2$

$$\frac{P}{\phi_c P_u} + \frac{8}{9}\left(\frac{M_{ax}}{\phi_b M_{ux}} + \frac{M_{ay}}{\phi_b M_{uy}}\right) \leq 1.0 \quad (3.10.15)$$

For $P/\phi_c P_u < 0.2$

$$\frac{P}{2\phi_c P_u} + \frac{M_{ax}}{\phi_b M_{ux}} + \frac{M_{ay}}{\phi_b M_{uy}} \leq 1.0 \quad (3.10.16)$$

where

P_u = ultimate axial compression capacity of the axially loaded column [= AISC/LRFD Column Curve Eq. (2.11.9) using the effective length of the column]

M_{ux}, M_{uy} = ultimate moment resisting capacity of the laterally unsupported beam about the x and y axes, respectively (see Chapter 5)

ϕ_c = column resistance factor (= 0.85)
ϕ_b = beam resistance factor (= 0.90)
P = design axial force

M_{ax}, M_{ay} = design moment for the member about the x and y axes, respectively, calculated as follows

$$M_a = B_1 M_{nt} + B_2 M_{lt} \quad (3.10.17)$$

in which

M_{nt} = moment in member assuming there is no lateral translation in the frame calculated by using first-order elastic analysis (see Fig. 3.34a)

M_{lt} = moment in member as a result of lateral translation of the frame only calculated by using first-order elastic analysis (see Fig. 3.34b)

$B_1 = P - \delta$ moment amplification factor (designated as A_F in the previous sections)

$$B_1 = \frac{C_m}{1 - \dfrac{P}{P_{ek}}} \geq 1 \quad (3.10.18)$$

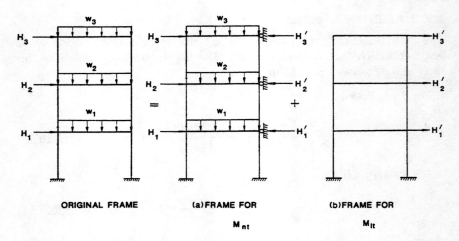

FIGURE 3.34 AISC/LRFD approach for calculating M_{nt} and M_{lt}

$B_2 = P - \Delta$ moment amplification factor, evaluated by

$$B_2 = \frac{1}{1 - \frac{\sum P \Delta_0}{\sum HL}} \qquad (3.10.19)$$

or alternatively

$$B_2 = \frac{1}{1 - \frac{\sum P}{\sum P_{ek}}} \qquad (3.10.20)$$

The terms in Eqs. (3.10.18) to (3.10.20) are defined as follows:

$C_m = 0.6 - 0.4 \, M_1/M_2$, same as defined previously in ASD, except that the limit condition $C_m \geq 0.4$ has been removed in LRFD. This limit was found to be overly conservative for $M_1/M_2 = 0.5$ to 1.0 when compared with the elastic C_m factor used in ASD with the exact elastic-plastic computer solutions.[14]

$P_{ek} = \pi^2 EI/(KL)^2$

$\sum P$ = axial loads on all columns in a story

Δ_0 = first-order translational deflection of the story under consideration

$\sum H$ = sum of all story horizontal forces producing Δ_0

L = story height

The $P - \Delta$ moment amplification factor B_2 expressed in Eq. (3.10.19) was developed based on the story stiffness concept.[15–17] By assuming that

3.10 Design Interaction Equations

(i) each story can behave independently of other stories, and (ii) the additional moments in the columns caused by the $P-\Delta$ effect is equivalent to that caused by a lateral force of $\sum P\Delta/L$, the sway stiffness of the story can be defined as

$$S_F = \frac{\text{horizontal force}}{\text{lateral displacement}}$$

$$= \frac{\sum H}{\Delta_0} = \frac{\sum H + \sum P\Delta/L}{\Delta} \qquad (3.10.21)$$

Solving Eq. (3.10.21) for Δ gives

$$\Delta = \left(\frac{1}{1-\frac{\sum P\Delta_0}{\sum HL}}\right)\Delta_0 \qquad (3.10.22)$$

If rigid connections and elastic behavior are assumed, the moment induced in the member as a result of sway will be proportional to the lateral deflection. Therefore, we can write the amplified sway moment M_{alt} as

$$M_{alt} = \left(\frac{1}{1-\frac{\sum P\Delta_0}{\sum HL}}\right)M_{lt} \qquad (3.10.23)$$

The alternate expression for B_2 expressed in Eq. (3.10.20) was developed based on the multiple-column buckling concept.[18] By assuming that when instability is to occur in a story, all columns in that story will become unstable simultaneously. As a result, a direct extension of Eq. (3.3.19) is justified by replacing the term P/P_{ek} by $\sum P/\sum P_{ek}$ where the summation is carried through all columns in a story. Thus

$$\Delta = \left(\frac{1}{1-\frac{\sum P}{\sum P_{ek}}}\right)\Delta_0 \qquad (3.10.24)$$

Using the same argument, that if rigid joints and elastic behavior are assumed, the sway moments are directly proportional to the lateral deflections of the story, we can now write the amplified sway moment as

$$M_{alt} = \left(\frac{1}{1-\frac{\sum P}{\sum P_{ek}}}\right)M_{lt} \qquad (3.10.25)$$

It should be mentioned that Eq. (3.10.17) is a rather conservative estimate of the maximum moment in the member. This is because the amplified moment resulting from the $P - \delta$ effect (i.e., the term $B_1 M_{nt}$) and the amplified moment resulting from the $P - \Delta$ effect (i.e., the term $B_2 M_{lt}$) do not necessarily coincide at the same location. For elastic behavior, the $P - \Delta$ effect usually magnifies the end moments. Nevertheless, because of the assumptions involved in developing the $P - \delta$ and $P - \Delta$ amplification factors, as well as the difficulties involved in locating the exact location of each of the magnified moments in the member, Eq. (3.10.17) gives a justifiable estimation of the design moment for the member.

Note that, unlike the ASD and PD interaction equations in which both the yielding and stability interactions equations are needed in the design process, only one interaction equation is needed if the LRFD approach is used. The applicable equation is determined by the term $P/\phi_c P_u$. If $P/\phi_c P_u \geq 0.2$, Eq. (3.10.15) is applicable, and if $P/\phi_c P_u < 0.2$, Eq. (3.10.16) is applicable. Another feature of the LRFD approach that is different from the ASD and PD approaches is that the $P - \delta$ and $P - \Delta$ moment magnification effect is treated independently, as is evident from Eq. (3.10.17). Recall that in the ASD or PD approach, if the member is subjected to sway, the factor C_m is taken as 0.85, therefore the moment magnification factor is $0.85/(1 - P/P_{ek})$ and this moment magnification factor is applied to the total first-order moment of the member regardless of whether it is caused by gravity load (M_{nt}) or lateral loads (M_{lt}).

In addition to Eqs. (3.10.15) and (3.10.16), the LRFD Specification also recommends a set of nonlinear interaction equations in its Appendix that are valid for nonsway members with end moments M_{ox} and M_{oy}. These equations are given as follows:

If yielding occurs,

$$\left(\frac{M_{ox}}{\phi_b M'_{px}}\right)^\zeta + \left(\frac{M_{oy}}{\phi_b M'_{py}}\right)^\zeta \leq 1.0 \qquad (3.10.26)$$

If stability controls

$$\left(\frac{C_{mx} M_{ox}}{\phi_b M'_{nx}}\right)^\eta + \left(\frac{C_{my} M_{oy}}{\phi_b M'_{ny}}\right)^\eta \leq 1.0 \qquad (3.10.27)$$

where

$$\zeta = 1.6 - \frac{P/P_y}{2[\ln(P/P_y)]} \quad \text{for} \qquad (3.10.28)$$

$$\eta = \begin{cases} 0.4 + \dfrac{P}{P_y} + \dfrac{b_f}{d} \geq 1.0 & \text{for} \quad b_f/d \geq 0.3 \\ 1 & \text{for} \quad b_f/d < 0.3 \end{cases} \qquad (3.10.29)$$

$\ln$ = natural logarithm
b_f = flange width, in inches
d = member depth, in inches

3.11 An Illustrative Example

$$M'_{px} = 1.2M_{px}[1 - (P/P_y)] \le M_{px} \qquad (3.10.30)$$

$$M'_{py} = 1.2M_{py}[1 - (P/P_y)^2] \le M_{py} \qquad (3.10.31)$$

$$M'_{nx} = M_{ux}[1 - (P/\phi_c P_u)][1 - (P/P_{ex})] \qquad (3.10.32)$$

$$M'_{ny} = M_{uy}[1 - (P/\phi_c P_u)][1 - (P/P_{ey})] \qquad (3.10.33)$$

The nonlinear interaction equations expressed in Eqs. (3.10.26) and (3.10.27) were developed by Tebedge and Chen[19] based on curve-fitting to theoretical elastic-plastic beam–column solutions.[3]

3.11 AN ILLUSTRATIVE EXAMPLE

Our discussion of the behavior of beam-columns in this chapter focuses primarily on isolated members. In reality, most structural members exist as parts of a framework and their behavior is therefore influenced by the behavior of other members of the frame. To illustrate some aspects of this interaction between the beams and columns in a frame, it is instructive to consider the following example.

Shown in Fig. 3.35 is a simple braced frame consisting of a beam and a column. The beam is loaded by a uniformly distributed load of w. After the full value of w is reached, the column is then loaded by a monotonically increasing concentric load of P until failure occurs. The behavior of the beam and the column will now be studied as P increases from zero to its ultimate value.

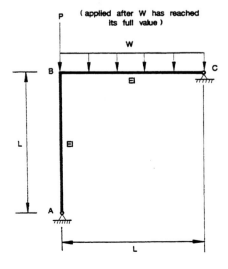

FIGURE 3.35 Two-member frame

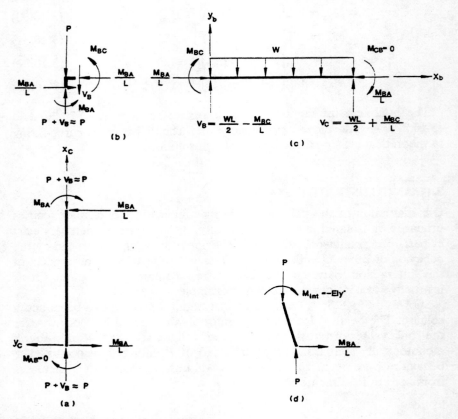

FIGURE 3.36 Free-body diagrams

In performing the analysis, the following assumptions are used:

1. The axial force in the beam is negligible.
2. The axial force in the column is represented by P.

These assumptions are illustrated in Fig. 3.36 in which the free-body diagrams of the beam and column are shown. Assumption 1 implies that the axial force M_{BA}/L induced in the beam by the column is negligible and assumption 2 implies that the additional axial force V_B induced in the column from the beam shear is negligible.

Column Analysis

As can be seen by referring to Fig. 3.36d, the differential equation of equilibrium for the column can be written as

$$-EIy_c'' - Py_c - \frac{M_{BA}}{L}x_c = 0 \qquad (3.11.1)$$

3.11 An Illustrative Example

Upon rearranging and using the notation $k^2 = P/EI$, Eq. (3.11.1) becomes

$$y_c'' + k^2 y_c = -\frac{M_{BA}}{LEI} x_c \qquad (3.11.2)$$

The general solution of Eq. (3.11.2) is

$$y_c = A \sin kx_c + B \cos kx_c - \frac{M_{BA}}{LEIk^2} x_c \qquad (3.11.3)$$

The constants A and B can be evaluated by using the boundary conditions

$$y_c(0) = 0 \qquad (3.11.4)$$
$$y_c(L) = 0 \qquad (3.11.5)$$

It can easily be shown that by using Eqs. (3.11.4) and (3.11.5)

$$B = 0 \qquad (3.11.6)$$

$$A = \frac{M_{BA}}{P \sin kL} \qquad (3.11.7)$$

Upon substituting the constants expressed in Eqs. (3.11.6) and (3.11.7) into Eq. (3.11.3), we obtain the equation for the deflected shape of the column as

$$y_c = \frac{M_{BA}}{P} \left(\frac{\sin kx_c}{\sin kL} - \frac{x_c}{L} \right) \qquad (3.11.8)$$

By successive differentiation, the equations for the slope, moment, and shear can be obtained as

$$y_c' = \frac{M_{BA}}{P} \left(\frac{k \cos kx_c}{\sin kL} - \frac{1}{L} \right) \qquad (3.11.9)$$

$$M_c = -EI y_c'' = \frac{M_{BA}}{\sin kL} \sin kx_c \qquad (3.11.10)$$

$$V_c = -EI y_c''' = \frac{M_{BA} k}{\sin kL} \cos kx_c \qquad (3.11.11)$$

The location of the *in-span* maximum moment in the column is obtained by setting the shear V_c equals zero.

$$V_c = \frac{M_{BA} k}{\sin kL} \cos kx_c = 0 \qquad (3.11.12)$$

Since the term $M_{BA} k / \sin kL$ is not zero, therefore, we must have

$$\cos kx_c = 0 \qquad (3.11.13)$$

The lowest value of kx_c satisfying the above equation is

$$kx_c = \frac{\pi}{2} \tag{3.11.14}$$

Substituting Eq. (3.11.14) into the moment equation (3.11.10) gives the value of the in-span maximum moment as

$$(M_c)_{max} = \frac{M_{BA}}{\sin kL} \tag{3.11.15}$$

In Eq. (3.11.15), M_{BA} is the column end moment at B whose value can be expressed as a function of the applied load w and P by consideration of joint equilibrium and compatibility at B as demonstrated in the following.

Beam Analysis

By neglecting the axial force in the beam, the slope-deflection equation for the beam can be written as

$$M_{BC} = \frac{EI}{L}(4\theta_B + 2\theta_C) - \frac{wL^2}{12} \tag{3.11.16}$$

$$M_{CB} = \frac{EI}{L}(2\theta_B + 4\theta_C) + \frac{wL^2}{12} \tag{3.11.17}$$

where θ_B and θ_C are the beam end rotations at B and C, respectively, and $wL^2/12$ is the fixed-end moment of a uniformly loaded beam.

Since the moment at C is zero, we have

$$M_{CB} = \frac{EI}{L}(2\theta_B + 4\theta_C) + \frac{wL^2}{12} = 0$$

from which

$$\theta_C = -\frac{wL^3}{48EI} - \frac{\theta_B}{2} \tag{3.11.18}$$

Substituting Eq. (3.11.18) into Eq. (3.11.16) gives

$$M_{BC} = \frac{EI}{L}(3\theta_B) - \frac{wL^2}{8} \tag{3.11.19}$$

and, upon rearranging,

$$\theta_B = \frac{M_{BC}L}{3EI} + \frac{wL^3}{24EI} \tag{3.11.20}$$

The location of the in-span maximum moment in the beam is obtained

3.11 An Illustrative Example

by setting the shear in the beam equal to zero:

$$V_b = \left(\frac{wL}{2} - \frac{M_{BC}}{L}\right) - wx_b = 0 \tag{3.11.21}$$

which gives

$$x_b = \frac{L}{2} - \frac{M_{BC}}{wL} \tag{3.11.22}$$

The value of the in-span maximum moment in the beam is then obtained by evaluating the moment at the distance x_b given by Eq. (3.11.22)

$$(M_b)_{max} = M \big|_{x_b = L/2 - M_{BC}/wL}$$

$$= \left[M_{BC} + \left(\frac{wl}{2} - \frac{M_{BC}}{L}\right)x_b - \frac{wx_b^2}{2}\right]_{x_b = (L/2) - M_{BC}/wL}$$

$$= \frac{M_{BC}^2}{2wL^2} + \frac{M_{BC}}{2} + \frac{wL^2}{8} \tag{3.11.23}$$

Joint Compatibility and Joint Equilibrium

The compatibility of joint B requires that

$$-y_c' \big|_{x_c = L} = \theta_B \tag{3.11.24}$$

The minus sign in Eq. (3.11.24) takes account of the fact that the column slope is negative at $x_c = L$, whereas θ_B is defined as positive when it rotates clockwise from the chord. Using Eqs. (3.11.9) and (3.11.20), Eq. (3.11.24) can be written as

$$-\frac{M_{BA}}{P}\left(\frac{k}{\tan kL} - \frac{1}{L}\right) = \frac{M_{BC}L}{3EI} + \frac{wL^3}{24EI} \tag{3.11.25}$$

For joint equilibrium (Fig. 3.36b), we must have

$$M_{BA} + M_{BC} = 0 \tag{3.11.26}$$

Solving Eqs. (3.11.25) and (3.11.26) simultaneously for M_{BA} and M_{BC} gives

$$M_{BA} = -M_{BC} = \frac{wL^2}{8}\left[\frac{k^2L^2}{3 - 3kL \cot kL + k^2L^2}\right] \tag{3.11.27}$$

The above expression for the joint moment can be used in Eqs. (3.11.15) and (3.11.23) for the maximum in-span moment in the column and beam, respectively.

In summary, the maximum in-span moment in the column is given by

$$(M_c)_{max} = \frac{M_{BA}}{\sin kL} \qquad (3.11.28)$$

and the maximum in-span moment in the beam is given by

$$(M_b)_{max} = \frac{M_{BC}^2}{2wL^2} + \frac{M_{BC}}{2} + \frac{wL^2}{8} \qquad (3.11.29)$$

where M_{BA} and M_{BC} are given by Eq. (3.11.27).

If the applied column force P is zero, it can easily be shown by using the L'Hospital rule in Eq. (3.11.27) or by a direct first-order analysis that the column and beam-end moments are given by

$$M_{BA} = -M_{BC} = \frac{wL^2}{16} \qquad (3.11.30)$$

The maximum in-span moment in the beam is given by

$$(M_b)_{max} = \frac{49}{512} wL^2 \qquad (3.11.31)$$

The maximum-column moment occurs at the end and is therefore equal to $wL^2/16$.

FIGURE 3.37 Behavior of the two-member frame

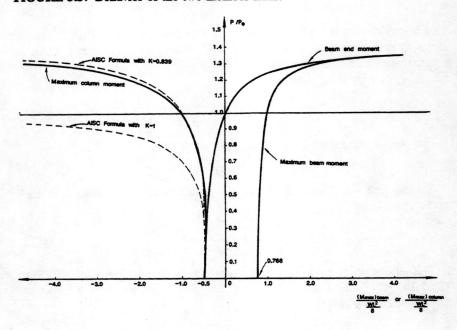

3.11 An Illustrative Example

Figure 3.37 shows a plot of the maximum-beam moment $(M_{max})_{beam}$, the maximum-column moment $(M_{max})_{column}$, and the beam-end moment M_{BC} ($= -M_{BA}$, the column-end moment) nondimensionalized by the quantity $wL^2/8$ as a function of the applied column force P nondimensionalized by the Euler load. The maximum beam moment is obtained as the larger value of the beam-end moment M_{BC} and the maximum in-span beam moment $(M_b)_{max}$. Similarly, the maximum-column moment is obtained as the larger value of the column-end moment M_{BA} and the maximum in-span column moment $(M_c)_{max}$. The sign convention used in the figure is that a positive-beam moment will cause tension on the bottom fiber of the beam and a positive-column moment will cause tension on the right-side fiber of the column.

Also shown in the figure are the maximum moments in the column as predicted by the AISC approach [Eq. (3.4.27)], that is $(M_{AB} = 0)$

$$(M_{max})_{column} = \left(\frac{C_m}{1-\frac{P}{P_e}}\right) M_{BA}$$

$$= \left[\frac{0.6 - 0.4(M_{AB}/M_{BA})}{1-\frac{P}{P_e}}\right] M_{BA}$$

$$= \left(\frac{0.6}{1-\frac{P}{P_e}}\right) M_{BA} \quad (3.11.32)$$

The lower dashed line is obtained by using an effective length factor $K=1$ in calculating P_e for the column while the upper dashed line is obtained by using $K = 0.839$ (from alignment chart discussed in Chapter 4) in calculating P_e.

A number of observations regarding the behavior of the two-member frame can be made from the figure.

1. As the applied column force P increases, the magnitude of the maximum-column moment and the maximum-beam moment both increase. Nevertheless, the locations of these maximum moments vary as a function of P. The change is apparent when one refers to Eqs. (3.11.14) and (3.11.22). The location of the maximum in-span column moment is given by

$$x_c = \frac{\pi}{2k} = \frac{L}{2\sqrt{\frac{P}{P_e}}} \quad (3.11.33)$$

The location of the maximum in-span beam moment is given by

$$x_b = \frac{L}{2} - \frac{M_{BC}}{wL} = \frac{L}{2} + \frac{L}{8}\left[\frac{\pi^2 P/P_e}{3 - 3\pi\sqrt{P/P_e}\cot(\pi\sqrt{P/P_e}) + (\pi^2 P/P_e)}\right]$$
(3.11.34)

In writing the above equation, the expression for M_{BC} given in Eq. (3.11.27) with $kL = \pi\sqrt{P/P_e}$ is used. It should be noted that the above expressions for x_c and x_b are valid only if the calculated value falls within the range 0 to L. If the calculated values fall outside this specific range, the location of the maximum moment is at the end rather than within the span of the member. Figure 3.38 shows a plot of the variation of x_c and x_b nondimensionalized by the length of the member L as a function of P/P_e. For the column, the location of the maximum moment shifts from the upper end to the middle of the

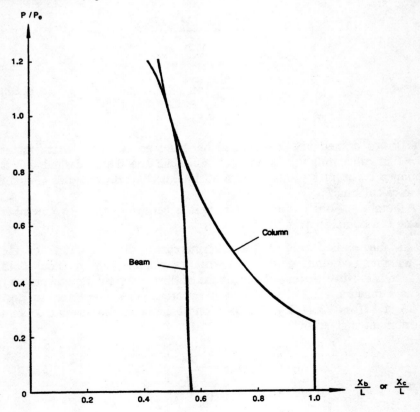

FIGURE 3.38 Variation of the location of the maximum beam and column moments with P/P_e.

3.11 An Illustrative Example

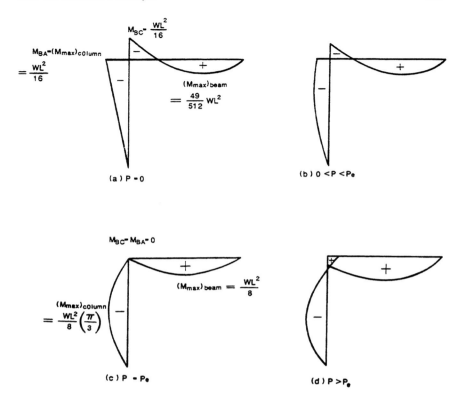

FIGURE 3.39 Moment redistribution in the frame

member as P increases from 0 to P_e. For the beam, the location of the maximum moment shifts from $x_b = 0.562L$ at $P = 0$ to $x_b = 0.5L$ at $P = P_e$.

2. The change in values in $(M_{max})_{beam}$, $(M_{max})_{column}$, and M_{BC} implies that the moment in the structure is being redistributed as the applied column force P increases. This change in moment distribution is revealed in Fig. 3.39 in which the bending moment diagrams for the frame at various values of P/P_e are shown. Note that there is a reversal in moment at the joint as P/P_e exceeds unity. In other words, when $P/P_e < 1$, the beam is *inducing* moment to the column (Fig. 3.40a), however, as $P/P_e > 1$, the beam is *restraining* the column against buckling (Fig. 3.40b). At $P/P_e = 1$, the beam end moment is zero, indicating that the beam is neither inducing moment to the column nor restraining it from buckling. It is also worth noting that as P/P_e exceeds unity, $(M_{max})_{beam}/(wL^2/8)$ will exceed unity, which implies that if the designer is to rely on the beam to restrain the

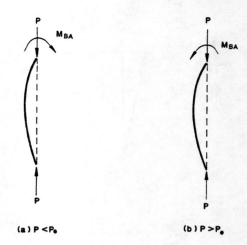

FIGURE 3.40 Beam-column interaction

column (that is, to design the column with a K-factor less than unity), the beam must be designed to carry a maximum in-span moment that exceeds $wL^2/8$ (that is, the maximum moment of a simply supported beam). This is in sharp contrast to the common notion that column-end moments do not change in a braced frame because the $P\delta$ moments are zero at the ends. This example clearly shows that this is not true. Consequently, second-order effects will change beam moments as in unbraced frames. When $P/P_e = 1.1$, the beam moment is $1.12 \times (wL^2/8)$, which will require a larger beam cross section.
3. The AISC formula for the maximum strength of a column [Eq. (3.4.27)] gives an excellent correlation to the exact result if an effective length $K = 0.839$ is used to compute the critical load in the magnification term (the upper dashed line in Fig. 3.37). If an effective length $K = 1$ is used, then the formula will underestimate the column moment (the lower dashed line in Fig. 3.37).

From observations 2 and 3, it can be concluded that for braced frames it is advisable to use an effective length factor $K = 1$ in the first term of the interaction equation [Eq. (3.10.4)] but not in the second term where we should use an effective length factor $K < 1$.

3.12 SUMMARY

The general governing equation of an elastic prismatic beam-column is a fourth-order linear differential equation relating the derivatives of the lateral displcement y, the axial force P, and the transverse load $w(x)$ in

3.12 Summary

the form

$$y^{IV} + k^2 y'' = \frac{w(x)}{EI} \qquad (3.12.1)$$

where $k^2 = P/EI$. The general solution to Eq. (3.12.1) is

$$y = A \sin kx + B \cos kx + Cx + D + f(x) \qquad (3.12.2)$$

If there are regions of constant transverse shear force V in the member, if may be more convenient to write the differential equation as

$$y''' + k^2 y' = -\frac{V}{EI} \qquad (3.12.3)$$

whose general solution is

$$y = A \sin kx + B \cos kx + C + f(x) \qquad (3.12.4)$$

Alternatively, one can draw a free-body diagram of a segment of beam-column and equate the external moment to the internal moment to obtain a second-order differential equation for a specific beam-column with lateral loads and end moments producing the primary bending moment $M(x)$ at some general location distance x from the left support

$$y'' + k^2 y = -\frac{M(x)}{EI} \qquad (3.12.5)$$

whose general solution is

$$y = A \sin kx + B \cos kx + f(x) \qquad (3.12.6)$$

In Eqs. (3.12.2), (3.12.4), and (3.12.6), $f(x)$ is the particular solution to the differential equations Eqs. (3.12.1), (3.12.3), and (3.12.5), respectively. The arbitrary constants (A, B, C, and D) can be obtained by enforcing boundary conditions of the member.

For design purposes, it is often necessary to determine the maximum deflection and maximum moment. The maximum deflection can be obtained by setting $y' = 0$ to solve for x and then backsubstitute this into the displacement function. The maximum moment can be obtained by setting $y''' = 0$, solving for x, and then backsubstituting into the moment expression. An exception to this is when the maximum moment occurs at the end(s) of a fixed-ended beam-column. In this case, the fixed-end moment(s) is (are) the maximum moment.

For simplicity and uniformity in a nonsway beam-column, it is often possible to approximate the value of the maximum moment by an expression in the simple form

$$M_{max} = A_F M_0 = \left(\frac{C_m}{1 - P/P_{ek}}\right) M_0 \qquad (3.12.7)$$

where

M_0 = maximum moment that would exist if the axial force in the member were absent (also referred to as the first-order moment)

A_F = $P - \delta$ moment amplification factor to reflect the effect of axial force on magnifying the primary moment in the member

C_m = defined as follows:

1. For members subjected to transverse loadings

$$C_m = 1 + \psi P/P_{ek} \qquad (3.12.8)$$

2. For members subjected to end moments only without transverse loadings

$$C_m = 0.6 - 0.4(M_A/M_B) \geq 0.4 \qquad (3.12.9)$$

The limiting condition $C_m \geq 0.4$ has been removed in the LRFD Specification.

$P_{ek} = \pi^2 EI/(KL)^2$

K = effective length factor

For members subjected to sidesway in an unbraced frame, in addition to the $P - \delta$ effect for an individual member (i.e., the effect of the axial force acting through the lateral displacement of the member relative to its chord), there is a $P - \Delta$ effect resulting from the frame sidesway action (i.e., the effect of the axial force acting through the sway of the member). Treatment of the $P - \Delta$ effect for a member in a frame is not as straightforward as the $P - \delta$ effect for an individual member only. The AISC/ASD and /PD Specifications account for both the $P - \delta$ and $P - \Delta$ effects indiscriminately by using $C_m = 0.85$ in Eq. (3.12.7) in design. However, the AISC/LRFD Specification accounts for these effects separately by first decomposing the first-order moment M_0 into a nonsway and sway component, designated as M_{nt} and M_{lt}, respectively. The nonsway component M_{nt} is multiplied by $P - \delta$ moment amplification factor $B_1(=A_F)$ to account for the $P - \delta$ effect, and the sway component M_{lt} is multiplied by a $P - \Delta$ moment amplification factor B_2 to account for the $P - \Delta$ effect. The maximum moment is then obtained as an algebraic sum of the two amplified moments

$$M_{max} = B_1 M_{nt} + B_2 M_{lt} \qquad (3.12.10)$$

This approach usually leads to conservative results, since the maximum secondary $P - \delta$ moment and the maximum secondary $P - \Delta$ moment do not necessarily coincide at the same location. Nevertheless, this is a more rational approach than the AISC/ASD and /PD approaches, in which the total first-order moment is magnified by the factor $0.85/(1 - P/P_{ek})$, because, in many cases, a larger percentage of the nonsway moment is not affected by the $P - \Delta$ effect.

The design of the beam-columns is facilitated by the use of interaction equations. These are equations that relate a combination of axial force and moments that will initiate the failure of a beam-column. They generally give good approximations to the more exact interaction curves developed on the basis of an inelastic analysis. Since inelastic beam-column analysis is rather complicated, interaction equations provide an attractive alternative for designers. The design of beam-columns in various interaction formats as provided by the current AISC Specifications is discussed in Sec. 3.10.

PROBLEMS

3.1 Use the design amplification factor for the lateral deflection y_{max} in Eq. (3.2.35) to derive the design moment amplification factor for the moment M_{max} in Eq. (3.2.41).

3.2 Use the four conditions (3.3.5) to (3.3.8) to determine the four constants A, B, C, and D as given in Eqs. (3.3.9) to (3.3.12).

3.3 Using L'Hospital's rule, show that the stability functions in Eqs. (3.7.15) and (3.7.16) reduce to $s_{ii} = 4$ and $s_{ij} = 2$ when $P = 0$.

3.4 Derive the expressions for the maximum deflection and maximum moments for the beam-columns shown in Figs. P3.4a,b. Using
  a. General Differential Equation
  b. Principle of Superposition
Determine the value of ψ if the expression

$$\left[\frac{1 + \psi(P/P_{ek})}{1 - (P/P_{ek})}\right] M_0$$

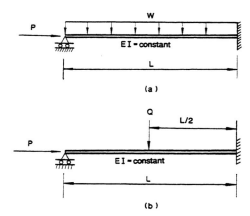

FIGURE P3.4

is used to approximate the fixed-end moment where $P_{ek} = \pi^2 EI/(KL)^2$ and M_0 = fixed-end moment that would exist in the member if P were absent (first-order moment).

3.5 Using the deflection function given in Eq. (3.3.13a, b) for the beam-column shown in Fig. P3.5a, formulate the expression for the deflection y for the beam-columns shown in Fig. P3.5b-d by the principle of superposition.

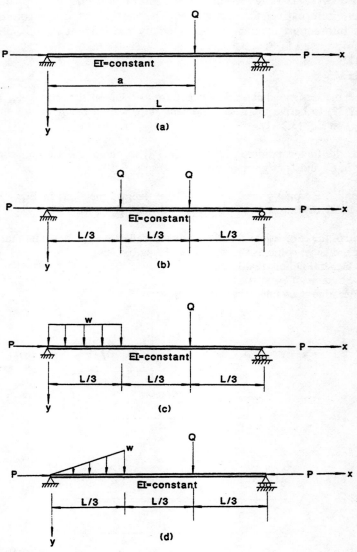

FIGURE P3.5

Problems

3.6 For the beam-column subjected to end-moments M_A and M_B as shown in Fig. P3.6, find the elastic maximum moment for $P/P_e = 0.4$ if
  a. $M_A/M_B = 0.4$
  b. $M_A/M_B = 0$
  c. $M_A/M_B = -0.4$
Where is the location of M_{max}?

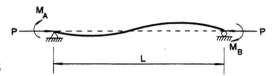

FIGURE P3.6

3.7 Find the fixed-end forces M_{FA} and M_{FB} for the beam-column shown in Fig. P3.7.

FIGURE P3.7

3.8 Find the design moments for the columns of the frame shown in Fig. P3.8 using the AISC's (a) ASD approach, (b) PD approach, and (c) LRFD approach.

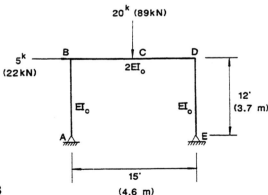

FIGURE P3.8

3.9 Plot the factors

$$\frac{1}{1-(f_a/F'_{ek})} \quad \text{in ASD}$$

$$\frac{1}{1-(P_u/P_{ek})} \quad \text{in LRFD}$$

versus P/P_e, where P is the axial force in the column and P_e is the critical load of the column for column DE of the frame shown in Fig. P3.8. What observation do you draw?

3.10 Find the fixed-ended moments M_{FA} and M_{FB} for the unsymmetrically loaded beam-column shown as Case 3 in Table 3.8.

3.11 Determine the exact C_m factor for the beam-column shown in Fig. 3.33(b).

3.12 Using the slope deflection Eqs. (3.8.1) and (3.8.2), determine the critical load of the frame shown in Fig. 3.33(a).

3.13 Find the design moments for the structure shown in Fig. P.3.13 using the LRFD method.

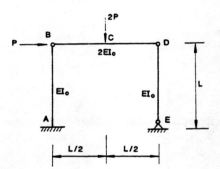

FIGURE P3.13

3.14 For the design of a beam-column in a steel frame, list the important differences between LRFD and ASD Specifications.

REFERENCES

1. Timoshenko, S. P., and Gere, J. M. Theory of Elastic Stability. Second edition. McGraw-Hill, New York, 1961.
2. Chen, W. F., and Atsuta, T. Theory of Beam-Columns: In-Plane Behavior and Design: Vol. 1. McGraw-Hill, New York, 1976.
3. Chen, W. F., and Atsuta, T. Theory of Beam-Columns: Space Behavior and Design; Vol. 2. McGraw-Hill, New York, 1977.
4. Ketter, R. L. Further studies of the strength of beam-columns. Journal of Structural Division. ASCE, 87(ST6): 135–152, 1961.
5. Massonnet, C. Stability considerations in the design of steel columns. Journal of Structural Division. ASCE, 85 (September): 75–111, 1959.
6. Austin, W. J. Strength and design of metal beam-columns. Journal of the Structural Division. ASCE. 87(ST4):1–32, 1961.
7. Specification for the Design, Fabrication and Erection of Structural Steel for Buildings. AISC, Chicago, IL, November 1978.

8. Load and Resistance Factor Design Specification for Structural Steel Buildings. AISC, Chicago, IL, October 1986.
9. Galambos, T. V., and Ketter, R. L. Columns under combined bending and thrust. Journal of Engineering Mechanics Division. ASCE, 85(2):1–30, 1959.
10. Jezek, K. Näherungsberechnung der Tragkraft exzentrisch gedrückter Stahlstäbe. Der Stahlbau. 8:89. 1935. Die Tragfähigkeit axial gedrückter und auf Biegung beanspruchter Stahlstäbe. Der Stahlbau. 9:12, 1936.
11. Chen, W. F. Approximate Solution of Beam-Columns. Journal of the Structural Division. ASCE, 97(ST2): 743–751, 1971.
12. Johnson, B. G., ed. SSRC Guide to Stability Design Criteria for Metal Structures. Third edition. John Wiley & Sons, New York, 1976.
13. Kanchanalai, T. The Design and Behavior of Beam-Columns in Unbraced Steel Frames. AISI Project No. 189, Report No. 2, Civil Engineering/Structures Research Lab, University of Texas at Austin, October, 1977.
14. Zhou, S. P., and Chen, W. F. On C_m factor in LRFD. Journal of Structural Engineering. ASCE, In press.
15. Rosenblueth, E. Slenderness effects in buildings. Journal of Structural Division, ASCE, 91(ST1): 229–252, 1965.
16. Stevens, L. K. Elastic stability of practical multistory frames. Proceedings of Civil Engineers. Vol. 36, London, England, 1967.
17. Cheong-Siat-Moy, F. Consideration of secondary effects in frame design. Journal of Structural Division. ASCE, 103(ST10): 2005–2019, 1972.
18. Yura, J. A. The Effective Length of Columns in Unbraced Frames. Engineering Journal. AISC. 8(2): pp. 37–42, 1971.
19. Tebedge, N., and Chen, W. F. Design criteria for H-Columns under biaxial bending. Journal of the Structural Division, ASCE, 104(ST9): 1355–1370, 1978.

Chapter 4

RIGID FRAMES

4.1 INTRODUCTION

In the preceding chapters, we have dealt only with *isolated* members with idealized end conditions (hinged, fixed, or guided). In reality, most structural members are connected to other members to form a framework. As a result, the behavior of these members will be affected by their adjacent members in the structure. For example, if a column in a framework buckles, its ends will rotate. This will cause rotations of adjacent members that are connected to the column, which in turn will cause deformations to other adjacent members. Thus, to determine the critical load of the column in a frame, it is necessary to investigate the stability of the frame as a whole.

If the frame is geometrically perfect and if the loadings are such that no primary bending moments are present in the members before buckling, then a frame's critical load can be obtained by an eigenvalue analysis done in a manner similar to that used for an individual member. Such a frame is shown in Fig. 4.1a, where the columns are perfectly straight and the loads are applied concentrically with the centroidal axes of the columns. The load-deflection behavior of the frame is shown in Fig. 4.2 as curve 1, and its critical load is designated as P_{cr} in the figure. Note that there is no bending deformations and so no bending moments in the members until P_{cr} is reached. Once the critical load P_{cr} is reached, a slight disturbance will induce large lateral deflections of the members.

If the columns are geometrically imperfect (Fig. 4.1b), or if the primary bending moments are present before buckling because of eccentricities of the applied loads (Fig. 4.1c), lateral deflections will occur as soon as the loads are applied. The *elastic* load-deflection behavior, as

4.1 Introduction

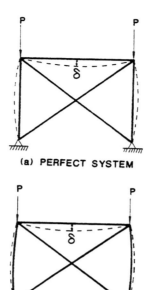

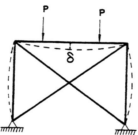

FIGURE 4.1 Simple braced portal frames

represented by curve 2 in Fig. 4.2, will be nonlinear because of the presence of secondary effects (P-δ and P-Δ effects), and the curve will approach its maximum or critical value P_{cr} asymptotically.

The elastic critical load P_{cr} can be reached only if the stresses in all members fall below the proportional or elastic limit of the stress–strain diagram of the material. Under this condition, failure of the frame is due solely to *elastic instability*. On the other extreme, if instability is excluded as a failure mode and material yielding or plasticity is the only factor accounted for in the failure analysis, failure of the frame will occur as a

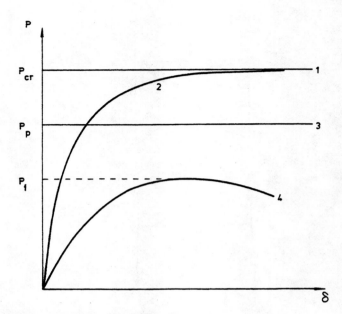

FIGURE 4.2 Load-deflection behavior of frames

result of the formation of a *collapse mechanism* when sufficient plastic hinges have developed in the structure. In this case, the *rigid plastic collapse load* P_p (Fig. 4.2, curve 3) rather than the elastic critical load P_{cr} will govern the limit state of the frame.

In many instances, the stability and plastic mechanism behavior of the frame will interact with each other and the true failure load P_f of the frame (Fig. 4.2, curve 4) is neither controlled by the elastic critical load, P_{cr}, nor the plastic mechanism load, P_p. To determine this failure load, P_f, a complete *elastic–plastic analysis* of the structure is often necessary. The rigorous analytical determination of the failure load P_f is generally very complex, and the amount of work involved does not justify its determination by the rigorous means for design applications. Fortunately, an approximate value of P_f can be obtained easily and directly, once the extreme values of P_{cr} and P_p are known. This approximate method will be presented in the later part of this chapter. The approximate determination of the failure load P_f is usually enough for design purposes.

In the first part of this chapter, we will present three methods for the determination of the elastic critical load P_{cr}: (1) the differential equation method, (2) the slope-deflection equation method, and (3) the matrix stiffness method. All these methods employ the concept of neutral equilibrium in which the critical load of the frame is obtained as the

4.2 Elastic Critical Loads by Differential Equation Method

eigenvalue of the system of equations generated by enforcing equilibrium and compatibility conditions of all members and joints in the structural system.

Toward the end of this chapter, we will give a simple method making use of the virtual work principle for the determination of the rigid plastic collapse load P_p. This will be followed by the presentation of a simple interaction equation making use of P_{cr} and P_p to estimate the true failure load P_f of the frame.

To conclude the chapter, we will discuss *the effective length factor K* as recommended by the AISC Specification for the design of members in a framed structure (references 7 and 8 in Chapter 3).

4.2 ELASTIC CRITICAL LOADS BY DIFFERENTIAL EQUATION METHOD

This section deals with the determination of the elastic critical load P_{cr} of frames. Since the critical load for a given frame is different depending on whether the frame is *braced* (sway prevented case) or *unbraced* (sway permitted case), we will discuss these two cases separately.

4.2.1 Sway-Prevented Case

Figure 4.3a shows a pin-ended portal frame braced against sidesway and loaded by two points loads P, one on each column. We will now evaluate the critical load of this frame. The subscripts b and c designate beam and column, respectively.

Because of symmetry, we need only to consider half of the structure. Referring to Fig. 4.3b, we see that the differential equation for the column is

$$EI_c y_c'' + Py_c = \frac{M_B}{L_c} x \qquad (4.2.1)$$

or

$$y_c'' + k_c^2 y_c = \frac{M_B}{EI_c} \frac{x}{L_c} \qquad (4.2.2)$$

where $k_c^2 = P/EI_c$.

The general solution is

$$y_c = A \sin k_c x + B \cos k_c x + \frac{M_B}{P} \frac{x}{L_c} \qquad (4.2.3)$$

Using the boundary conditions of

$$y_c(0) = 0, \quad y_c(L_c) = 0 \qquad (4.2.4)$$

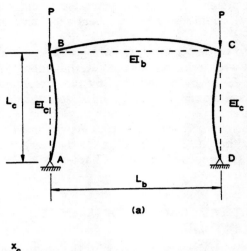

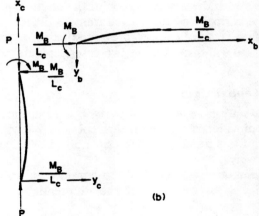

FIGURE 4.3 Nonsway buckling of a pinned-based portal frame

results in

$$A = -\frac{M_B}{P \sin k_c L_c} \quad (4.2.5)$$

$$B = 0 \quad (4.2.6)$$

Therefore, the deflection function for the column, Eq. (4.2.3), can be written as

$$y_c = \frac{M_B}{P}\left(\frac{x_c}{L_c} - \frac{\sin k_c x_c}{\sin k_c L_c}\right) \quad (4.2.7)$$

4.2 Elastic Critical Loads by Differential Equation Method

from which

$$y'_c = \frac{M_B}{P}\left(\frac{1}{L_c} - \frac{k_c \cos k_c x_c}{\sin k_c L_c}\right) \quad (4.2.8)$$

and

$$y'_c(L_c) = \frac{M_B}{P}\left(\frac{1}{L_c} - \frac{k_c}{\tan k_c L_c}\right) \quad (4.2.9)$$

For the beam, the effect of axial force on the behavior of the member is usually negligible (Fig. 4.3b). Therefore, the differential equation has the simple form

$$EI_b y''_b = M_B \quad (4.2.10)$$

or

$$y''_b = \frac{M_B}{EI_b} \quad (4.2.11)$$

The general solution is

$$y_b = Cx_b + D + \frac{M_B}{EI_b}\frac{x_b^2}{2} \quad (4.2.12)$$

Using the boundary conditions of

$$y_b(0) = 0, \quad y_b(L_b) = 0 \quad (4.2.13)$$

results in

$$C = \frac{-M_B L_b}{2EI_b} \quad (4.2.14)$$

$$D = 0 \quad (4.2.15)$$

Therefore, the deflection function for the beam, Eq. (4.2.12), can be written as

$$y_b = \frac{-M_B}{2EI_b}(L_b x_b - x_b^2) \quad (4.2.16)$$

from which

$$y'_b = \frac{-M_B}{2EI_b}(L_b - 2x_b) \quad (4.2.17)$$

$$y'_b(0) = \frac{-M_B L_b}{2EI_b} \quad (4.2.18)$$

The joint compatibility requires

$$y'_c(L_c) = y'_b(0) \quad (4.2.19)$$

Upon substituting Eqs. (4.2.9) and (4.2.18) into the joint compatibility equation (4.2.19), we have

$$\frac{M_B}{P}\left(\frac{1}{L_c} - \frac{k_c}{\tan k_c L_c}\right) = \frac{-M_B L_b}{2EI_b} \tag{4.2.20}$$

or, rearranging and denoting $k_b^2 = P/EI_b$,

$$\left(\frac{k_b^2 L_b}{2} + \frac{1}{L_c} - \frac{k_c}{\tan k_c L_c}\right) M_B = 0 \tag{4.2.21}$$

At bifurcation, M_B increases without bound. For Eq. (4.2.21) to be valid, the term in the parenthesis must be zero, i.e.,

$$\frac{k_b^2 L_b}{2} + \frac{1}{L_c} - \frac{k_c}{\tan k_c L_c} = 0 \tag{4.2.22}$$

or

$$k_b^2 L_b L_c \tan k_c L_c + 2 \tan k_c L_c - 2 k_c L_c = 0 \tag{4.2.23}$$

Equation (4.2.23) is the *characteristic equation* of the frame buckled in the *nonsway mode*. The eigenvalue determined from this equation is the critical load of the frame.

For simplicity, if we take $L_b = L_c = L$ and $k_b = k_c = k$, we can write the characteristic equation as

$$(kL)^2 \tan kL + 2 \tan kL - 2kL = 0 \tag{4.2.24}$$

By trial and error or by graphical means, the solution of Eq. (4.2.24) is found to be

$$kL = 3.59 \tag{4.2.25}$$

Since

$$kL = (\sqrt{P/EI})L \tag{4.2.26}$$

we have

$$(\sqrt{P/EI})L = 3.59 \tag{4.2.27}$$

or

$$P_{cr} = (3.59)^2 \frac{EI}{L^2} = 12.9 \frac{EI}{L^2} \tag{4.2.28}$$

Before we proceed to the sway-permitted case, we shall examine the two extreme cases of P_{cr} as expressed in Eq. (4.2.28). On one extreme, if the bending stiffness of the beam approaches zero (Fig. 4.4a), the two columns will behave like a hinged-hinged member and the critical load is $\pi^2 EI/L^2 (= 9.87 EI/L^2)$. On the other extreme, if the stiffness of the beam approaches infinity (Fig. 4.4b), the columns will behave like a hinged-fixed member and the critical load is $20.1 EI/L^2$. Thus, in the present case the lower bound for P_{cr} is $9.87 EI/L^2$ and the upper bound is $20.1 EI/L^2$. The actual value for P_{cr} for the frame with a finite beam stiffness should fall between these two extreme values as Eq. (4.2.28) does.

4.2 Elastic Critical Loads by Differential Equation Method

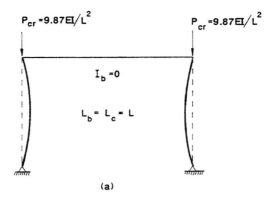

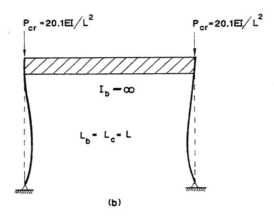

FIGURE 4.4 Extreme values of P_{cr}

4.2.2 Sway-Permitted Case

If the frame shown in Fig. 4.3a is not braced against sidesway, it may buckle in a *sway mode* at a lower buckling load level, as shown in Fig. 4.5a. Because of antisymmetry, only half of the structure needs to be considered (Fig. 4.5b). Note that the horizontal reaction H at the column base is zero, since there is no external horizontal force acting on the frame. Assuming that the applied load P is much greater than the beam shear force $2M_B/L_b$, it follows that the axial force in the column can be approximately taken as P. As a result, the differential equation for the column can be written as

$$EI_c y_c'' + P y_c = 0 \tag{4.2.29}$$

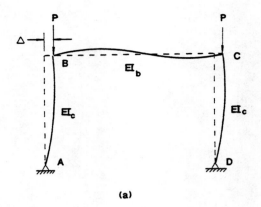

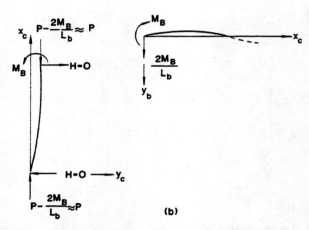

FIGURE 4.5 Sway buckling of a pinned-based portal frame

Introducing the notation $k_c^2 = P/EI_c$, we can write Eq. (4.2.29) as

$$y_c'' + k_c^2 y_c = 0 \qquad (4.2.30)$$

The general solution is

$$y_c = A \sin k_c x_c + B \cos k_c x_c \qquad (4.2.31)$$

Using the boundary conditions

$$y_c(0) = 0, \qquad y(L_c) = \Delta \qquad (4.2.32)$$

the two constants can be evaluated as

$$A = \frac{\Delta}{\sin k_c L_c} \qquad (4.2.33)$$

$$B = 0 \qquad (4.2.34)$$

4.2 Elastic Critical Loads by Differential Equation Method

Therefore, the column's lateral deflection can be expressed in terms of the sway deflection Δ of the frame as

$$y_c = \frac{\Delta}{\sin k_c L_c} \sin k_c x_c \qquad (4.2.35)$$

From equilibrium consideration of the column, we have

$$M_B = P\Delta \qquad (4.2.36)$$

from which we obtain the sway deflection

$$\Delta = \frac{M_B}{P} \qquad (4.2.37)$$

Upon substituting Eq. (4.2.37) into Eq. (4.2.35), we obtain the deflection function of the column as

$$y_c = \frac{M_B}{P \sin k_c L_c} \sin k_c x_c \qquad (4.2.38)$$

from which

$$y' = \frac{k_c M_B}{P \sin k_c L_c} \cos k_c x_c \qquad (4.2.39)$$

and

$$y'(L_c) = \frac{k_c M_B}{P \tan k_c L_c} \qquad (4.2.40)$$

The differential equation for the beam is

$$EI_b y_b'' = \frac{2M_B}{L_b} x_b - M_B \qquad (4.2.41)$$

or

$$y_b'' = \frac{M_B}{EI_b} \left(\frac{2x_b}{L_b} - 1 \right) \qquad (4.2.42)$$

The general solution is

$$y_b = C x_b + D + \frac{M_B}{EI_b} \left(\frac{x_b^3}{3L_b} - \frac{x_b^2}{2} \right) \qquad (4.2.43)$$

Using the boundary conditions

$$y_b(0) = 0, \quad y_b\left(\frac{L_b}{2}\right) = 0 \qquad (4.2.44)$$

the two constants can be evaluated as

$$C = \frac{M_B L_b}{6 EI_b} \qquad (4.2.45)$$

$$D = 0 \qquad (4.2.46)$$

Substituting Eqs. (4.2.45) and (4.2.46) into Eq. (4.2.43) gives the deflection function of the beam as

$$y_b = \frac{M_B L_b}{6EI_b} x_b + \frac{M_B}{EI_b}\left(\frac{x_b^3}{3L_b} - \frac{x_b^2}{2}\right) \quad (4.2.47)$$

from which

$$y_b' = \frac{M_B L_b}{6EI_b} + \frac{M_B}{EI_b}\left(\frac{x_b^2}{L_b} - x_b\right) \quad (4.2.48)$$

and

$$y_b'(0) = \frac{M_B L_b}{6EI_b} \quad (4.2.49)$$

Joint compatibility at B requires that

$$y_c'(L_c) = y_b'(0) \quad (4.2.50)$$

or, using Eq. (4.2.40) and Eq. (4.2.49), we obtain

$$\frac{k_c M_B}{P \tan k_c L_c} = \frac{M_B L_b}{6EI_b} \quad (4.2.51)$$

or

$$\left(\frac{k_c}{P \tan k_c L_c} - \frac{L_b}{6EI_b}\right) M_B = 0 \quad (4.2.52)$$

At the bifurcation load, M_B increases without bound. To ensure the validity of Eq. (4.2.52), we must have

$$\frac{k_c}{P \tan k_c L_c} - \frac{L_b}{6EI_b} = 0 \quad (4.2.53)$$

or

$$6k_c L_c - k_b^2 L_b L_c \tan k_c L_c = 0 \quad (4.2.54)$$

where $k_b^2 = P/EI_b$.

Equation (4.2.54) is the characteristic equation of the frame buckled in the sway mode. For the special case for which $L_b = L_c = L$ and $k_b = k_c = k$, Eq. (4.2.54) reduces to

$$6kL - (kL)^2 \tan kL = 0 \quad (4.2.55)$$

from which kL can be evaluated as

$$kL = (\sqrt{P/EI})L = 1.35 \quad (4.2.56)$$

and so

$$P_{cr} = (1.35)^2 \frac{EI}{L^2} = 1.82 \frac{EI}{L^2} \quad (4.2.57)$$

Again, we shall examine the two extreme cases of P_{cr} as expressed in Eq.

4.2 Elastic Critical Loads by Differential Equation Method

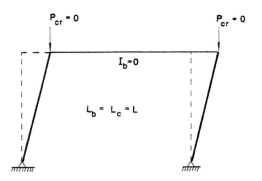

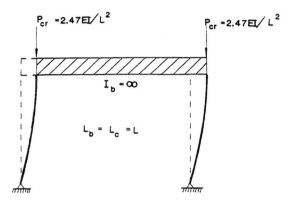

FIGURE 4.6 Extreme values of P_{cr}

(4.2.57). If the bending stiffness of the beam approaches zero (Fig. 4.6a), the two columns do not possess any sway stiffness and the critical load is zero. If the bending stiffness of the beam approaches infinity, the two columns act like a hinged-guided member and the critical load is $2.47EI/L^2$. The critical load as expressed in Eq. (4.2.57) does fall between these two extreme cases of 0 and $2.47EI/L^2$.

Note that P_{cr} for the sway-permitted case is much less than that of the sway-prevented case; the frame will undoubtedly buckle on the sway mode if no physical constraint is provided to prevent the frame from sidesway buckling.

The differential equation method described above can be extended to a more complex frame. A differential equation is written for each and every member of the frame. The arbitrary constants of the general solution to each differential equation solution are then evaluated using the boundary conditions and joint compatibility conditions of each member and each joint from which the charactersitic equation of the frame can be obtained. The eigenvalue of this characteristic equation will give the critical load of the frame.

Although the differential equation method can, in theory, be used to determine P_{cr} for all types of frames, the actual implementation of the method for the solution of a given frame is rather complex, especially when it is applied to frames of more than one story and one bay. Fortunately, there is a simpler method available, which makes use of the slope-deflection equations[1] developed in the previous chapter. This will be discussed in the following section.

4.3 ELASTIC CRITICAL LOADS BY SLOPE-DEFLECTION EQUATION METHOD

In the slope-deflection equation method, the slope-deflection equations developed in Sections 3.7 and 3.8 are written for each and every member of the frame. These equations are then related to one another by enforcing moment equilibrium at the joints (for braced frames) or by enforcing moment equilibrium at the joints and story shear equilibrium for every story of the frame (for unbraced frames). The characteristic equation is obtained by setting the determinant of the coefficient matrix of the resulting set of equilibrium equations equal to zero. The critical load is then obtained as the eigenvalue of the characteristic equation. To demonstrate the use of this approach, we will reanalyze the braced and unbraced frames shown in Figs. 4.3a and 4.5a using the slope-deflection equation approach. Note that because of symmetry and antisymmetry, only half of the structure needs to be considered in the analysis.

4.3.1 Sway-Prevented Case

For this case, there is no relative lateral translation between the ends of the column; therefore, by using Eq. (3.8.5), the slope-deflection equation for the column (Fig. 4.7a) can be written as

$$M_{BA} = \frac{EI_c}{L_c}\left(s_{iic} - \frac{s_{ijc}^2}{s_{iic}}\right)\theta_B \qquad (4.3.1)$$

in which the subscript c denotes the column, and in which expressions for s_{ii} and s_{ij} are given in Eqs. (3.7.15) and (3.7.16), respectively.

Since the beam is bent in a single curvature, we use the slope-deflection equation (3.8.15) for the beam

$$M_{BC} = \frac{EI_b}{L_b}(s_{iib} - s_{ijb})\theta_B \qquad (4.3.2)$$

in which the subscript b denotes the beam.

4.3 Elastic Critical Loads by Slope-Deflection Equation Method

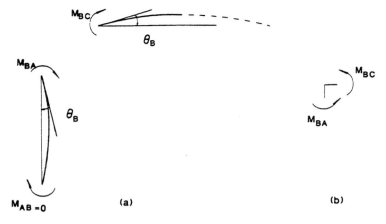

FIGURE 4.7 Slope-deflection equation approach for P_{cr} of nonsway buckling of simple portal frame

Neglecting the effect of axial force on the bending stiffness of the beam, we can set $s_{iib} = 4$ and $s_{ijb} = 2$, so that Eq. (4.3.2) becomes

$$M_{BC} = \frac{2EI_b}{L_b} \theta_B \quad (4.3.3)$$

From joint equilibrium (Fig. 4.7b), we must have

$$M_{BA} + M_{BC} = 0 \quad (4.3.4)$$

Using Eqs. (4.3.1) and (4.3.3), the joint equilibrium condition expressed in Eq. (4.3.4) can be written as

$$\frac{EI_c}{L_c}\left(s_{iic} - \frac{s_{ijc}^2}{s_{iic}}\right)\theta_B + \frac{2EI_b}{L_b}\theta_B = 0 \quad (4.3.5)$$

or

$$\left[\frac{EI_c}{L_c}\left(s_{iic} - \frac{s_{ijc}^2}{s_{iic}}\right) + \frac{2EI_b}{L_b}\right]\theta_B = 0 \quad (4.3.6)$$

Since at bifurcation, $\theta_B \neq 0$, we must have

$$\frac{EI_c}{L_c}\left(s_{iic} - \frac{s_{ijc}^2}{s_{iic}}\right) + \frac{2EI_b}{L_b} = 0 \quad (4.3.7)$$

Equation (4.3.7) is the characteristic equation of the frame. For the special case for which $I_b = I_c = I$ and $L_b = L_c = L$, Eq. (4.3.7) becomes

$$\frac{EI}{L}\left[s_{iic} - \frac{s_{ijc}^2}{s_{iic}} + 2\right] = 0 \quad (4.3.8)$$

or

$$s_{iic} - \frac{s_{ijc}^2}{s_{iic}} + 2 = 0 \tag{4.3.9}$$

By trial and error and using Table 3.7, the value of kL that satisfies Eq. (4.3.9) is found to be

$$kL = (\sqrt{P/EI})L = 3.59 \tag{4.3.10}$$

from which the critical load

$$P_{cr} = 12.9 \frac{EI}{L^2} \tag{4.3.11}$$

is obtained. This load is the same as before using the differential equation approach.

4.3.2 Sway-Permitted Case

Referring to Fig. 4.8a, we see that the slope-deflection equations (3.8.1) and (3.8.2) for the swayed column are

$$M_{AB} = \frac{EI_c}{L_c}\left[s_{iic}\theta_A + s_{ijc}\theta_B - (s_{iic} + s_{ijc})\frac{\Delta}{L_c}\right] = 0 \tag{4.3.12}$$

$$M_{BA} = \frac{EI_c}{L_c}\left[s_{ijc}\theta_A + s_{iic}\theta_B - (s_{iic} + s_{ijc})\frac{\Delta}{L_c}\right] \tag{4.3.13}$$

Solving Eq. (4.3.12) for θ_A and substituting θ_A into Eq. (4.2.13), we obtain

$$M_{BA} = \frac{EI_c}{L_c}\left[\left(s_{iic} - \frac{s_{ijc}^2}{s_{iic}}\right)\theta_B - \left(s_{iic} - \frac{s_{ijc}^2}{s_{iic}}\right)\frac{\Delta}{L_c}\right] \tag{4.3.14}$$

Since the beam is bent in double curvature, we use the slope-deflection equation (3.8.17) for the beam

$$M_{BC} = \frac{EI_b}{L_b}(s_{iib} + s_{ijb})\theta_B \tag{4.3.15}$$

Because there is no axial force in the beam, we set $s_{iib} = 4$ and $s_{ijb} = 2$, or

$$M_{BC} = \frac{6EI_b}{L_b}\theta_B \tag{4.3.16}$$

From joint equilibrium (Fig. 4.8b), we know

$$M_{BA} + M_{BC} = 0 \tag{4.3.17}$$

4.3 Elastic Critical Loads by Slope-Deflection Equation Method

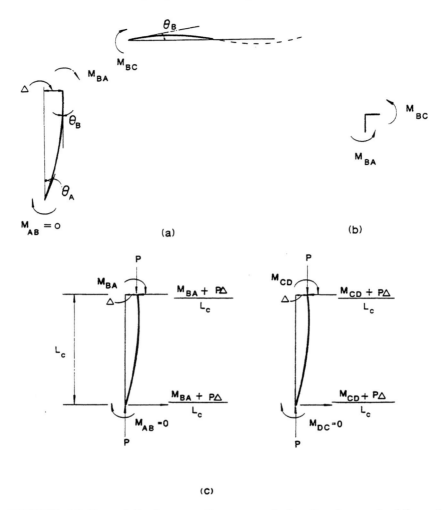

FIGURE 4.8 Slope-deflection equation approach for P_{cr} of sway buckling of simple portal frame

Using Eqs. (4.3.14) and (4.3.16), the joint equilibrium condition (4.3.17) becomes

$$\frac{EI_c}{L_c}\left[\left(s_{iic} - \frac{s_{ijc}^2}{s_{iic}}\right)\theta_B - \left(s_{iic} - \frac{s_{ijc}^2}{s_{iic}}\right)\frac{\Delta}{L_c}\right] + \frac{6EI_b}{L_b}\theta_B = 0 \qquad (4.3.18)$$

or

$$\left(s_{iic} - \frac{s_{ijc}^2}{s_{iic}} + 6\frac{I_b L_c}{I_c L_b}\right)\theta_B - \left(s_{iic} - \frac{s_{ijc}^2}{s_{iic}}\right)\frac{\Delta}{L_c} = 0 \qquad (4.3.19)$$

From story shear equilibrium (Fig. 4.8c), we have

$$\frac{M_{AB} + M_{BA} + P\Delta}{L_c} + \frac{M_{CD} + M_{DC} + P\Delta}{L_c} = 0 \quad (4.3.20)$$

Realizing that

$$M_{AB} = M_{DC} = 0 \quad \text{(hinged)} \quad (4.3.21)$$

and

$$M_{CD} = M_{BA} \quad \text{(antisymmetry)} \quad (4.3.22)$$

we can write the story-shear equilibrium equation (4.3.20) as

$$\frac{2M_{BA} + 2P\Delta}{L_c} = 0 \quad (4.3.23)$$

or

$$\frac{M_{BA} + P\Delta}{L_c} = 0 \quad (4.3.24)$$

Using Eq. (4.3.14) for M_{BA} in Eq. (4.3.24), we can write

$$\frac{EI_c}{L_c^2}\left[\left(s_{iic} - \frac{s_{ijc}^2}{s_{iic}}\right)\theta_B - \left(s_{iic} - \frac{s_{ijc}^2}{s_{iic}}\right)\frac{\Delta}{L_c}\right] + \frac{P\Delta}{L_c} = 0 \quad (4.3.25)$$

or

$$\left(s_{iic} - \frac{s_{ijc}^2}{s_{iic}}\right)\theta_B - \left(s_{iic} - \frac{s_{ijc}^2}{s_{iic}} - k_c^2 L_c^2\right)\frac{\Delta}{L_c} = 0 \quad (4.3.26)$$

Equations (4.3.19) and (4.3.26) are the two equilibrium equations of the frame, they can be written in matrix form

$$\begin{bmatrix} S + 6\frac{I_b L_c}{I_c L_b} & -S \\ S & -S + (k_c L_c)^2 \end{bmatrix} \begin{pmatrix} \theta_B \\ \frac{\Delta}{L_c} \end{pmatrix} = \begin{pmatrix} 0 \\ 0 \end{pmatrix} \quad (4.3.27)$$

where

$$S = s_{iic} - \frac{s_{ijc}^2}{s_{iic}}$$

Note that the coefficient matrix in Eq. (4.3.27) can be made symmetric by multiplying Eq. (4.3.26) by minus one. If we do this, and also let $I_b = I_c = I$ and $L_b = L_c = L$, Eq. (4.3.27) becomes

$$\begin{bmatrix} S + 6 & -S \\ -S & S - (kL)^2 \end{bmatrix} \begin{pmatrix} \theta_B \\ \frac{\Delta}{L} \end{pmatrix} = \begin{pmatrix} 0 \\ 0 \end{pmatrix} \quad (4.3.28)$$

At bifurcation, both θ_B and Δ increase without bound. For Eq.

(4.3.28) to be valid, we must set

$$\det \begin{vmatrix} S+6 & -S \\ -S & S-(kL)^2 \end{vmatrix} = 0 \qquad (4.3.29)$$

Equation (4.3.29) is the characteristic equation of the frame. By trial and error and by using Table 3.7, the value kL can be found to be

$$kL = (\sqrt{P/EI})L = 1.35 \qquad (4.3.30)$$

from which the critical load can be solved

$$P_{cr} = 1.82 \frac{EI}{L^2} \qquad (4.3.31)$$

Note the correspondence of Eq. (4.3.31) obtained using the slope-deflection method with Eq. (4.2.57) obtained previously using the differential equation method.

The slope-deflection equation method, as in the differential equation method, can in theory, be extended to evaluate P_{cr} for all types of frames. The resulting coefficient matrix obtained by enforcing joint (and story-shear) equilibrium will be an $n \times n$ matrix in which n is the number of independent degrees of freedom of the frame. However, if n is large, it is cumbersome to obtain a solution. In the next section, the slope-deflection equation method will be generalized; the resulting formulation we will see is called the *matrix stiffness* method.[2,3] This procedure to obtain solutions for large frames can be greatly enhanced by the use of computers.

4.4 ELASTIC CRITICAL LOADS BY MATRIX STIFFNESS METHOD

In the matrix stiffness method, the element stiffness matrix that relates the element end forces to end displacements is first formulated for each and every member of the frame. These element stiffness matrices are then assembled into the structure stiffness matrix that relates the structure nodal force to the structure nodal displacements. At bifurcation, the stiffness of the structure vanishes. Therefore, by setting the determinant of the structure stiffness matrix to zero, the critical load of the frame can be obtained.

4.4.1 Element Stiffness Formulation

We shall begin our discussion of the matrix stiffness method by developing the element stiffness matrix from the slope-deflection equation. Figure 4.9a shows the sign convention for the positive directions of

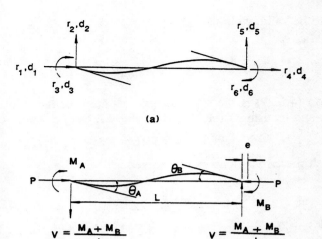

FIGURE 4.9 Element end forces and displacements notations

element end forces and end displacements of a frame member. The end forces and end displacements used in the slope-deflection equation are shown in Fig. 4.9b. By comparing the two figures, we can easily express the following equilibrium and kinematic relationships.

Equilibrium

$$r_1 = P \tag{4.4.1}$$

$$r_2 = -V = -\frac{M_A + M_B}{L} \tag{4.4.2}$$

$$r_3 = M_A \tag{4.4.3}$$

$$r_4 = -P \tag{4.4.4}$$

$$r_5 = \frac{M_A + M_B}{L} \tag{4.4.5}$$

$$r_6 = M_B \tag{4.4.6}$$

Kinematic

$$e = -(d_4 - d_1) \tag{4.4.7}$$

$$\theta_A = d_3 + \left(\frac{d_5 - d_2}{L}\right) \tag{4.4.8}$$

$$\theta_B = d_6 + \left(\frac{d_5 - d_2}{L}\right) \tag{4.4.9}$$

4.4 Elastic Critical Loads by Matrix Stiffness Method

Equations (4.4.1) to (4.4.6) can be written in matrix form as

$$\begin{pmatrix} r_1 \\ r_2 \\ r_3 \\ r_4 \\ r_5 \\ r_6 \end{pmatrix} = \begin{bmatrix} 1 & 0 & 0 \\ 0 & -\dfrac{1}{L} & -\dfrac{1}{L} \\ 0 & 1 & 0 \\ -1 & 0 & 0 \\ 0 & \dfrac{1}{L} & \dfrac{1}{L} \\ 0 & 0 & 1 \end{bmatrix} \begin{pmatrix} P \\ M_A \\ M_B \end{pmatrix} \quad (4.4.10)$$

Similarly, Eqs. (4.7) to (4.4.9) can be written in matrix form as

$$\begin{pmatrix} e \\ \theta_A \\ \theta_B \end{pmatrix} = \begin{bmatrix} 1 & 0 & 0 & -1 & 0 & 0 \\ 0 & -\dfrac{1}{L} & 1 & 0 & \dfrac{1}{L} & 0 \\ 0 & -\dfrac{1}{L} & 0 & 0 & \dfrac{1}{L} & 1 \end{bmatrix} \begin{pmatrix} d_1 \\ d_2 \\ d_3 \\ d_4 \\ d_5 \\ d_6 \end{pmatrix} \quad (4.4.11)$$

Equation (4.4.10) and Eq. (4.4.11) can be related by recognizing that

$$P = \frac{EA}{L} e \quad (4.4.12)$$

$$M_A = \frac{EI}{L}(s_{ii}\theta_A + s_{ij}\theta_B) \quad (4.4.13)$$

$$M_B = \frac{EI}{L}(s_{ij}\theta_A + s_{ii}\theta_B) \quad (4.4.14)$$

Equation (4.4.12) relates the axial force P to the axial displacement e of the member, Eqs. (4.4.13) and (4.4.14) are the slope-deflection equations of the member, and s_{ii}, s_{ij} are the stability functions. In writing Eq. (4.4.12), it is tacitly assumed that the effect of member shortening due to the bending curvature is negligible. This assumption is satisfactory for most practical purposes.

Putting Eqs. (4.4.12) to (4.4.14) in matrix form, we have

$$\begin{pmatrix} P \\ M_A \\ M_B \end{pmatrix} = \frac{EI}{L} \begin{bmatrix} \dfrac{A}{I} & 0 & 0 \\ 0 & s_{ii} & s_{ij} \\ 0 & s_{ij} & s_{ii} \end{bmatrix} \begin{pmatrix} e \\ \theta_A \\ \theta_B \end{pmatrix} \quad (4.4.15)$$

Substituting Eq. (4.4.15) into Eq. (4.4.10), and then substituting Eq. (4.4.11) into the resulting equation, we can relate the element end forces (r_1 to r_6) with the element end displacements (d_1 to d_6) as

$$\begin{pmatrix} r_1 \\ r_2 \\ r_3 \\ r_4 \\ r_5 \\ r_6 \end{pmatrix}_{ns} = \frac{EI}{L} \begin{bmatrix} \frac{A}{I} & 0 & 0 & -\frac{A}{I} & 0 & 0 \\ & \frac{2(s_{ii}+s_{ij})}{L^2} & -\frac{(s_{ii}+s_{ij})}{L} & 0 & \frac{-2(s_{ii}+s_{ij})}{L^2} & -\frac{(s_{ii}+s_{ij})}{L} \\ & & s_{ii} & 0 & \frac{s_{ii}+s_{ij}}{L} & s_{ij} \\ & \text{sym.} & & \frac{A}{I} & 0 & 0 \\ & & & & \frac{2(s_{ii}+s_{ij})}{L^2} & \frac{(s_{ii}+s_{ij})}{L} \\ & & & & & s_{ii} \end{bmatrix}_{ns} \begin{pmatrix} d_1 \\ d_2 \\ d_3 \\ d_4 \\ d_5 \\ d_6 \end{pmatrix}$$

(4.4.16)

Symbolically, Eq. (4.4.16) can be written as

$$\mathbf{r}_{ns} = \mathbf{k}_{ns}\mathbf{d} \qquad (4.4.17)$$

where the subscript ns is used here to indicate that there is no sidesway in the member. If the member is permitted to sway as shown in Fig. 4.10, an additional shear force equal to $P\Delta/L$ will be induced in the member due to the swaying of the member by an amount Δ given by

$$\Delta = d_2 - d_5 \qquad (4.4.18)$$

We can relate this additional shear force due to member sway to the member end displacement as

$$\begin{pmatrix} r_1 \\ r_2 \\ r_3 \\ r_4 \\ r_5 \\ r_6 \end{pmatrix}_s = \begin{bmatrix} 0 & 0 & 0 & 0 & 0 & 0 \\ & -\frac{P}{L} & 0 & 0 & \frac{P}{L} & 0 \\ & & 0 & 0 & 0 & 0 \\ & & & 0 & 0 & 0 \\ & \text{sym.} & & & -\frac{P}{L} & 0 \\ & & & & & 0 \end{bmatrix}_s \begin{pmatrix} d_1 \\ d_2 \\ d_3 \\ d_4 \\ d_5 \\ d_6 \end{pmatrix} \qquad (4.4.19)$$

or symbolically

$$\mathbf{r}_s = \mathbf{k}_s\mathbf{d} \qquad (4.4.20)$$

4.4 Elastic Critical Loads by Matrix Stiffness Method

where the subscript s is used to indicate the quantities due to sidesway of the member.

By combining Eq. (4.4.17) and Eq. (4.4.20), we obtain the general beam-column element force-displacement relationship as

$$\mathbf{r} = \mathbf{k}\mathbf{d} \qquad (4.4.21)$$

where

$$\mathbf{r} = \mathbf{r}_{ns} + \mathbf{r}_s \qquad (4.4.22a)$$

$$\mathbf{k} = \mathbf{k}_{ns} + \mathbf{k}_s \qquad (4.4.22b)$$

$$\mathbf{k} = \frac{EI}{L} \begin{bmatrix} \frac{A}{I} & 0 & 0 & -\frac{A}{I} & 0 & 0 \\ & \frac{2(s_{ii}+s_{ij})-(kL)^2}{L^2} & -\frac{(s_{ii}+s_{ij})}{L} & 0 & \frac{-2(s_{ii}+s_{ij})+(kL)^2}{L^2} & -\frac{(s_{ii}+s_{ij})}{L} \\ & & s_{ii} & 0 & \frac{s_{ii}+s_{ij}}{L} & s_{ij} \\ & & & \frac{A}{I} & 0 & 0 \\ & \text{sym.} & & & \frac{2(s_{ii}+s_{ij})-(kL)^2}{L^2} & \frac{(s_{ii}+s_{ij})}{L} \\ & & & & & s_{ii} \end{bmatrix}$$

$$(4.4.23)$$

Substituting the expressions for the stability functions (s_{ii}, s_{ij}) in Eq. (4.4.23) and simplifying, we obtain

$$\mathbf{k} = \frac{EI}{L} \begin{bmatrix} \frac{A}{I} & 0 & 0 & -\frac{A}{I} & 0 & 0 \\ & \frac{12}{L^2}\phi_1 & \frac{-6}{L}\phi_2 & 0 & \frac{-12}{L^2}\phi_1 & \frac{-6}{L}\phi_2 \\ & & 4\phi_3 & 0 & \frac{6}{L}\phi_2 & 2\phi_4 \\ & & & \frac{A}{I} & 0 & 0 \\ & \text{sym.} & & & \frac{12}{L^2}\phi_1 & \frac{6}{L}\phi_2 \\ & & & & & 4\phi_3 \end{bmatrix} \qquad (4.4.24)$$

The expressions for ϕ_1, ϕ_2, ϕ_3, and ϕ_4 are given in Table 4.1. Note that as P approaches zero, the functions ϕ_1, ϕ_2, ϕ_3, and ϕ_4 become

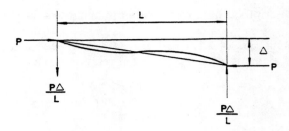

FIGURE 4.10 Additional shear due to swaying of the member

indefinite. However, by using the L'Hospital's rule, it can be shown that these functions will approach unity and Eq. (4.4.24) reduces to the first-order (linear) element stiffness matrix for a frame member.

Also shown in Table 4.1 are the ϕ_i functions expressed in the form of a power series by using the following series expansion for the trigonometric functions:

For compression

$$\sin kL = kL - \frac{(kL)^3}{6} + \frac{(kL)^5}{120} + \cdots \quad (4.4.25a)$$

$$\cos kL = 1 - \frac{kL}{2} + \frac{(kL)^4}{24} + \cdots \quad (4.4.25b)$$

For tension

$$\sinh kL = kL + \frac{(kL)^3}{6} + \frac{(kL)^5}{120} + \cdots \quad (4.4.26a)$$

$$\cosh kL = 1 + \frac{kL}{2} + \frac{(kL)^4}{24} + \cdots \quad (4.4.26b)$$

It has been shown[4] that these power series expressions are convenient and efficient to use in a computer-aided analysis because no numerical difficulties will arise even if the axial force P is small. In addition, the expressions in the series are the same regardless of whether P is tensile or compressive. For most cases, the series will converge to a high degree of accuracy if $n = 10$ is used.

If the axial force in the member is small, Eq. (4.4.24) can be simplified by using a Taylor series expansion for the ϕ_i's. If we retain only the first two terms in the Taylor series, it can be shown that the resulting stiffness

Table 4.1 Expressions for ϕ_1, ϕ_2, ϕ_3, and ϕ_4

ϕ	P		
	Compressive	Zero	Tensile
ϕ_1	$\dfrac{(kL)^3 \sin kL}{12\phi_c}$	1	$\dfrac{(kL)^3 \sinh kL}{12\phi_t}$
ϕ_2	$\dfrac{(kL)^2(1 - \cos kL)}{6\phi_c}$	1	$\dfrac{(kL)^2(\cosh kL - 1)}{6\phi_t}$
ϕ_3	$\dfrac{(kL)(\sin kL - kL \cos kL)}{4\phi_c}$	1	$\dfrac{(kL)(kL \cosh kL - \sinh kL)}{4\phi_t}$
ϕ_4	$\dfrac{(kL)(kL - \sin kL)}{2\phi_c}$	1	$\dfrac{(kL)(\sinh kL - kL)}{2\phi_t}$

where

$\phi_c = 2 - 2\cos kL - kL \sin kL \qquad \phi_t = 2 - 2\cosh kL + kL \sinh kL$

Alternatively, the ϕ_i functions can be expressed in the form of power series, as in reference 4:

$$\phi_1 = \frac{1 + \sum_{n=1}^{\infty} \dfrac{1}{(2n+1)!}[\mp(kL)^2]^n}{12\phi}$$

$$\phi_2 = \frac{\tfrac{1}{2} + \sum_{n=1}^{\infty} \dfrac{1}{(2n+2)!}[\mp(kL)^2]^n}{6\phi}$$

$$\phi_3 = \frac{\tfrac{1}{3} + \sum_{n=1}^{\infty} \dfrac{2(n+1)}{(2n+3)!}[\mp(kL)^2]^n}{4\phi}$$

$$\phi_4 = \frac{\tfrac{1}{6} + \sum_{n=1}^{\infty} \dfrac{1}{(2n+3)!}[\mp(kL)^2]^n}{2\phi}$$

where

$$\phi = \tfrac{1}{12} + \sum_{n=1}^{\infty} \frac{2(n+1)}{(2n+4)!}[\mp(kL)^2]^n$$

Use the minus sign if the axial force is compressive.
Use the plus sign if the axial force is tensile.

matrix that is valid for small axial force is given by

$$\mathbf{k} = \frac{EI}{L} \begin{bmatrix} \frac{A}{I} & 0 & 0 & -\frac{A}{I} & 0 & 0 \\ & \frac{12}{L^2} & \frac{-6}{L} & 0 & \frac{-12}{L^2} & \frac{-6}{L} \\ & & 4 & 0 & \frac{6}{L} & 2 \\ & & & \frac{A}{I} & 0 & 0 \\ & & & & \frac{12}{L^2} & \frac{6}{L} \\ & & & & & 4 \end{bmatrix} \mp \frac{P}{L} \begin{bmatrix} 0 & 0 & 0 & 0 & 0 & 0 \\ & \frac{6}{5} & \frac{-L}{10} & 0 & \frac{-6}{5} & \frac{-L}{10} \\ & & \frac{2L^2}{15} & 0 & \frac{L}{10} & \frac{-L^2}{30} \\ & & & 0 & 0 & 0 \\ & & & & \frac{6}{5} & \frac{L}{10} \\ & & & & & \frac{2L^2}{15} \end{bmatrix}$$

(4.4.27)

in which the negative sign preceding the second matrix corresponds to a compressive axial force and the positive sign corresponds to a tensile axial force.

Symbolically, Eq. (4.4.27) can be written as

$$\mathbf{k} = \mathbf{k}_0 + \mathbf{k}_G \qquad (4.4.28)$$

where $\mathbf{k}_0$ is the first-order (linear) elastic stiffness matrix and $\mathbf{k}_G$ is the geometric stiffness matrix (sometimes referred to as the initial stress stiffness matrix), which accounts for the effect of the axial force P on the bending stiffness of the member.

The following example will be used to demonstrate the procedure of using the stiffness matrix method to obtain the critical load of frames.

4.4.2 Sway Buckling of a Pinned-Base Portal Frame

The matrix stiffness method is applied here to determine the critical load P_{cr} for the frame shown in Fig. 4.5a. Because of symmetry, we consider only one half of the structure in the analysis. This is shown in Fig. 4.11a together with the structural nodal forces and displacements. To reduce the number of degrees of freedom of the structure, we assume that all members are inextensible (i.e., the change in length due to axial force is neglected). As a result, only four degrees of freedom, are labeled: three rotational degrees of freedom, D_1, D_2, and D_3, and one translational degree of freedom, D_4. The corresponding structural nodal forces, $R_1, \ldots, R_4$, are also shown in Fig. 4.11a. The directions of these

4.4 Elastic Critical Loads by Matrix Stiffness Method

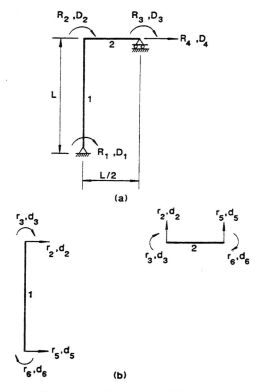

FIGURE 4.11 Structure and member forces and displacements notations

rotations, translations, and forces are shown in their positive sense in the figure.

Because of the assumption of inextensional behavior, the axial force-axial displacement relationship expressed in Eq. (4.4.12) is not valid anymore. As a consequence, the 6×6 element stiffness matrix relating the element end forces to the element end displacements will be reduced to a 4×4 matrix as

$$\mathbf{k} = \frac{EI}{L} \begin{bmatrix} \frac{12}{L^2} & -\frac{6}{L} & -\frac{12}{L^2} & -\frac{6}{L} \\ & 4 & \frac{6}{L} & 2 \\ & & \frac{12}{L^2} & \frac{6}{L} \\ \text{sym.} & & & 4 \end{bmatrix} \mp \frac{P}{L} \begin{bmatrix} \frac{6}{5} & -\frac{L}{10} & -\frac{6}{5} & -\frac{L}{10} \\ & \frac{2L^2}{15} & \frac{L}{10} & -\frac{L^2}{30} \\ & & \frac{6}{5} & \frac{L}{10} \\ \text{sym.} & & & \frac{2L^2}{15} \end{bmatrix}$$

(4.4.29)

This stiffness matrix relates the four end forces (r_2, r_3, r_5, and r_6) to the four end displacements (d_2, d_3, d_5, and d_6) of an inextensible member. Note that the element stiffness matrix for an inextensible member [Eq. (4.4.29)] is obtained simply by deleting the first and fourth rows and the first and fourth columns from the element stiffness matrix for an extensible member [Eq. (4.4.27)].

Figure 4.11b shows the four degrees of freedom (d_2, d_3, d_5, and d_6) and the corresponding end forces (r_2, r_3, r_5, and r_6) associated with each member of the structure. Again, the directions are shown in their positive sense in the figure.

By using Eq. (4.4.29), the element stiffness matrix for the column (element 1) can be written as

$$\mathbf{k}_1 = \frac{EI}{L}\begin{bmatrix} \frac{12}{L^2} & \frac{-6}{L} & \frac{-12}{L^2} & \frac{-6}{L} \\ & 4 & \frac{6}{L} & 2 \\ & & \frac{12}{L^2} & \frac{6}{L} \\ \text{sym.} & & & 4 \end{bmatrix} - \frac{P}{L}\begin{bmatrix} \frac{6}{5} & \frac{-L}{10} & \frac{-6}{5} & \frac{-L}{10} \\ & \frac{2L^2}{15} & \frac{L}{10} & \frac{-L^2}{30} \\ & & \frac{6}{5} & \frac{L}{10} \\ \text{sym.} & & & \frac{2L^2}{15} \end{bmatrix} \quad (4.4.30)$$

and the element stiffness matrix for beam with $P = 0$ and $L/2$ for L (element 2) can be written as

$$\mathbf{k}_2 = 2\frac{EI}{L}\begin{bmatrix} \frac{48}{L^2} & \frac{-12}{L} & \frac{-48}{L^2} & \frac{-12}{L} \\ & 4 & \frac{12}{L} & 2 \\ & & \frac{48}{L^2} & \frac{12}{L} \\ \text{sym.} & & & 4 \end{bmatrix} \quad (4.4.31)$$

The structure stiffness matrix can be obtained by assembling these element stiffness matrices. The process of assemblage is described in detail in most matrix structural analysis textbooks.[5-7] So we will discuss it only very briefly here.

For each element, the element end displacements are first related to the structure nodal displacements by consideration of joint compatibility. It can easily be seen from Fig. 4.11 that for element 1, this kinematic

4.4 Elastic Critical Loads by Matrix Stiffness Method

relationship is

$$\begin{pmatrix} d_2 \\ d_3 \\ d_5 \\ d_6 \end{pmatrix}_1 = \begin{bmatrix} 0 & 0 & 0 & 1 \\ 0 & 1 & 0 & 0 \\ 0 & 0 & 0 & 0 \\ 1 & 0 & 0 & 0 \end{bmatrix}_1 \begin{pmatrix} D_1 \\ D_2 \\ D_3 \\ D_4 \end{pmatrix} \quad (4.4.32)$$

For element 2, the kinematic relationship is

$$\begin{pmatrix} d_2 \\ d_3 \\ d_5 \\ d_6 \end{pmatrix}_2 = \begin{bmatrix} 0 & 0 & 0 & 0 \\ 0 & 1 & 0 & 0 \\ 0 & 0 & 0 & 0 \\ 0 & 0 & 1 & 0 \end{bmatrix}_2 \begin{pmatrix} D_1 \\ D_2 \\ D_3 \\ D_4 \end{pmatrix} \quad (4.4.33)$$

Symbolically, Eqs. (4.4.32) and (4.4.33) can be written respectively as

$$\mathbf{d}_1 = \mathbf{T}_1 \mathbf{D} \quad (4.4.34)$$
$$\mathbf{d}_2 = \mathbf{T}_2 \mathbf{D} \quad (4.4.35)$$

On the other hand, the portion of the structure nodal forces resisted by element 1 is

$$\begin{pmatrix} R_1 \\ R_2 \\ R_3 \\ R_4 \end{pmatrix}_1 = \begin{bmatrix} 0 & 0 & 0 & 1 \\ 0 & 1 & 0 & 0 \\ 0 & 0 & 0 & 0 \\ 1 & 0 & 0 & 0 \end{bmatrix} \begin{pmatrix} r_2 \\ r_3 \\ r_5 \\ r_6 \end{pmatrix}_1 \quad (4.4.36)$$

and the portion of the structural nodal force resisted by element 2 is

$$\begin{pmatrix} R_1 \\ R_2 \\ R_3 \\ R_4 \end{pmatrix}_2 = \begin{bmatrix} 0 & 0 & 0 & 0 \\ 0 & 1 & 0 & 0 \\ 0 & 0 & 0 & 1 \\ 0 & 0 & 0 & 0 \end{bmatrix} \begin{pmatrix} r_2 \\ r_3 \\ r_5 \\ r_6 \end{pmatrix}_2 \quad (4.4.37)$$

By comparing Eq. (4.4.36) with Eq. (4.4.32) and Eq. (4.4.37) with Eq. (4.4.33), it can be seen that the matrix relating the structure nodal forces R's to the element end forces r's is the transposition of the matrix relating the element end displacements d's to the structure nodal displacements D's. This observation is not a coincidence, but represents a theory in structural analysis known as the *contragradient* law.[7]

In view of the above observation, Eqs. (4.4.36) and (4.4.37) can be written symbolically as

$$\mathbf{R}_1 = \mathbf{T}_1^T \mathbf{r}_1 \qquad (4.4.38)$$

$$\mathbf{R}_2 = \mathbf{T}_2^T \mathbf{r}_2 \qquad (4.4.39)$$

From consideration of joint equilibrium, we can write

$$\mathbf{R} = \mathbf{R}_1 + \mathbf{R}_2 \qquad (4.4.40)$$

Substituting the member equilibrium relationships Eqs. (4.4.38) and (4.4.39) into Eq. (4.4.40) gives

$$\mathbf{R} = \mathbf{T}_1^T \mathbf{r}_1 + \mathbf{T}_2^T \mathbf{r}_2 \qquad (4.4.41)$$

Since, from Eq. (4.4.21) the element force-displacement relationship for elements 1 and 2 can be written, respectively, as

$$\mathbf{r}_1 = \mathbf{k}_1 \mathbf{d}_1 \qquad (4.4.42)$$

and

$$\mathbf{r}_2 = \mathbf{k}_2 \mathbf{d}_2 \qquad (4.4.43)$$

we can write Eq. (4.4.41) as

$$\mathbf{R} = \mathbf{T}_1^T \mathbf{k}_1 \mathbf{d}_1 + \mathbf{T}_2^T \mathbf{k}_2 \mathbf{d}_2 \qquad (4.4.44)$$

Now, using the member kinematic relationships, Eqs. (4.4.34) and (4.4.35), we can write Eq. (4.4.44) as

$$\mathbf{R} = \mathbf{T}_1^T \mathbf{k}_1 \mathbf{T}_1 \mathbf{D} + \mathbf{T}_2^T \mathbf{k}_2 \mathbf{T}_2 \mathbf{D}$$
$$= (\mathbf{T}_1^T \mathbf{k}_1 \mathbf{T}_1 + \mathbf{T}_2^T \mathbf{k}_2 \mathbf{T}_2) \mathbf{D} \qquad (4.4.45)$$

or

$$\mathbf{R} = \mathbf{KD} \qquad (4.4.46)$$

where

$$\mathbf{K} = \mathbf{T}_1^T \mathbf{k}_1 \mathbf{T}_1 + \mathbf{T}_2^T \mathbf{k}_2 \mathbf{T}_2 \qquad (4.4.47)$$

is the structure stiffness matrix.

The process shown above is referred to as assemblage and it involves the process of transforming and putting together element stiffness matrices to form the structure stiffness matrix. In general, if these are n elements in the structure, the structure stiffness matrix can be obtained as

$$\mathbf{K} = \sum_{i=1}^{n} \mathbf{T}_i^T \mathbf{k}_i \mathbf{T}_i \qquad (4.4.48)$$

Now, referring back to the example problem, upon substituting the matrices $\mathbf{T}_1$, $\mathbf{T}_2$, $\mathbf{k}_1$, $\mathbf{k}_2$ into the structure stiffness matrix Eq. (4.4.47) and carrying out the matrix products, we see that the structure stiffness matrix

4.4 Elastic Critical Loads by Matrix Stiffness Method

can be written as

$$\mathbf{K} = \frac{EI}{L} \begin{bmatrix} 4 & 2 & 0 & \frac{-6}{L} \\ & 12 & 4 & \frac{-6}{L} \\ & & 8 & 0 \\ \text{sym.} & & & \frac{12}{L^2} \end{bmatrix} - \frac{P}{L} \begin{bmatrix} \frac{2L^2}{15} & \frac{-L^2}{30} & 0 & \frac{-L}{10} \\ & \frac{2L^2}{15} & 0 & \frac{-L}{10} \\ & & 0 & 0 \\ \text{sym.} & & & \frac{6}{5} \end{bmatrix} \quad (4.4.49)$$

Denoting

$$\lambda = \frac{PL^2}{30EI} = \frac{(kL)^2}{30} \quad (4.4.50)$$

Eq. (4.4.49) can be written as

$$\mathbf{K} = \frac{EI}{L} \begin{bmatrix} 4 - 4\lambda & 2 + \lambda & 0 & \frac{-6 + 3\lambda}{L} \\ & 12 - 4\lambda & 4 & \frac{-6 + 3\lambda}{L} \\ & & 8 & 0 \\ \text{sym.} & & & \frac{12 - 36\lambda}{L^2} \end{bmatrix} \quad (4.4.51)$$

At bifurcation, the determinant of the stiffness matrix must vanish. Thus, by setting

$$\det |\mathbf{K}| = 0 \quad (4.4.52)$$

we obtain a polynomial in λ. The smallest root satisfying this equation is $\lambda = 0.061$, and from Eq. (4.4.50)

$$P_{cr} = 30\lambda \frac{EI}{L^2} = 1.83 \frac{EI}{L^2} \quad (4.4.53)$$

The slight discrepancy of Eq. (4.4.53) compared to the value obtained previously by the differential equation method or the slope-deflection equation method is due to the round-off error, and this error was introduced earlier as a result of the approximation from Eq. (4.4.24) to Eq. (4.4.27).

At first glance, it seems that there is much more work involved in the stiffness matrix approach than that of the differential equation or the slope-deflection equation approaches. However, it should be noted that the steps shown above can easily be programmed in a digital computer, and so P_{cr} can be obtained quite conveniently for any type of frame.

4.5 SECOND-ORDER ELASTIC ANALYSIS

In the preceding sections, we determined the load that corresponds to a state of bifurcation of equilibrium of a perfect frame by an eigenvalue analysis. In an eigenvalue analysis, the system is assumed to be perfect. There will be no lateral deflections in the members until the load reaches the critical load P_{cr}. At the critical load P_{cr}, the original configuration of the frame ceases to be stable and with a slight disturbance, the lateral deflections of the members begin to increase without bound as indicated by curve 1 in Fig. 4.2. However, if the system is not perfect, lateral deflections will occur as soon as the load is applied, as shown by curve 2 in Fig. 4.2. For an elastic frame, curve 2 will approach curve 1 asymptotically. To trace this curve, a complete load-deflection analysis of the frame is necessary. A second-order elastic analysis will generate just such load-deflection response of the frame.

In a second-order analysis, such secondary effects as the $P - \delta$ and $P - \Delta$ effects, which we discussed previously in Chapter 3, can be incorporated directly into the analysis procedure. As a result, the use of $P - \delta$ and $P - \Delta$ moment magnification factors (denoted as B_1 and B_2 in Chapter 3) are not necessary.

Because for a second-order analysis the equilibrium equations are formulated with respect to the deformed geometry of the structure, which is not known in advance and is constantly changing with the applied loads, it is necessary to employ an iterative technique to obtain solutions. In a numerical implementation, one of the most popular solution techniques is the incremental load approach. In this approach, the applied load is divided into increments and applied incrementally to the structure. The deformed configurations of the structure at the end of each cycle of calculation is used as the basis for the formulation of equilibrium equations for the next cycle. At a particular cycle of calculation, the structure is assumed to behave linearly. In effect, the nonlinear response of the structure as a result of geometric changes is approximated by a series of linear analyses, the geometry of the structure used in the analysis for a specific cycle is the deformed geometry of the structure corresponding to the previous cycle of calculation. Because of the linearization process, equilibrium may be violated and the external force may not always balance the internal force. This unbalanced force must be reapplied to the structure and the process repeated until equilibrium is satisfied.

For a second-order elastic frame analysis, the iteration process is

4.5 Second-Order Elastic Analysis

summarized in the following steps (in the following discussion, a subscript refers to the load step and a superscript refers to the cycle of calculation within each load step):

1. First, discretize the frame into a number of beam-column elements.
2. Next, formulate the element stiffness matrix **k** for each and every element. The element stiffness matrix is given in Eq. (4.4.24), or in its approximate form, Eq. (4.4.27). (P can be set equal to zero in these equations for the first cycle of calculations.)
3. Assemble all these element stiffness matrices to form the structure stiffness matrix **K**.
4. Solve for the incremental displacement vector using

$$\Delta \mathbf{R}_i = \mathbf{K}_i^1 \, \Delta \mathbf{D}_i^1 \qquad (4.5.1)$$

   from which

$$\Delta \mathbf{D}_i^1 = (\mathbf{K}_i^1)^{-1} \, \Delta \mathbf{R}_i \qquad (4.5.2)$$

   where

   $\Delta \mathbf{R}_i$ = prescribed incremental load vector of the i load step
   $\mathbf{K}_i^1$ = structure secant stiffness matrix at the beginning of i load step
   $\Delta \mathbf{D}_i^1$ = incremental structure nodal displacement vector at i load step.

5. Update the structure nodal displacement vector from

$$\mathbf{D}_i^1 = \mathbf{D}_i + \Delta \mathbf{D}_i^1 \qquad (4.5.3)$$

   where

   $\mathbf{D}_i^1$ = structure nodal displacement vector at the end of the first cycle of calculation at the i load step
   $\mathbf{D}_i$ = structure nodal displacement vector at the beginning of the i load step
   $\Delta \mathbf{D}_i^1$ = incremental structure nodal displacement vector evaluated at Step 4.

6. Extract the element end displacement vector $\mathbf{d}_i$ from $\mathbf{D}_i^1$ for each and every element in the structure.
7. For each element, evaluate the element axial displacement e and element end rotations θ_A, θ_B from Eqs. (4.4.7) to (4.4.9).
8. For each element, evaluate element axial force P and element end moments M_A, M_B from Eqs. (4.14.12) to (4.4.14).
9. For each element, evaluate element end forces from Eq. (4.4.10).
10. Form the structure internal force vector $\mathbf{R}_i^1$ at the end of the first cycle of calculation by assembling the element end forces evaluated in Step 9 for all the elements.

11. Calculate the external force vector from
$$\mathbf{R}_{i+1} = \mathbf{R}_i + \Delta \mathbf{R}_i \qquad (4.5.4)$$
12. Evaluate the unbalanced force $\Delta \mathbf{Q}_i^1$ at the end of the cycle from
$$\Delta \mathbf{Q}_i^1 = \mathbf{R}_{i+1} - \mathbf{R}_i^1 \qquad (4.5.5)$$
13. Using the current value of axial force P, update the element stiffness matrix $\mathbf{k}$ for each and every element. Assemble $\mathbf{k}$ for all the elements to form an updated secant structure stiffness matrix $\mathbf{K}_i^2$. Evaluate the incremental displacement vector $\Delta \mathbf{D}_i^2$ from
$$\Delta \mathbf{D}_i^2 = (\mathbf{K}_i^2)^{-1} \Delta \mathbf{Q}_i^1 \qquad (4.5.6)$$
where $\Delta \mathbf{Q}_i^1$ is the unbalanced force vector calculated in the previous cycle of calculation.
14. Update the structure nodal displacement vector from
$$\mathbf{D}_i^2 = \mathbf{D}_i + \sum_{k=1}^{2} \Delta \mathbf{D}_i^k \qquad (4.5.7)$$
15. Extract the element end displacement vector $\mathbf{d}_i$ from $\mathbf{D}_i^2$ calculated in Eq. (4.5.7) for each and every element. Update e, θ_A, and θ_B and, hence, P, M_A, and M_B as done in Steps 7 and 8 for all elements in the frame.
16. Update the element end forces for all the elements and form the new structure internal force vector $\mathbf{R}_i^2$.
17. Evaluate the new unbalanced force $\Delta \mathbf{Q}_i^2$ from
$$\Delta \mathbf{Q}_i^2 = \mathbf{R}_{i+1} - \mathbf{R}_i^2 \qquad (4.5.8)$$
18. Repeat Steps 13 through 17 as many times as possible until convergence. Convergence is said to have been attained if the unbalanced force $\Delta \mathbf{Q}_i^j$, where the superscript j refers to the j cycle of calculation, falls within a prescribed tolerance.
19. After convergence the structure nodal displacement at the end of the i load step is obtained by
$$\mathbf{D}_{i+1} = \mathbf{D}_i^n = \mathbf{D}_i + \sum_{k=1}^{n} \Delta \mathbf{D}_i^k \qquad (4.5.9)$$
20. Prescribe another load increment and repeat Step 2 to 19.

A schematic representation of the above procedure is shown in Fig. 4.12a,b for a one degree of freedom structure. In performing the above procedure, the complete load-deflection response of the frame can be traced, and the stability limit point is obtained as the peak point of this load-deflection curve.

As the stability limit point is approached in the analysis, convergence

4.5 Second-Order Elastic Analysis

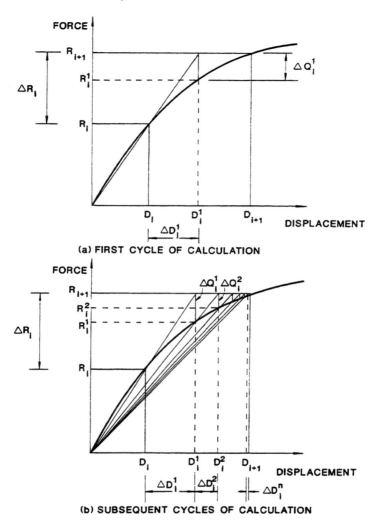

FIGURE 4.12 Iteration technique for second-order elastic frame analysis

of the solution may be slow. To facilitate convergence, a smaller load increment should be used.

The numerical procedure described above can be programmed in a computer. By using the computer to perform a second-order analysis, the design moments for the members can be obtained directly. Comparative studies using second-order elastic analysis and first-order elastic analysis in conjunction with B_1, B_2 moment amplification factors described on Chapter 3 have been made.[4,8] It was demonstrated that for rigidly

4.6 PLASTIC COLLAPSE LOADS

In the preceding discussions, the frame is assumed to behave elastically throughout the entire stage of loading up to failure. The failure of the frame is a result of instability when the stiffness of the frame vanishes. Consequently, the elastic critical load P_{cr} is the maximum load for an *elastic* frame. On the other extreme, if we exclude the instability effect but consider only the plastic yielding of the material, failure of the frame will be controlled by the formation of a *plastic collapse mechanism*. The plastic collapse mechanism load P_p is the maximum load-carrying capacity of the frame. A plastic collapse mechanism will form when there are sufficient number of plastic hinges developed in the structure to render it statically unstable. Before we proceed to the discussion of the method of determining the collapse load P_p, it would be pertinent here to briefly explain the basic concept of plastic hinge and plastic collapse mechanism in a simple plastic theory.

4.6.1 Plastic Hinge

If a simple tension test is performed on a structural steel specimen, its stress–strain diagram will be as seen in Fig. 4.13: there a definitive knee

FIGURE 4.13 Uniaxial stress–strain behavior of structural steel

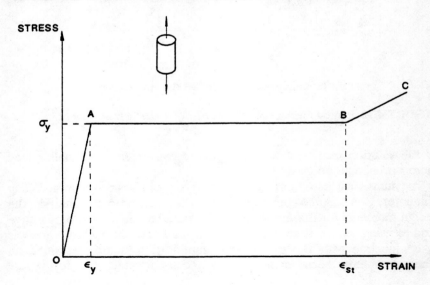

4.6 Plastic Collapse Loads

at A marks the *yield point* of the material. The stress level that corresponds to point A is the yield stress of the material. If the stress is below the yield stress σ_y, the material behaves elastically as shown by line OA. After the yield stress has been reached, the strain can increase greatly without any further increase in stress as indicated by line AB. When the strain has reached $\varepsilon_{st} \simeq 12\varepsilon_y$, further increase in strain will bring about a further increase in stress as a result of *strain-hardening*, as indicated by line BC. For simplicity, the effect of strain-hardening is usually not considered in a simple plastic design analysis. Neglecting this effect will obviously lead to a conservative design. When material can sustain a large deformation without fracture, this is known as *ductility*. It is this unique property of structural steel that makes plastic design possible.

When a member is subjected to pure bending, and if the usual assumption of plane sections before bending remain plane after bending is made, a series of stress and strain distributions across the section corresponding to an increasing bending moment can be sketched, as we do in Fig. 4.14a–d. The corresponding moment-curvature relationship is shown in Fig. 4.15. The points a, b, c, and d on the figure correspond to the various stages shown in Fig. 4.14. Before the stress in any fiber in the cross section reaches the yield stress σ_y (Fig. 4.14a), the section behaves elastically (line OA in Fig. 4.15). When the extreme fibers of the cross section just reach σ_y (Fig. 4.14b), the corresponding moment is referred to as the *yield moment* M_y and is denoted by point A in Fig. 4.15. Further increase in moment above the yield moment will increase the curvature of the cross section at a faster rate. As the load continues to increase, yielding of fibers will spread and penetrate toward those fibers located closer to the neutral axis of the cross section (Fig. 4.14c). This process of successive yielding of fibers towards the neutral axis of the cross section is referred to as *plastification*. Note that because of the stress–strain behavior of steel, the stresses of the yielded fibers remain at σ_y. When plastification of the fibers across the cross section is completed (Fig. 4.14d), the cross section cannot carry any additional moment. The moment that corresponds to the full plastification is referred to as the *plastic moment* M_p denoted by point B in Fig. 4.15. It is clear from Fig. 4.15 that the yield moment M_y is not the maximum moment capacity of the cross section. Rather, the maximum moment capacity is the plastic moment M_p. The exact shape of the M-Φ curve from M_y to M_p (curve AB in Fig. 4.15), as well as their relative magnitudes, are different for different cross-sectional shapes. For example, for a rectangular cross section, $M_p/M_y = 1.5$, and for hot-rolled, wide-flange sections, it is $M_p/M_y \simeq 1.12$ if bent about the strong axis, and $M_p/M_y \simeq 1.5$ if bent about the weak axis.

Figure 4.15 shows also that the cross section can sustain a constant M_p

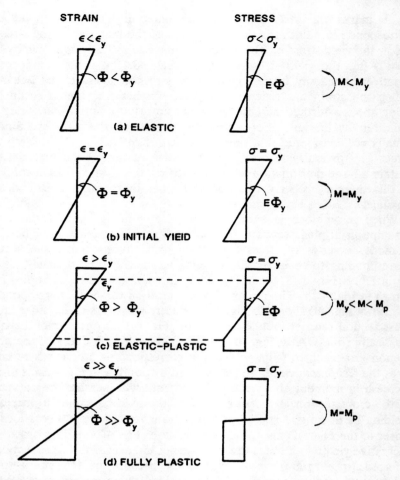

FIGURE 4.14 Strain and stress diagrams

through a large rotation capacity, as indicated by the horizontal line BC, precisely because of the large ductility of steel. In other words, the cross section behaves like a real hinge, but carrying a constant moment capacity M_p. Hence the word *plastic hinge* is used to indicate such a property for a steel cross section.

4.6.2 Plastic Collapse Load P_p by Hinge-by-Hinge Method

Consider a propped cantilever of length L subjected to a concentrated load P acting at midspan (Fig. 4.16a). The structure is statically

4.6 Plastic Collapse Loads

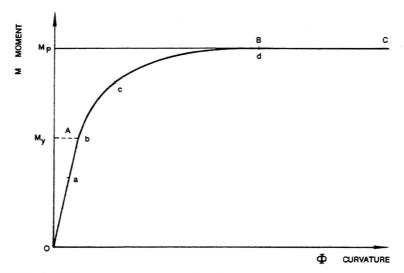

FIGURE 4.15 Moment-curvature relationship

indeterminate to the first degree. Thus, if a plastic hinge is developed in the beam at the maximum moment location, the beam will become statically determinate. If, in addition, a second plastic hinge is developed at a subsequent critical location, the beam will become statically unstable and a collapse mechanism will develop for the beam. In the following, we shall denote the stage of loading from beginning to the formation of the first plastic hinge as *load stage 1* (Fig. 4.16b), and the *additional* loading beyond load stage 1 as *load stage 2* (Fig. 4.16e). We use the subscripts 1 and 2 to distinguish these two load stages.

Load Stage 1

Before the formation of the first plastic hinge, the beam behaves elastically; the elastic moment diagram under the applied load P_1 is shown in Fig. 4.16c. Since the moment at the fixed end is larger than the moment at midspan, the first plastic hinge will form at the fixed end (point A). The load that corresponds to the formation of the first plastic hinge can be found by equating $3P_1 L/16$ to M_p.

$$\frac{3P_1 L}{16} = M_p \tag{4.6.1}$$

from which, we obtain the first plastic hinge load

$$P_1^* = \frac{16 M_p}{3L} \tag{4.6.2}$$

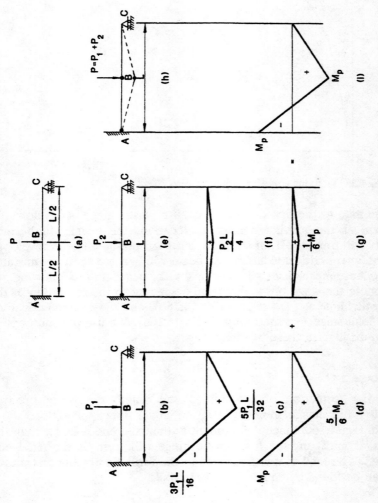

FIGURE 4.16 P_p of a propped cantilever

4.6 Plastic Collapse Loads

The moment at midspan (point B), when the first hinge is just formed, is

$$\frac{5P_1L}{32} = \frac{5(16M_p/3L)L}{32} = \frac{5}{6}M_p \qquad (4.6.3)$$

Figure 4.16d shows the moment diagram at the end of load stage 1.

Load Stage 2

After the formation of the first plastic hinge at A, the propped cantilever becomes a simply supported beam with a constant moment M_p at A to carry the load P_1^*. When the additional load P_2 is applied, the beam behaves as a simply supported member (Fig. 4.16e); Figure 4.16f shows the corresponding moment diagram due to P_2. The load P_2 is now added to the first plastic hinge load P_1^* and the resulting maximum moment at midspan under the combined load $P_1^* + P_2$ is $5M_p/6 + P_2L/4$. The second plastic hinge will form at midspan, when this moment reaches the plastic moment capacity M_p, that is, when

$$\frac{5M_p}{6} + \frac{P_2L}{4} = M_p \qquad (4.6.4)$$

from which we obtain the second plastic hinge load

$$P_2^* = \frac{2}{3}\frac{M_p}{L} \qquad (4.6.5)$$

At the formation of second plastic hinge, the structure becomes statically unstable and a plastic collapse mechanism, such as shown in Fig. 4.16h, emerges.

The collapse load P_p, that brings the beam to its collapse state, is the sum of P_1^* and P_2^*

$$\begin{aligned} P_p &= P_1^* + P_2^* \\ &= \frac{16M_p}{3L} + \frac{2}{3}\frac{M_p}{L} \\ &= \frac{6M_p}{L} \end{aligned} \qquad (4.6.6)$$

and the moment diagram at the collapse state is shown in Fig. 4.16i obtained by superposing the moment diagrams of Fig. 4.16d and g.

The procedure described above is called the *hinge-by-hinge analysis*. It is essentially a sequence of elastic analyses with additional plastic hinges introduced during the course of loading. The method can be programmed in a computer and thus extended to cases in which the structure and loadings are more complicated.[9] In the hinge-by-hinge analysis, the sequence of formation of plastic hinges is traced. In general, if the structure is statically indeterminate to the nth degree, then the formation of $(n+1)$ plastic hinges will be necessary for the structure to reach its collapse state.

4.6.3 Plastic Collapse Load by Mechanism Method

To determine the plastic collapse P_p in a more direct manner, a simpler method, known as the *mechanism method,* will be presented here. This method is based on the *upper bound theorem* of plastic analysis. The theorem states that a load computed on the basis of an assumed failure or collapse mechanism will always be greater than or at most equal to the true collapse load. Thus, in using this theorem, all possible collapse mechanisms of the structure are identified and the load corresponding to each of these mechanisms is evaluated. The mechanism that gives the lowest value of P_p will be the collapse mechanism. Strictly speaking, to ascertain the mechanism so chosen will give the lowest value of P_p, a *moment check* is often necessary to ensure that the moment everywhere in the structure is less than or at most equal to M_p. For an assumed mechanism that is not the true collapse mechanism, this moment check cannot generally be made. The application of the mechanism method to obtain P_p is greatly facilitated by the use of the virtual work equations.

P_p for a Propped Cantilever

To demonstrate the use of the virtual work equations in the mechanism method to obtain P_p in a direct manner, consider the same propped cantilever shown in Fig. 4.16a as reproduced in Fig. 4.17a. The collapse

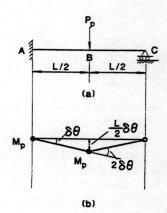

FIGURE 4.17 Collapse mechanism of a propped cantilever

4.6 Plastic Collapse Loads

load P_p can be determined directly from the assumed collapse mechanism as shown in Fig. 4.17b. The calculations can be made in the following manner: Assuming the plastic hinge at A undergoes a virtual rotation of $\delta\theta$, it is readily seen from the geometry of the collapse mechanism that the plastic hinge at B will undergo a virtual rotation of $2\delta\theta$, and the vertical drop of point B from its original position is $L\delta\theta/2$. Note that small displacement assumption is used in evaluating this vertical drop.

The external virtual work done during this virtual displacement is equal to the applied load P times the distance it travels, i.e.,

$$\delta W_{ext} = P(L\delta\theta/2) \tag{4.6.7}$$

and the virtual strain energy stored in the structure during this virtual displacement is equal to the sum of the plastic moment times the hinge rotation at points A and B, i.e.,

$$\delta U_{int} = M_p(\delta\theta) + M_p(2\delta\theta) \tag{4.6.8}$$

Note, when writing Eq. (4.6.8) it is tacitly assumed that *all* deformations are concentrated in the plastic hinges. As a result, no virtual strain energy is stored anywhere else but in locations of plastic hinges during the virtual displacements.

Equating Eq. (4.6.7) to Eq. (4.6.8), we have

$$P(L\delta\theta/2) = M_p(\delta\theta) + M_p(2\delta\theta) \tag{4.6.9}$$

from which we obtain the collapse load

$$P_p = \frac{6M_p}{L} \tag{4.6.10}$$

which is the same as Eq. (4.6.6), which was obtained previously by the hinge-by-hinge analysis.

The mechanism method can be extended to obtain P_p for a framed structure. This is described in the following example.

P_p for a Pinned-Based Portal Frame

Figure 4.18a shows a pinned-based portal frame loaded by a vertical force of $1.5P$ at midspan of the beam and a horizontal force of P at point B. We will now calculate the collapse load of the frame using the mechanism method.

Unlike the previous example of the propped cantilever in which only one collapse mechanism was identified, three possible collapse mechanisms for the portal frame can be identified. They are shown in Fig. 4.18b–d. The mechanism shown in Fig. 4.18b is called the *beam mechanism*; in Fig. 4.18c, it is called the *sway mechanism*; and in Fig. 4.18d, we have the *combined mechanism*, which contains the features of

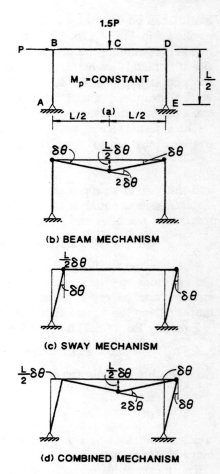

FIGURE 4.18 Collapse mechanisms of a simple portal frame

both the beam and sway mechanisms. The corresponding virtual displacements are also indicated in the figures. To calculate P_p, we write the virtual work equations for all these mechanisms.

$$\delta W_{ext} = \delta U_{int} \tag{4.6.11}$$

Beam Mechanism

$$(1.5P)(L\delta\theta/2) = M_p(\delta\theta) + M_p(2\delta\theta) + M_p(\delta\theta) \tag{4.6.12}$$

from which we obtain

$$P_1 = \frac{16}{3}\frac{M_p}{L} \tag{4.6.13}$$

4.6 Plastic Collapse Loads

Sway Mechanism

$$P(L\delta\theta/2) = M_p(\delta\theta) + M_p(\delta\theta) \tag{4.6.14}$$

from which we obtain

$$P_2 = 4\frac{M_p}{L} \tag{4.6.15}$$

Combined Mechanism

$$(1.5P)(L\delta\theta/2) + P(L\delta\theta/2) = M_p(2\delta\theta) + M_p(2\delta\theta) \tag{4.6.16}$$

from which we obtain

$$P_3 = \frac{16}{5}\frac{M_p}{L} \tag{4.6.17}$$

Since the lowest value is P_3, from the upper bound theorem, we therefore choose P_3 as the collapse load. However, to ensure that P_3 is the true collapse load, we need to perform a moment check on the structure. By drawing free-body diagrams of the frame (Fig. 4.19a), it is

FIGURE 4.19 Moment check

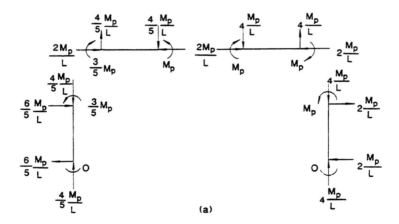

(a)

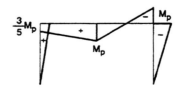

(b)

readily seen that the condition $M \le M_p$ is satisfied everywhere in the frame (Fig. 4.19b). Therefore, we can conclude from the lower bound theorem of plastic analysis[10] that $P_p = P_3 = 16M_p/5L$ is the collapse load and the combined mechanism is the collapse mechanism of the frame.

For multistory multibay frames, a number of possible collapse mechanisms will exist. It may be difficult for the analyst to envision all these possible mechanisms. However, various mechanisms can be constructed systematically from the two basic mechanisms, namely the beam and the sway mechanisms, by a process known as *combination of mechanism*. This method is described in detail in reference 10 and will not be presented here. Alternatively, one can obtain P_p for such frames by the hinge-by-hinge method with the aid of a computer.[9]

Generally speaking, the plastic collapse load P_p will give a reasonable estimate of the failure load P_f of the frame only if the effect of instability is small and can be ignored, as in, for example, single-story frames that consist of stocky members. For multistory frames in which stability is important, the rigid-plastic collapse load will not be representative of the failure load P_f. In reality, failure of frames is a result of both instability and plasticity effects. Thus, neither the critical load P_{cr}, evaluated by considering elastic instability effect only, nor the rigid-plastic collapse load P_p, evaluated by considering plasticity effect only, will represent the failure load P_f of the frame. To obtain a precise value of P_f, a rigorous analysis, such as a complete elasto-plastic analysis of the structure, may be necessary. This type of analysis is rather complex and costly and inevitably required the use of a computer. However, for design purposes it is more desirable if P_f can be obtained by a simpler means. In the following section, a simple method to estimate P_f is discussed.

4.7 MERCHANT–RANKINE INTERACTION EQUATION

We have pointed out in the previous sections that neither P_{cr} nor P_p will represent P_f of most frameworks. They represent only two extreme cases in which only the instability effect or only the plasticity effect is considered in the analysis. In reality, the effects of instability and plasticity interact with each other. The exact interaction is rather complex, and so approximate interaction equations involving relatively simple calculations are desirable. One such interaction equation has been proposed by Horne and Merchant.[11] It has the simple form

$$\frac{P_f}{P_{cr}} + \frac{P_f}{P_p} = 1 \qquad (4.7.1)$$

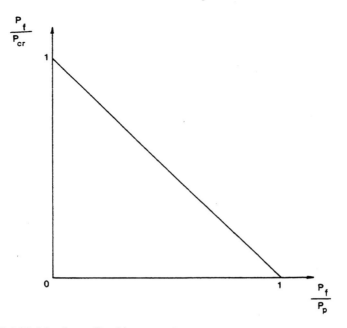

FIGURE 4.20 Merchant–Rankine equation

where

P_{cr} = elastic critical load of the frame
P_p = plastic collapse load of the frame
P_f = failure load of the frame

Equation (4.7.1) is called the *Merchant–Rankine equation*. This equation is plotted in Fig. 4.20. As can be seen, P_{cr} and P_p represent end points of a straight line interaction equation for the failure load P_f. It has been demonstrated[11] that the failure load P_f obtained from Eq. (4.7.1) is usually conservative and reasonably accurate for design purposes.

4.8 EFFECTIVE LENGTH FACTORS OF FRAMED MEMBERS

In the design of rigid frames, it is common practice to isolate each member from the frame and design it as an individual beam-column, using the beam-column interaction equations discussed in the previous chapter. But, as mentioned previously, the behavior of a framed member is affected by all its adjacent members in the frame. As a result, the influence of other members on the particular member in question must be taken into account in a design. One convenient way to include this interaction effect is to use the concept of *effective length factor K*. We discussed the concept of effective length factor for isolated column with idealized end conditions thoroughly in Chapter 2. Here we will discuss the effective length factor for a framed member. The determination of the effective length factor K for a framed member is more involved than that for an isolated member, because the stiffness of all adjacent members, as well as the rigidities of the connections, must be included in estimating the rotational restraint at the ends of the member in question. In theory, the effective length factor K for a framed column should be determined from a stability analysis of the entire structure. However, for design purposes this procedure is impractical and the use of a simpler procedure is much more desirable. One such procedure was proposed by Julian and Lawrence.[12] Their procedure was recommended by the AISC, and we will therefore discuss it in what follows. Since the behavior of a framed column will be different, depending on whether the frame is braced (sidesway prevented) or unbraced (sidesway permitted), we will discuss each case separately.

4.2.1 Braced Frame

The model used for the determination of K for a framed column braced against sidesway is shown in Fig. 4.21a. The column in question is denoted by c2 in the figure. The assumptions used for the model are:

1. All members are prismatic and behave elastically.
2. The axial forces in the beams are negligible.
3. All columns in a story buckle simultaneously.
4. At a joint, the restraining moment provided by the beams is distributed among the columns in proportion to their stiffnesses.
5. At buckling, the rotations at the near and far ends of the girders are equal and opposite (i.e., the girders are bent in single curvature).

Using the slope-deflection equations (3.7.13) and (3.7.14) for the columns and (3.8.15) and (3.8.16) for the beams, we have

For Column 1

$$(M_A)_{c1} = \left(\frac{EI}{L}\right)_{c1} [s_{ii}\theta_A + s_{ij}\theta_B] \qquad (4.8.1)$$

For Column 2

$$(M_A)_{c2} = \left(\frac{EI}{L}\right)_{c2} [s_{ii}\theta_A + s_{ij}\theta_B] \qquad (4.8.2)$$

$$(M_B)_{c2} = \left(\frac{EI}{L}\right)_{c2} [s_{ij}\theta_A + s_{ii}\theta_B] \qquad (4.8.3)$$

4.8 Effective Length Factors of Framed Members

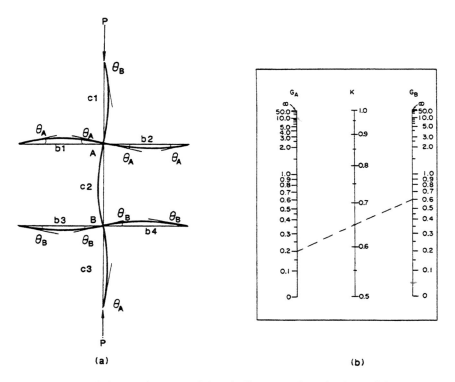

FIGURE 4.21 Subassemblage model and alignment chart for braced frame

For Column 3
$$(M_B)_{c3} = \left(\frac{EI}{L}\right)_{c3}[s_{ij}\theta_A + s_{ii}\theta_B] \quad (4.8.4)$$

For Beam 1
$$(M_A)_{b1} = \left(\frac{EI}{L}\right)_{b1}[4\theta_A - 2\theta_A] = \left(\frac{EI}{L}\right)_{b1}(2\theta_A) \quad (4.8.5)$$

For Beam 2
$$(M_A)_{b2} = \left(\frac{EI}{L}\right)_{b2}[4\theta_A - 2\theta_A] = \left(\frac{EI}{L}\right)_{b2}(2\theta_A) \quad (4.8.6)$$

For Beam 3
$$(M_B)_{b3} = \left(\frac{EI}{L}\right)_{b3}[4\theta_B - 2\theta_B] = \left(\frac{EI}{L}\right)_{b3}(2\theta_B) \quad (4.8.7)$$

For Beam 4

$$(M_B)_{b4} = \left(\frac{EI}{L}\right)_{b4}[4\theta_B - 2\theta_B] = \left(\frac{EI}{L}\right)_{b4}(2\theta_B) \qquad (4.8.8)$$

Note that because of Assumption 2, we can use $s_{ii} = 4$ and $s_{ij} = 2$ for the beams.

For joint equilibrium at A, we must have

$$(M_A)_{c1} + (M_A)_{c2} + (M_A)_{b1} + (M_A)_{b2} = 0 \qquad (4.8.9)$$

from which we obtain

$$(M_A)_{c2} = -(M_A)_{b1} - (M_A)_{b2} - (M_A)_{c1} \qquad (4.8.10)$$

Substituting Eqs. (4.8.5), (4.8.6), and (4.8.1) for $(M_A)_{b1}$, $(M_A)_{b2}$, and $(M_A)_{c1}$, respectively, into Eq. (4.8.10), we have

$$(M_A)_{c2} = -2\left[\left(\frac{EI}{L}\right)_{b1} + \left(\frac{EI}{L}\right)_{b2}\right]\theta_A - \left(\frac{EI}{L}\right)_{c1}[s_{ii}\theta_A + s_{ij}\theta_B] \qquad (4.8.11)$$

From Eq. (4.8.2), we can write

$$[s_{ii}\theta_A + s_{ij}\theta_B] = \frac{(M_A)_{c2}}{\left(\frac{EI}{L}\right)_{c2}} \qquad (4.8.12)$$

Finally, by substituting Eq. (4.8.12) into Eq. (4.8.11), and rearranging, we obtain

$$(M_A)_{c2} = -2\left(\frac{EI}{L}\right)_{c2}\frac{\sum_A\left(\frac{EI}{L}\right)_b}{\sum_A\left(\frac{EI}{L}\right)_c}\theta_A \qquad (4.8.13)$$

where we have used the notation

$$\sum_A\left(\frac{EI}{L}\right)_b = \left(\frac{EI}{L}\right)_{b1} + \left(\frac{EI}{L}\right)_{b2} \qquad (4.8.14)$$

$$\sum_A\left(\frac{EI}{L}\right)_c = \left(\frac{EI}{L}\right)_{c1} + \left(\frac{EI}{L}\right)_{c2} \qquad (4.8.15)$$

Following the same procedure by considering equilibrium at joint B, the moment at the B end of column 2 can be written as

$$(M_B)_{c2} = -2\left(\frac{EI}{L}\right)_{c2}\frac{\sum_B\left(\frac{EI}{L}\right)_b}{\sum_B\left(\frac{EI}{L}\right)_c}\theta_B \qquad (4.8.16)$$

4.8 Effective Length Factors of Framed Members

where we use the same notation

$$\sum_{B}\left(\frac{EI}{L}\right)_b = \left(\frac{EI}{L}\right)_{b3} + \left(\frac{EI}{L}\right)_{b4} \qquad (4.8.17)$$

$$\sum_{B}\left(\frac{EI}{L}\right)_c = \left(\frac{EI}{L}\right)_{c2} + \left(\frac{EI}{L}\right)_{c3} \qquad (4.8.18)$$

Eliminating $(M_A)_{c2}$ from Eqs. (4.8.2) and (4.8.13), and $(M_B)_{c2}$ from Eqs. (4.8.3) and (4.8.16), we can obtain the following equations

$$\left(s_{ii} + 2\frac{\sum_A\left(\frac{EI}{L}\right)_b}{\sum_A\left(\frac{EI}{L}\right)_c}\right)\theta_A + s_{ij}\theta_B = 0 \qquad (4.8.19)$$

$$s_{ij}\theta_A + \left(s_{ii} + 2\frac{\sum_B\left(\frac{EI}{L}\right)_b}{\sum_B\left(\frac{EI}{L}\right)_c}\right)\theta_B = 0 \qquad (4.8.20)$$

Denoting

$$G_A = \frac{\sum_A\left(\frac{EI}{L}\right)_c}{\sum_A\left(\frac{EI}{L}\right)_b} = \frac{\text{sum of column stiffnesses meeting at joint A}}{\text{sum of beam stiffnesses meeting at joint A}} \qquad (4.8.21)$$

and

$$G_B = \frac{\sum_B\left(\frac{EI}{L}\right)_c}{\sum_B\left(\frac{EI}{L}\right)_b} = \frac{\text{sum of column stiffness meeting at joint B}}{\text{sum of beam stiffness meeting at joint B}} \qquad (4.8.22)$$

The equilibrium equations (4.8.19) and (4.8.20) can be written in matrix form as

$$\begin{bmatrix} s_{ii} + \frac{2}{G_A} & s_{ij} \\ s_{ij} & s_{ii} + \frac{2}{G_B} \end{bmatrix} \begin{pmatrix} \theta_A \\ \theta_B \end{pmatrix} = \begin{pmatrix} 0 \\ 0 \end{pmatrix} \qquad (4.8.23)$$

At bifurcation, we must have the determinant of the coefficient matrix vanished.

$$\det \begin{vmatrix} s_{ii} + \frac{2}{G_A} & s_{ij} \\ s_{ij} & s_{ii} + \frac{2}{G_B} \end{vmatrix} = 0 \qquad (4.8.24)$$

Using the expressions for s_{ii} and s_{ij} [(Eqs. (3.7.15) and (3.7.16)] in Eq. (4.8.24) and realizing that

$$kL = (\sqrt{P/EI})L = \pi\sqrt{P/P_e} = \pi/K \tag{4.8.25}$$

Equation (4.8.24) can be simplified to give

$$\frac{G_A G_B}{4}\left(\frac{\pi}{K}\right)^2 + \left(\frac{G_A + G_B}{2}\right)\left(1 - \frac{\pi/K}{\tan(\pi/K)}\right) + \frac{2\tan(\pi/2K)}{\pi/K} - 1 = 0 \tag{4.8.26}$$

Equation (4.8.26) can be expressed in a nomograph form as shown in Fig. 4.21b (see reference 12). To obtain the effective length factor K of column AB, one needs only to evaluate the relative stiffness factors G_A and G_B expressed in Eqs. (4.8.21) and (4.8.22), respectively, at its two ends. A straight line joining the two G values will cut the middle line which gives the value of K. For example, if $G_A = 0.2$, $G_B = 0.6$, the K value for the column is 0.65 (see dotted line). Note that the range of K is from 0.5 to 1 for a braced frame.

4.8.2 Unbraced Frame

The model for the determination of K for a framed column subjected to sidesway is shown in Fig. 4.22a. The column in question is denoted by c2 in the figure. The assumptions used for this model are the same as that used for the model of the braced frame, except that for this model assumption 5 is modified to the following: at buckling, the rotations at the near and far ends of the girders are equal in magnitude and direction (i.e., the girders are bent in double curvature).

Again, making use of the slope-deflection equations (3.8.1) and (3.8.2) for the columns and (3.8.17) and (3.8.18) for the beams, we can write

For Column 1

$$(M_A)_{c1} = \left(\frac{EI}{L}\right)_{c1}\left[s_{ii}\theta_A + s_{ij}\theta_B - (s_{ii} + s_{ij})\frac{\Delta}{L_{c1}}\right] \tag{4.8.27}$$

For Column 2

$$(M_A)_{c2} = \left(\frac{EI}{L}\right)_{c2}\left[s_{ii}\theta_A + s_{ij}\theta_B - (s_{ii} + s_{ij})\frac{\Delta}{L_{c2}}\right] \tag{4.8.28}$$

$$(M_B)_{c2} = \left(\frac{EI}{L}\right)_{c2}\left[s_{ij}\theta_A + s_{ii}\theta_B - (s_{ii} + s_{ij})\frac{\Delta}{L_{c2}}\right] \tag{4.8.29}$$

For Column 3

$$(M_B)_{c3} = \left(\frac{EI}{L}\right)_{c3}\left[s_{ij}\theta_A + s_{ii}\theta_B - (s_{ii} + s_{ij})\frac{\Delta}{L_{c3}}\right] \tag{4.8.30}$$

4.8 Effective Length Factors of Framed Members

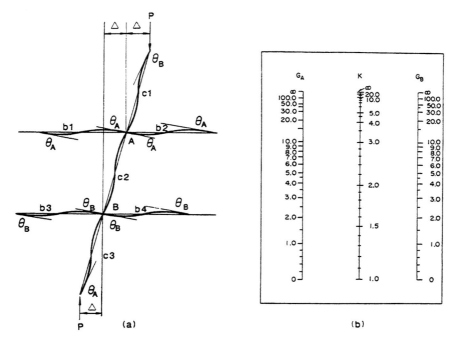

FIGURE 4.22 Subassemblage model and alignment chart for unbraced frame

For Beam 1
$$(M_A)_{b1} = \left(\frac{EI}{L}\right)_{b1} [4\theta_A + 2\theta_A] = \left(\frac{EI}{L}\right)_{b1} (6\theta_A) \quad (4.8.31)$$

For Beam 2
$$(M_A)_{b2} = \left(\frac{EI}{L}\right)_{b2} [4\theta_A + 2\theta_A] = \left(\frac{EI}{L}\right)_{b2} (6\theta_A) \quad (4.8.32)$$

For Beam 3
$$(M_B)_{b3} = \left(\frac{EI}{L}\right)_{b3} [4\theta_B + 2\theta_B] = \left(\frac{EI}{L}\right)_{b3} (6\theta_B) \quad (4.8.33)$$

For Beam 4
$$(M_B)_{b4} = \left(\frac{EI}{L}\right)_{b4} [4\theta_B + 2\theta_B] = \left(\frac{EI}{L}\right)_{b4} (6\theta_B) \quad (4.8.34)$$

For joint equilibrium at A, we must have
$$(M_A)_{c1} + (M_A)_{c2} + (M_A)_{b1} + (M_A)_{b2} = 0 \quad (4.8.35)$$

from which
$$(M_A)_{c2} = -(M_A)_{b1} - (M_A)_{b2} - (M_A)_{c1} \tag{4.8.36}$$

Substituting Eqs. (4.8.31), (4.8.32), and (4.8.27) for $(M_A)_{b1}$, $(M_A)_{b2}$, and $(M_A)_{c1}$ into Eq. (4.8.36), we have

$$(M_A)_{c2} = -6\left[\left(\frac{EI}{L}\right)_{b1} + \left(\frac{EI}{L}\right)_{b2}\right]\theta_A$$
$$-\left(\frac{EI}{L}\right)_{c1}\left[s_{ii}\theta_A + s_{ij}\theta_B - (s_{ii} + s_{ij})\frac{\Delta}{L_{c1}}\right] \tag{4.8.37}$$

From Eq. (4.8.28), we can write
$$s_{ii}\theta_A + s_{ij}\theta_B - (s_{ii} + s_{ij})\frac{\Delta}{L_{c2}} = \frac{(M_A)_{c2}}{\left(\frac{EI}{L}\right)_{c2}} \tag{4.8.38}$$

If $L_{c1} = L_{c2}$, Eq. (4.8.38) can be substituted into Eq. (4.8.37). The result is

$$(M_A)_{c2} = -6\left(\frac{EI}{L}\right)_{c2} \frac{\sum_A \left(\frac{EI}{L}\right)_b}{\sum_A \left(\frac{EI}{L}\right)_c} \theta_A \tag{4.8.39}$$

where we have used the notation

$$\sum_A \left(\frac{EI}{L}\right)_b = \left(\frac{EI}{L}\right)_{b1} + \left(\frac{EI}{L}\right)_{b2} \tag{4.8.40}$$

$$\sum_A \left(\frac{EI}{L}\right)_c = \left(\frac{EI}{L}\right)_{c1} + \left(\frac{EI}{L}\right)_{c2} \tag{4.8.41}$$

Following the same procedure, we see that by considering equilibrium at joint B, the moment at the B end of column 2 can be written as

$$(M_B)_{c2} = -6\left(\frac{EI}{L}\right)_{c2} \frac{\sum_B \left(\frac{EI}{L}\right)_b}{\sum_B \left(\frac{EI}{L}\right)_c} \theta_B \tag{4.8.42}$$

where we use the same notations

$$\sum_B \left(\frac{EI}{L}\right)_b = \left(\frac{EI}{L}\right)_{b3} + \left(\frac{EI}{L}\right)_{b4} \tag{4.8.43}$$

$$\sum_B \left(\frac{EI}{L}\right)_c = \left(\frac{EI}{L}\right)_{c2} + \left(\frac{EI}{L}\right)_{c3} \tag{4.8.44}$$

4.8 Effective Length Factors of Framed Members

Eliminating $(M_A)_{c2}$ from Eqs. (4.8.28) and (4.8.39), and $(M_B)_{c2}$ from Eqs. (4.8.29) and (4.8.42), we can obtain the following equations

$$\left(s_{ii} + 6\frac{\sum_A\left(\frac{EI}{L}\right)_b}{\sum_A\left(\frac{EI}{L}\right)_c}\right)\theta_A + s_{ij}\theta_B - (s_{ii} + s_{ij})\frac{\Delta}{L_{c2}} = 0 \quad (4.8.45)$$

and

$$s_{ij}\theta_A + \left(s_{ii} + 6\frac{\sum_B\left(\frac{EI}{L}\right)_b}{\sum_B\left(\frac{EI}{L}\right)_c}\right)\theta_B - (s_{ii} + s_{ij})\frac{\Delta}{L_{c2}} = 0 \quad (4.8.46)$$

Equations (4.8.45) and (4.8.46) are obtained by considering joint equilibrium at A and B, respectively. A third equation can be obtained by considering member equilibrium of column 2.

$$(M_A)_{c2} + (M_B)_{c2} + P\Delta - VL_{c2} = 0 \quad (4.8.47)$$

Since there is no external horizontal force acting, $V = 0$ and Eq. (4.8.47) becomes

$$(M_A)_{c2} + (M_B)_{c2} + P\Delta = 0 \quad (4.8.48)$$

Substituting Eqs. (4.8.39) and (4.8.42) into Eq. (4.8.48) and realizing that $P\Delta = \left(\frac{EI}{L}\right)_{c2}(kL)^2_{c2}\frac{\Delta}{L_{c2}}$, we can write Eq. (4.8.48) as

$$-6\frac{\sum_A\left(\frac{EI}{L}\right)_b}{\sum_A\left(\frac{EI}{L}\right)_c}\theta_A - 6\frac{\sum_B\left(\frac{EI}{L}\right)_b}{\sum_B\left(\frac{EI}{L}\right)_c}\theta_B + (kL)^2_{c2}\frac{\Delta}{L_{c2}} = 0 \quad (4.8.49)$$

The equilibrium equations (4.8.45), (4.8.46), and (4.8.49) can be put into matrix form as

$$\begin{bmatrix} s_{ii} + \dfrac{6}{G_A} & s_{ij} & -(s_{ii} + s_{ij}) \\ s_{ij} & s_{ii} + \dfrac{6}{G_B} & -(s_{ii} + s_{ij}) \\ -\dfrac{6}{G_A} & -\dfrac{6}{G_B} & (kL)^2_{c2} \end{bmatrix} \begin{pmatrix} \theta_A \\ \theta_B \\ \dfrac{\Delta}{L_{c2}} \end{pmatrix} = \begin{pmatrix} 0 \\ 0 \\ 0 \end{pmatrix} \quad (4.8.50)$$

where G_A and G_B, are defined in Eqs. (4.8.21) and (4.8.22), respectively.

At bifurcation, we must have

$$\det \begin{vmatrix} s_{ii} + \dfrac{6}{G_A} & s_{ij} & -(s_{ii}+s_{ij}) \\ s_{ij} & s_{ii} + \dfrac{6}{G_B} & -(s_{ii}+s_{ij}) \\ -\dfrac{6}{G_A} & -\dfrac{6}{G_B} & (kL)^2_{c2} \end{vmatrix} = 0 \qquad (4.8.51)$$

Using the expressions for s_{ii} and s_{ij} [Eqs. (3.7.15) and (3.7.16)] in Eq. (4.8.51) and realizing that $kL = \pi/K$, Eq. (4.8.51) can be simplified to

$$\frac{G_A G_B (\pi/K)^2 - 36}{6(G_A + G_B)} - \frac{(\pi/K)}{\tan(\pi/K)} = 0 \qquad (4.8.52)$$

Equation (4.8.52) is expressed in a nomograph form as shown in Fig. 4.22b (see reference 12). Note that K is always greater than or equal to unity.

Equations (4.8.26) and (4.8.52) express length factor K of a framed column as a function of the end restraint factors G_A and G_B. In the present development, all members are assumed to behave elastically. However, in many cases the magnitude of axial load in the column is such that inelasticity may set in when buckling occurs. To account for inelasticity in the column, Yura (see reference 13) suggested that the end restraint parameters be modified to

$$G_{\text{inelastic}} = G_{\text{elastic}}\left(\frac{E_t}{E}\right) \qquad (4.8.53)$$

where

E_t = tangent modulus of the material
G_{elastic} = end restraint parameter assuming all members are elastic
$G_{\text{inelastic}}$ = end restraint parameter accounting for inelasticity in the column

The K-factor evaluated using $G_{\text{inelastic}}$ will be smaller than that evaluated using G_{elastic}. The reason for this is that the apparent end restraint from the beams will be greater for an inelastic than an elastic column because of a reduction in bending stiffness of an inelastic column.

An alternative approach to determine the effective length factor K for framed columns based on the stiffness distribution approach was proposed by Wood.[14]

4.9 ILLUSTRATIVE EXAMPLES

Example 4.1. Two-Member Frame
For the structure shown in Fig. 4.23, determine the effective length factor K for column AB using

(a) the slope-deflection equation
(b) the nomograph

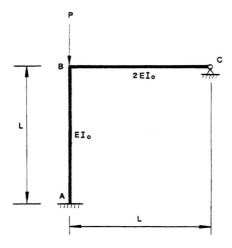

FIGURE 4.23 Two-member frame

SOLUTION

(a) *Slope-Deflection Equation*
Referring to Fig. 4.24 and using the slope-deflection equations, we see that we have

Column AB

$$M_{AB} = \frac{EI_0}{L}(s_{iic}\theta_A + s_{ijc}\theta_B) = \frac{EI_0}{L}(s_{ijc}\theta_B) \quad (4.9.1)$$

$$M_{BA} = \frac{EI_0}{L}(s_{ijc}\theta_A + s_{iic}\theta_B) = \frac{EI_0}{L}(s_{iic}\theta_B) \quad (4.9.2)$$

Beam BC

$$M_{BC} = \frac{2EI_0}{L}(s_{iib}\theta_B + s_{ijb}\theta_C) \simeq \frac{2EI_0}{L}(4\theta_B + 2\theta_C) \quad (4.9.3)$$

$$M_{CB} = \frac{2EI_0}{L}(s_{ijb}\theta_B + s_{iib}\theta_C) \simeq \frac{2EI_0}{L}(2\theta_B + 4\theta_C) = 0 \quad (4.9.4)$$

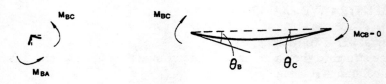

FIGURE 4.24 Beam and column moments

The equation $M_{CB} = 0$ leads to
$$\theta_C = -\tfrac{1}{2}\theta_B \tag{4.9.5}$$
therefore, the end moment M_{BC} reduces to
$$M_{BC} = \frac{2EI_0}{L}(3\theta_B) \tag{4.9.6}$$

Joint Equilibrium
$$M_{BA} + M_{BC} = 0 \tag{4.9.7}$$
Therefore
$$\frac{EI_0}{L}(s_{iic}\theta_B) + \frac{2EI_0}{L}(3\theta_B) = 0 \tag{4.9.8}$$
$$\frac{EI_0}{L}\theta_B(s_{iic} + 6) = 0 \tag{4.9.9}$$
$$s_{iic} = -6 \tag{4.9.10}$$

Using Table 3.7, we obtain
$$kL = (\sqrt{P/EI})L = 5.5405 \tag{4.9.11}$$

4.9 Illustrative Examples

from which

$$P = P_{cr} = \frac{30.7EI}{L^2} \quad \text{and} \quad K = \sqrt{\frac{P_e}{P_{cr}}} = \sqrt{\frac{\pi^2}{30.7}} = 0.57 \quad (4.9.12)$$

(b) *Nomograph*

$$G_A = 0 \quad \text{(fixed-end)} \quad (4.9.13)$$

$$G_B = \frac{(EI_0/L)}{2EI_0/L} = \frac{1}{2} \quad (4.9.14)$$

The G_B value calculated above is valid only if $\theta_C = -\theta_B$ (see Fig. 4.21). In our case $\theta_C \neq -\theta_B$, and so G_B must be adjusted. The adjustment can be made by realizing that for $\theta_C = -\theta_B$,

$$M_{BC} = \frac{2EI_0}{L}(2\theta_B) = \frac{4EI_0}{L}\theta_B \quad (4.9.15)$$

and for our case

$$M_{BC} = \frac{6EI_0}{L}\theta_B \quad (4.9.16)$$

Thus, the apparent stiffness of the beam with its far end hinged is $6/4 = 1.5$ times greater. Consequently, to adjust G_B we need only to divide it by the factor 1.5

$$(G_B)_{\text{adjusted}} = \frac{G_B}{1.5} = 0.333 \quad (4.9.17)$$

Using the nomograph shown in Fig. 4.21 with $G_A = 0$ and $(G_B)_{\text{adjusted}} = 0.333$, we obtain

$$K = 0.57 \quad (4.9.18)$$

Example 4.2. Simple Portal Frame
Determine the design moments for the simple frame shown in Fig. 4.25 using the LRFD method.

SOLUTION: In the LRFD method, a first-order analysis is performed on the structure. The secondary effects are taken care of by the use of the member stability $(P - \delta)$ and frame stability $(P - \Delta)$ moment magnification factors. These factors are designated as B_1 [Eq. (3.10.18)] and B_2 [Eqs. (3.10.19) or (3.10.20)], respectively.

First-Order Analysis
Figure 4.26a shows the decomposition of the frame into a nonsway and sway component. The corresponding nonsway and sway moments are shown in Fig. 4.26b and c, respectively.

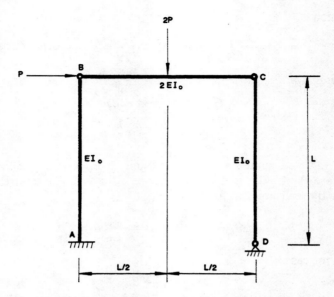

FIGURE 4.25 Simple frame under vertical and lateral loads

FIGURE 4.26 Nonsway and sway moments

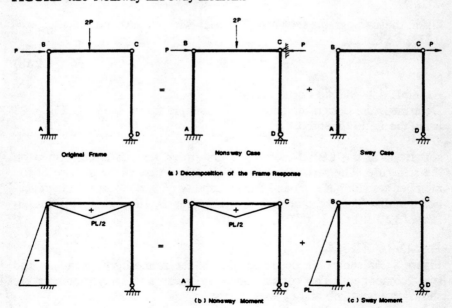

4.9 Illustrative Examples

Column AB

The design moment for column AB is determined from Eq. (3.10.17). The magnitudes for M_{nt} and M_{lt} can easily be determined from Fig. 4.26b as 0 and PL, respectively. To determine the B_1 and B_2 factors, we need to evaluate the effective length K for the column. From the nomograph in Fig. 4.22, K was found to be 2. However, since the right column is pinned at both ends, it cannot resist any sidesway motion. As a consequence, all the resistance to frame instability effect comes from the left column. For this situation, the column that does not provide any sidesway resistance is said to *lean on* the column, which provides the sidesway resistance. Since not all of the columns are effective in resisting sidesway, the effective length of the column providing the sidesway resistance must be modified. Le Messurier[15] suggested a formula for this modification

$$K_i = \sqrt{\frac{\pi^2 EI}{L^2 P_i}\left(\frac{\Sigma P}{\Sigma P_{ek}}\right)} \tag{4.9.19}$$

where

K_i = modified effective length of the column providing the sidesway resistance
P_i = axial force in the column providing the sidesway resistance
ΣP = axial loads on all columns in a story
ΣP_{ek} = Euler loads of all columns in a story providing sidesway resistance for the frame evaluating using the effective length obtained from the nomograph.

For our case, the modified effective length for column AB is

$$K_{AB} = \sqrt{\frac{\pi^2 EI}{L^2 P}\left(\frac{2P}{P_e/K^2}\right)}$$
$$= \sqrt{2K^2} = 2\sqrt{2} \tag{4.9.20}$$

Thus, from Eq. (3.10.18)

$$B_1 = \frac{C_m}{1 - \frac{P}{P_{ek}}} \geq 1$$

$$= \frac{0.6}{1 - \frac{P}{P_e/(0.8)^2}} \geq 1$$

$$= \frac{0.6}{1 - 0.64\frac{P}{P_e}} \geq 1 \tag{4.9.21}$$

and, from Eq. (3.10.20)

$$B_2 = \frac{1}{1 - \frac{\sum P}{\sum P_{ek}}} = \frac{1}{1 - \frac{2P}{P_e/(2\sqrt{2})^2}}$$

$$= \frac{1}{1 - 16\frac{P}{P_e}} \qquad (4.9.22)$$

The design moment is, from Eq. (3.10.17)

$$M_a = B_1 M_{nt} + B_2 M_{lt}$$

$$= \frac{0.6}{1 - 0.64\frac{P}{P_e}}(0) + \frac{1}{1 - 16\frac{P}{P_e}}(PL)$$

$$= \frac{PL}{1 - 16\frac{P}{P_e}} \qquad (4.9.23)$$

Column CD
This column is pinned at both ends, and so both M_{nt} and M_{lt} are zero. As a consequence, the design moment is zero and, therefore, this column is designed as a centrally loaded member.

Example 4.3. Two-Bay Frame
Check the adequacy of column CD of the frame shown in Fig. 4.27 using the LRFD approach. The frame is braced against out-of-plane bending at story height of every column and at midheight of the exterior columns. Assume the loadings to be

$$D = 0.9 \text{ kips/ft}$$
$$L = 1.6 \text{ kips/ft}$$
$$W = 0.8 \text{ kips/ft}$$

and column CD can develop its full plastic moment capacity.

SOLUTION: For the given loadings, it can be seen from Table 1.1 that the pertinent load combinations are

1. $1.4D$
2. $1.2D + 1.6L$
3. $1.2D + 0.5L$
4. $1.2D + 1.3W + 0.5L$
5. $0.9D - 1.3W$

4.9 Illustrative Examples

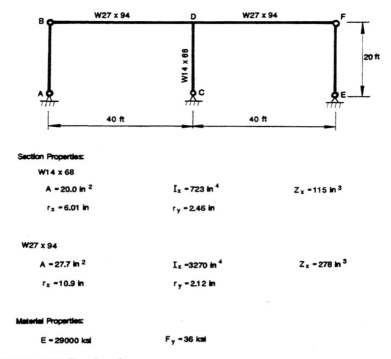

FIGURE 4.27 Two-bay frame

For gravity loadings, it is obvious that load combination (2) is the most severe load case and for combined gravity and lateral loadings, it can be seen that load combination (4) is the most severe load case. As a result, we need only to check the frame for load combinations (2) and (4).

Load Combination (2): $1.2D + 1.6L$

Figure 4.28 shows the result of a first-order analysis of the frame. The first-order moment in column CD is zero, therefore, the second and third terms of the interaction equations [Eqs. (3.10.15) and (3.10.16)] vanish. We need only to check the first term.

Determine P_u

Strong axis bending

Under the section properties given in Fig. 4.27, we have

$$G_C = \infty \quad \text{(pinned-end)}$$

$$G_D = \frac{\sum (EI/L)_c}{\sum (EI/L)_b} = \frac{\sum (I/L)_c}{\sum (I/L)_b}$$

$$= \frac{723/20}{(2)(3270)/40} = 0.221 \qquad (4.9.24)$$

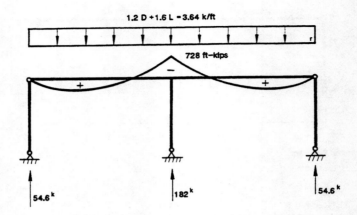

FIGURE 4.28 First-order analysis for load combination 1.2D + 1.6L

Although the theoretical G value for pinned-end is infinity, for design purposes it is customary to use $G = 10$ to account for the fact that an ideal pinned-ended condition does not exist. Therefore

$$(G_C)_{\text{adjusted}} = 10$$

Also, since the far end of the beams are hinged, G_D needs to be modified. Since the frame does not sway under the present load combination, we have $(G_D)_{\text{adjusted}} = \dfrac{G_D}{1.5}$.

Using the nomograph in Fig. 4.21 with the adjusted values for the G's, the effective length of column CD was found to be

$$K_x = 0.73$$

and the slenderness parameter can be calculated from

$$\begin{aligned}
\lambda_{cx} &= \frac{K_x L}{\pi r_x} \sqrt{\frac{F_y}{E}} \\
&= \frac{(0.73)(20)(12)}{\pi(6.01)} \sqrt{\frac{36}{29,000}} \\
&= 0.327
\end{aligned} \qquad (4.9.25)$$

Weak axis bending
For weak axis bending, $K_y = 1$ and so

$$\lambda_{cy} = \frac{K_y L}{\pi r_y} \sqrt{\frac{F_y}{E}} = \frac{(1)(20)(12)}{\pi(2.46)} \sqrt{\frac{36}{29,000}}$$
$$= 1.09 \qquad (4.9.26)$$

4.9 Illustrative Examples

Upon comparing λ_{cx} and λ_{cy}, it can be concluded that weak axis bending controls. Using Eq. (2.11.9), P_u was found to be

$$P_u = \exp[-0.419\lambda_{cy}^2]P_y$$
$$= \exp[-0.419(1.09)^2](20.0)(36) = 438 \text{ kips} \qquad (4.9.27)$$

Check interaction equation

$$\frac{P}{\phi_c P_u} = \frac{182}{(0.85)(438)} = 0.489 > 0.2 \qquad (4.9.28)$$

Use the Expression (3.10.15)

$$\frac{P}{\phi_c P_u} + \frac{8}{9}\left(\frac{M_{ax}}{\phi_b M_{ux}} + \frac{M_{ay}}{\phi_b M_{uy}}\right) = 0.489 + 0 + 0 = 0.489 \qquad (4.9.29)$$

which is <1; therefore all right.

Load Combination (4): $1.2D + 1.3W + 0.5L$

Figure 4.29 shows the results of a first-order analysis of the nonsway and sway components of the frame. Notice that the wind load is assumed to distribute to the windward and leeward sides of the frame in a 7:3 ratio. This is because in addition to exerting a positive pressure on a windward wall, the wind can simultaneously create a suction on the leeward wall. Moreover, an uplift force on the roof could be created. However this uplift force is not considered in this example because it has a beneficial effect on the frame. In other words, the loadings to be considered here represent the most unfavorable condition for the frame.

Determine P_u

Strong axis bending

$$G_C = \infty \quad \text{(pinned-end)} \qquad (4.9.30)$$
$$G_D = 0.221 \qquad (4.9.31)$$

To account for the fact that the pinned-end is not ideal, use

$$(G_C)_{\text{adjusted}} = 10 \qquad (4.9.32)$$

To account for the far end of the beam being hinged and for the fact that the frame sways under the present load combination, the apparent beam stiffness is only one-half of the beam stiffness used in calculating G_D (see Prob. 4.6). Therefore

$$(G_D)_{\text{adjusted}} = \frac{G_D}{1/2} = 0.442 \qquad (4.9.33)$$

and, from the nomograph shown in Fig. 4.22b, we have

$$K_x = 1.73 \qquad (4.9.34)$$

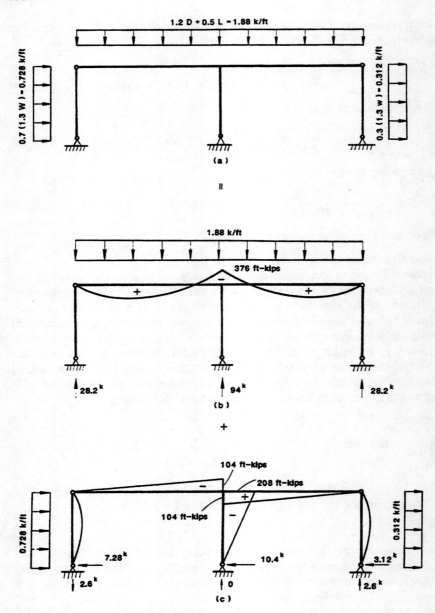

FIGURE 4.29 First-order analysis for load combination $1.2D + 1.3W + 0.5L$

Since the exterior pinned-ended columns lean on column CD, K_x must be adjusted. Using the method proposed in reference 15 as cited in the commentary of LRFD, the modified effective length is given by

$$(K_x)_{\text{adjusted}} = \sqrt{\frac{\pi^2 EI}{L^2 P_{CD}} \left(\frac{\sum P}{\sum P_{ek}}\right)}$$

$$= \sqrt{\frac{\pi^2 EI}{L^2(94)} \left(\frac{150.4}{P_e/K_x^2 + 0 + 0}\right)}$$

$$= \sqrt{1.6 K_x^2} = 2.19 \qquad (4.9.35)$$

and so

$$\lambda_{cx} = \frac{(K_x)_{\text{adjusted}} L}{\pi r_x} \sqrt{\frac{F_y}{E}}$$

$$= \frac{(2.19)(20)(12)}{\pi(6.01)} \sqrt{\frac{36}{29,000}} = 0.981 \qquad (4.9.36)$$

Weak axis bending

$$K_y = 1 \qquad (4.9.37)$$

$$\lambda_{cy} = \frac{K_y L}{\pi r_y} \sqrt{\frac{F_y}{E}}$$

$$= \frac{(1)(20)(12)}{\pi(2.46)} \sqrt{\frac{36}{29,000}} = 1.09 \qquad (4.9.38)$$

Since $\lambda_{cy} > \lambda_{cx}$, therefore P_u should be determined based on weak axis buckling

$$P_u = \exp[-0.419 \lambda_{cy}^2] P_y = 438 \text{ kips}$$

Determine M_{ax}

B_1 factor

$$B_1 = \frac{C_m}{1 - \dfrac{P}{P_{ek}}} \geq 1$$

$$= \frac{0.6}{1 - \dfrac{94}{6742}} = 0.608 \qquad (4.9.39)$$

which is <1, therefore we use $B_1 = 1$ ($P_{ek} = \Pi^2 EI_x/(K_x L)^2 = 6742$ kips)

B_2 *factor*

$$B_2 = \frac{1}{1 - \frac{\sum P}{\sum P_{ek}}}$$

$$= \frac{1}{1 - \frac{150.4}{749}} = 1.25 \qquad (4.9.40)$$

$$M_{ax} = B_1 M_{nt} + B_2 M_{lt}$$
$$= (1)(0) + 1.25(208)$$
$$= 260 \text{ ft} - \text{kip} = 3120 \text{ in} - \text{kip} \qquad (4.9.41)$$

Determine M_{ux}

$$M_{ux} = M_{px} = Z_x F_y = (115)(36) = 4140 \text{ in} - \text{kip} \qquad (4.9.42)$$

Check interaction equation

$$\frac{P}{\phi_c P_u} = \frac{94}{(0.85)(438)} = 0.252 > 0.2 \qquad (4.9.43)$$

Therefore, use Eq. (3.10.15)

$$\frac{P}{\phi_c P_u} + \frac{8}{9}\left(\frac{M_{ax}}{\phi_b M_{ux}} + \frac{M_{ay}}{\phi_b M_{uy}}\right) \leq 1.0 \qquad (4.9.44)$$

Since the frame is braced against out-of-plane bending, it follows that $M_{ay} = 0$. The first two terms have the value

$$\frac{94}{(0.85)(438)} + \frac{8}{9}\left[\frac{3120}{(0.90)(4140)}\right]$$
$$= 0.997 \qquad (4.9.45)$$

which is <1; therefore, it is all right.

4.10 SUMMARY

In this chapter, we focussed on the study of the behavior of frameworks. In particular, two reference loads for frames were discussed. They are the elastic critical load P_{cr} and the plastic collapse load P_p. The elastic critical load can be obtained by using one of the following methods: (1) the differential equation method, (2) the slope-deflection equation method, and (3) the matrix stiffness method. In obtaining P_{cr}, only the effect of instability is considered. However, if only the effect of plasticity or yielding of the material is considered, one can obtain the plastic collapse load P_p by using either the hinge-by-hinge method or the mechanism method.

In reality, the collapse loads P_f of most frameworks are neither P_{cr} nor P_p, because collapse in most cases is a result of an interaction of the effects of instability and plasticity. The exact interaction relation is rather complex. However, for design purposes, the Merchant–Rankine interaction equation provides a simple but reasonably accurate method to estimate P_f.

For the design of members in a frame, the K-factor approach provides a convenient means to account for the end restraint effects of other members on the behavior of the member in question. The two nomographs or alignment charts for braced and unbraced frames have been developed to aid designers to obtain K-factors for columns that are component parts of a frame.

PROBLEMS

4.1 Find P_{cr} for the frame in Fig. P4.1 using
   a. differential equation approach
   b. slope-deflection approach
   c. matrix stiffness approach

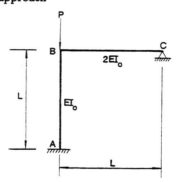

FIGURE P4.1

4.2 Find P_{cr} for the frames in Fig. P4.2a-b for both sway-prevented and sway-permitted cases. What conclusion can be drawn from the results?

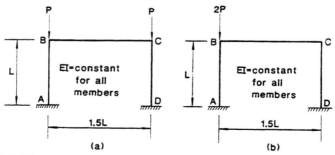

FIGURE P4.2

4.3 For the frame in Fig. P4.3,
   a. sketch the buckled shape of the frame
   b. establish upper and lower bounds for the critical load
   c. find the critical load for $a = 1$

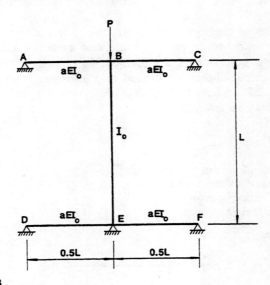

FIGURE P4.3

4.4 Find P_p for the frame in Fig. P4.4 using the mechanism method.

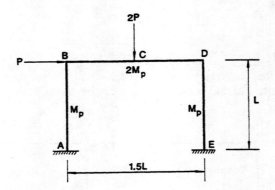

FIGURE P4.4

4.5 For the structure in Fig. P4.5, find the effective length factor K for the column BD using
   a. slope-deflection equation
   b. nomograph

4.6 Discuss how the G factors for the nomograph can be adjusted to account for the cases where the far ends of the beams are all
- **a.** pinned
- **b.** fixed

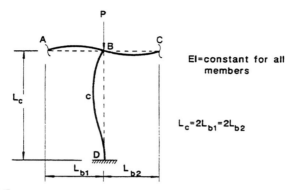

FIGURE P4.5

REFERENCES

1. Winter, G., Hsu, P. T., Koo, B., and Loh, M. H. Buckling of Trusses and Rigid Frames. Cornell University Engineering Experimental Station Bulletin, No. 36, Ithaca, N.Y., 1948.
2. Hartz, B. J. Matrix formulation of structural stability problems. Journal of the Structural Division. ASCE, Vol. 91, No. ST6, December 1965, pp. 141–158.
3. Halldorsson, O., and Wang, C. K. Stability analysis of frameworks by matrix methods. Journal of the Structural Division. ASCE, Vol. 94, No. ST7, July 1968, pp. 1745–1760.
4. Goto, Y., and Chen, W. F. On Second-Order Elastic Analysis for Design. Session on Design of Columns in Frames, ASCE, Structural Engineering Congress, New Orleans, September 15–18, 1986. See also, On the computer-based design analysis for the flexibly jointed frames. Journal of Constructional Steel Research, Vol. 7, No. 2, 1987.
5. Przemieniecki, J. S. Theory of Matrix Structural Analysis. McGraw-Hill, New York, 1968.
6. Weaver, W., Jr., and Gere, J. M. Matrix Analysis of Framed Structures. 2nd edition, Van Nostrand, New York, 1980.
7. Meyers, V. J. Matrix Analysis of Structures. Harper & Row, New York, 1984.
8. White, D. W., and McGuire, W. Methods of Analysis in LRFD. Structural Engineering Congress, ASCE, Chicago, September 16–18, 1985.
9. Wang, C. K. General computer program for limit analysis. Journal of the Structural Division. ASCE, Vol. 89, No. ST6, December 1963, pp. 101–117.
10. Baker, J., and Heyman, J. Plastic Design of Frames. Vol. I: Fundamentals. Cambridge University Press, Cambridge, 1969.

11. Horne, M. R., and Merchant, W. The Stability of Frames. Pergamon Press, New York, 1965.
12. Julian, O. G., and Lawrence, L. S. Notes on J and L Nomograms for Determination of Effective Lengths. Unpublished report.
13. Yura, J. A. The effective length of columns in unbraced frames, AISC Engineering Journal, 8(2), April, 1971, pp. 37–42.
14. Wood, R. H. Effective lengths of columns in multistory buildings. Structural Engineers. Vol. 52, Nos. 7, 8, 9, July, August, September, 1974, pp. 235–244, 295–302, 341–346.
15. LeMessuier, W. J. A practical method of second order analysis, part 2 — rigid frames, AISC Engineering Journal, 14(2), April, 1977, pp. 49–67.

General References

Bleich, F. Buckling Strength of Metal Structures. McGraw-Hill, New York, 1952.

Chajes, A. Principles of Structural Stability Theory. Prentice Hall, Englewood Cliffs, New Jersey, 1974.

Timoshenko, S. P., and Gere, J. M. Theory of Elastic Stability. 2nd edition. McGraw-Hill, New York, 1961.

Chapter 5

BEAMS

5.1 INTRODUCTION

Beams are structural members that resist the applied loads primarily by bending and shearing actions. When a slender beam is under the action of bending loads acting in the plane of the weak axis of the cross section (or in the plane of the web of a wide-flange section), in-plane bending about the strong axis will occur at the commencement of the loadings. However, if sufficient lateral bracing is not provided to the compression flange, out-of-plane bending and twisting of the cross section will occur when the applied loads reach a certain limit. For a geometrically perfect elastic beam, the limit of the applied loads at which lateral instability commences is called the *elastic lateral torsional buckling load*. The value of the lateral torsional buckling load is influenced by a number of factors. Among the important ones are the cross-sectional shape, the unbraced length and the support conditions of the beam, the type and position of the applied loads along the member axis, and the location of the applied loads with respect to the centroidal axis of the cross section.

It it well-known that beams of thin-walled *open* cross sections composed of slender component plates, such as I-sections, channel sections, and Z-sections, are particularly susceptible to lateral torsional buckling. This is because the torsional rigidities of such cross sections are very low and so their resistance to torsional instability is very limited. The effects of unbraced length and end conditions on the lateral torsional buckling load of the beam are rather evident. The longer the unbraced length and the less restraint the support can deliver to the beam, the lower the critical lateral buckling load will be. Although it is quite obvious that different types of loads applied at different locations of the

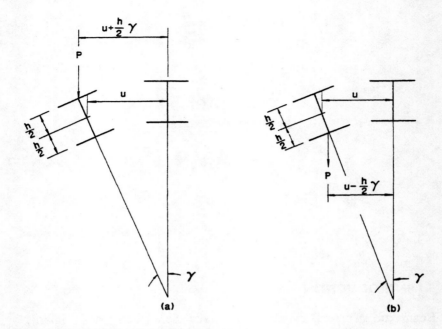

FIGURE 5.1 Effect of location of loading

beam will give different values for the various critical lateral buckling loads, the reason behind the importance of the position of the applied loads with respect to the centroidal axis of the cross section requires some explanation. For instance, consider a concentrated force acting on the top flange of an I-beam (Fig. 5.1a). As lateral torsional buckling occurs, the cross section will rotate and deflect laterally from its original position. It can be seen from the figure that the applied force has a destabilizing effect, since it enhances the rotation of the cross section from its original undeflected position. However, if the same load is applied at the bottom flange (Fig. 5.1b), it has a stabilizing effect, since it reduces the rotation of the cross section. Consequently, the critical load corresponding to Fig. 5.1a will be lower than that corresponding to Fig. 5.1b.

The analysis of lateral torsional buckling behavior of beams is considerably more complex than that of in-plane buckling behavior of columns discussed in Chapter 2 because the lateral buckling problem is intrinsically three-dimensional. The problem is further complicated because the lateral (out-of-plane) deflection and twisting are coupled, so this coupling effect must be considered in the analysis.

5.2 Uniform Torsion of Thin-Walled Open Sections

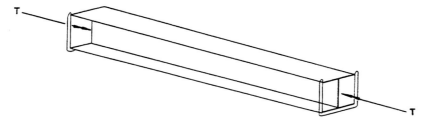

FIGURE 5.2 Simply supported I-beam subjected to twisting moment

5.2 UNIFORM TORSION OF THIN-WALLED OPEN SECTIONS

When an equal and opposite torque T is applied to the ends of a simply supported beam with a thin-walled open section, such as an I-section (Fig. 5.2), the twisting moment along the length of the members is constant and the beam is said to be under a *uniform torsion*. Under the action of the torque, warping of the cross section will occur. This is illustrated in Fig. 5.3. It shows how plane sections of the cross section no longer remain plane as a result of the uneven axial deformation that takes place over the entire cross section. If the applied torque is constant and all cross sections are free to warp, then the warping deformation in the beam is the same for all cross sections and takes place freely without inducing any axial strain on the longitudinal fibers.

For the simply supported beam shown in Fig. 5.2, in which warping of all the cross sections is unrestrained, the applied torque (twisting moment) is resisted solely by shear stresses developed in the cross section. These stresses act parallel to the edge of the component plates of the cross section, as shown in Fig. 5.4. The distribution of these shear stresses is the same for all thin-walled, open cross sections. It is usually assumed that the shear stress at any point acts parallel to the tangent to the midline of the cross section. The magnitude of these shear stresses will be proportional to the distance from the midline of the component plate. These shear stresses are called *St. Venant shear stresses* and the

FIGURE 5.3 Warping of I-section under uniform twisting moment

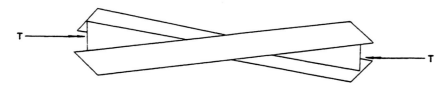

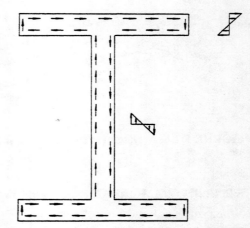

FIGURE 5.4 St. Venant shear stress distribution in an I-section

torsion that is associated with these shear stresses is referred to as *St. Venant torsion* T_{sv}.

From mechanics of materials, the angle of twist γ over a length L caused by the St. Venant torsion is given by

$$\frac{\gamma}{L} = \frac{T_{sv}}{GJ} \qquad (5.2.1)$$

where

γ/L = angle of twist per unit length
T_{sv} = St. Venant torsion
G = shear modulus
J = torsional constant of the cross section

For a thin-walled, open section of constant thickness t, the torsional constant can be expressed[1] by

$$J = \tfrac{1}{3}bt^3 \qquad (5.2.2)$$

where b is the length of the midline of the cross section and t is the thickness of the cross section. If the cross section is made up of n slender component plates, each with midline length b_i and thickness t_i, the torsional constant can be assumed to be

$$J = \sum_{i=1}^{n} \tfrac{1}{3} b_i t_i^3 \qquad (5.2.3)$$

Table 5.1 gives the expression of J for a doubly symmetric I-section. It should be mentioned that the above expressions for J are valid only if b/t is larger than 10. If b/t is smaller than 10, a correction factor must be used in calculating J (see Problem 5.3).

5.3 Non-Uniform Torsion of Thin-Walled, Open Cross Sections

Table 5.1 Torsional Constant and Warping Constant for a Doubly Symmetric I-Section

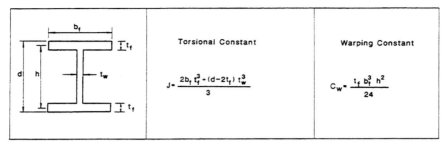

For the purpose of analysis, it is more convenient to express Eq. (5.2.1) in the form of the rate of twist as

$$\frac{d\gamma}{dz} = \frac{T_{sv}}{GJ} \qquad (5.2.4)$$

in which z is the coordinate axis along the length of the beam. Note that the rate of twist will be constant for a prismatic member subjected to a uniform torque.

Upon rearranging, Eq. (5.2.4) can be written as

$$T_{sv} = GJ \frac{d\gamma}{dz} \qquad (5.2.5)$$

The St. Venant torsion expressed in Eq. (5.2.5) is also referred to as *uniform* or *pure torsion*.

5.3 NON-UNIFORM TORSION OF THIN-WALLED, OPEN CROSS SECTIONS

Consider a cantilever beam subjected to a torque applied at the free end (Fig. 5.5). At the free end the cross section is free to warp, so the applied torque is resisted solely by St. Venant torsion. At the fixed end, however, warping is prevented. As a result, in addition to St. Venant torsion, there exists another type of torsion known as *warping restraint torsion* in the cross section. If the cross section is prevented from warping, axial strain and so axial stresses must be induced in the cross section in addition to the shear stresses. These induced axial stresses are in self-balance because no external axial force is applied to the beam. For an I-section, the axial stresses developed at the fixed end of the beam are illustrated in Fig. 5.5. The resultant of these axial stresses in the two flanges constitutes a pair of equal moments called the *bi-moment* M_f acting oppositely in each of these two planes of the flanges (Fig. 5.5).

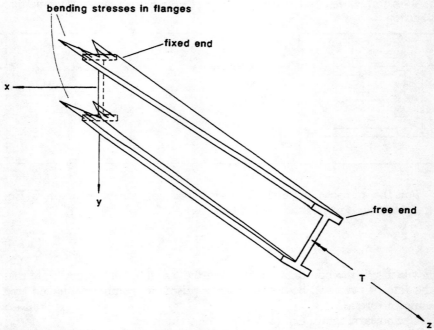

FIGURE 5.5 Cantilever beam subjected to a twisting moment

The development of these bending moments, or bi-moment, in the flanges in the cross section becomes evident if one refers to Fig. 5.6. Since warping is prevented at the fixed end, the two flanges of the beam must bend in opposite directions as the cross section rotates under the action of the applied torque. The bending of the flanges will thus induce

FIGURE 5.6 Bending of flanges due to warping restraint at the fixed end

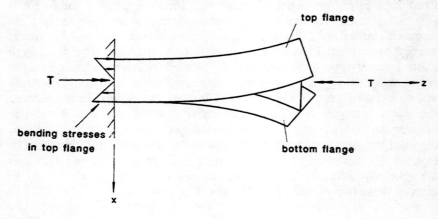

5.3 Non-Uniform Torsion of Thin-Walled, Open Cross Sections

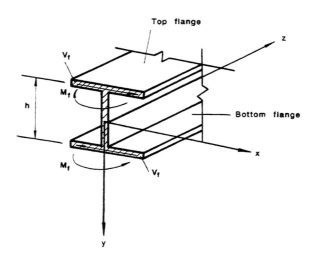

FIGURE 5.7 Moment and shear developed at the fixed-end cross section of an I-section

bending moments M_f at the fixed end (Fig. 5.7). The bending moment M_f in either the top or the bottom flange can be expressed in terms of lateral displacement u_f by the usual beam moment-curvature relationship as

$$M_f = EI_f \frac{d^2 u_f}{dz^2} \qquad (5.3.1)$$

where E is the modulus of elasticity, I_f the moment of inertia of one flange about the y axis of the cross section, and u_f the lateral displacement of the flange, as shown in Fig. 5.8.

Associated with the bending moment M_f in one flange is the shear force V_f given by the usual beam theory

$$V_f = -\frac{dM_f}{dz} = -EI_f \frac{d^3 u}{dz^3} \qquad (5.3.2)$$

The shear forces V_f are present in both flanges of the I-section. They are equal in magnitude but act in opposite directions, as shown in Fig. 5.7. This pair of shear forces constitute a couple acting on the cross section. The resulting torsion, which is referred to as the *warping restraint torsion* or *non-uniform torsion*, is given by

$$T_w = V_f h \qquad (5.3.3)$$

where h is the distance between the lines of action of the shear forces.

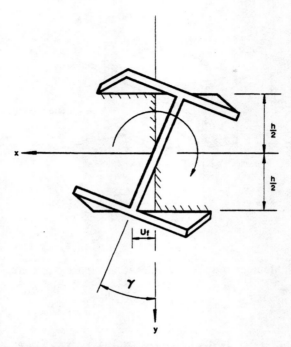

FIGURE 5.8 Rotational and translational relationship

In view of Eq. (5.3.2), Eq. (5.3.3) can be written as

$$T_w = -EI_f h \frac{d^3 u_f}{dz^3} \qquad (5.3.4)$$

From Fig. 5.8, it can be seen that the rotation γ of the cross section is related to the lateral deflection u_f by

$$u_f = \frac{h}{2} \gamma \qquad (5.3.5)$$

from which

$$\frac{d^3 u_f}{dz^3} = \frac{h}{2} \frac{d^3 \gamma}{dz^3} \qquad (5.3.6)$$

Upon substituting Eq. (5.3.6) into Eq. (5.3.4), we can obtain the warping restraint torsion as

$$T_w = -EI_f \frac{h^2}{2} \frac{d^3 \gamma}{dz^3} = -EC_w \frac{d^3 \gamma}{dz^3} \qquad (5.3.7)$$

5.3 Non-Uniform Torsion of Thin-Walled, Open Cross Sections

where

$$C_w = \frac{I_f h^2}{2} \quad (5.3.8)$$

is called the *warping constant* of the I-section (Table 5.1). It should be noted that the warping constant is different for different cross sections. A general expression for the warping constant is found in reference 2; it is

$$C_w = \int_0^s (\bar{w}_s - w_s)^2 t \, ds \quad (5.3.9)$$

in which $w_s = \int_0^s r \, ds$, r equals the distance from the shear center of the cross section to the tangent at any point around the cross section, the equation

$$\bar{w}_s = \frac{1}{s} \int_0^s w_s \, ds$$

equals the average value of w_s over the entire cross section, t is the thickness of the thin-walled element, and s is the length of the midline of the entire cross section. From Eq. (5.3.9), it can be seen that the warping constant C_w will be zero for those thin-walled, open cross sections for which all component plates intersect at a common point, such as the angle, the tee, and the cruciform sections (see Fig. 5.9a). This is because

FIGURE 5.9 Cross sections with $C_w = 0$

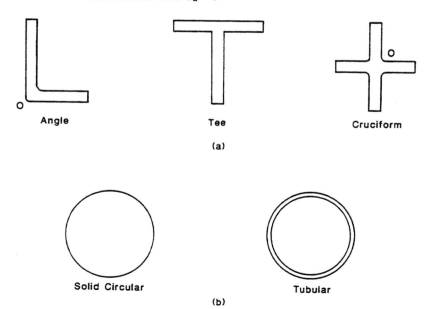

the shear centers 0 for these sections are located at the point of intersection of the component plates. So w_s and $\bar{w}_s$ will be zero. Physically, this means that all these sections do not warp when subjected to twisting moments. Another type of cross sections for which no warping will occur is the axisymmetric sections, such as the solid or circular tubes (see Fig. 5.9b). Here, those sections that are originally plane will remain plane after the twisting moment is applied. For sections other than those shown in Fig. 5.9, warping will generally occur when a twisting moment is applied. Warping for narrow rectangular sections and box sections composed of narrow rectangular elements are usually negligible, and so C_w may be taken as zero for these sections. If warping is restrained, the applied twisting moment will be resisted by both St. Venant torsion and warping restraint torsion.

$$T = T_{sv} + T_w \qquad (5.3.10)$$

In view of Eq. (5.2.5) and Eq. (5.3.7), we have

$$T = GJ\frac{d\gamma}{dz} - EC_w\frac{d^3\gamma}{dz^3} \qquad (5.3.11)$$

Equation (5.3.11) represents the internal twisting moment that will develop in the cross section when the member is twisted. The first term represents the resistance of the cross section to twist and the second term represents the resistance of the cross section to warp. Thus, for the I-section shown in Fig. 5.5, the applied twisting moment is resisted solely by St. Venant torsion at the free end ($z = L$) where the cross section is free to warp. However, further away from the free end, warping is partially restrained, so both St. Venant and warping restraint torsion will be present. The proportion of the applied twisting moment transmitted by these two forms of torsional resistance varies. As we move toward the fixed end, a greater and greater share of the applied torque will be resisted by the warping restraint torsion. At the fixed end ($z = 0$), warping is totally restrained and the applied torque will be transmitted completely by the warping restraint torsion.

From the above discussion, it can be seen that St. Venant torsion is always present when a member is subjected to twisting and rotates. On the other hand, warping restraint torsion will develop if a cross section is prevented from warping when it is being twisted. Warping restraint torsion will also develop in the cross sections if the twisting moment is not uniform along the length of the member. This is because under a non-uniform torsion, different cross sections will warp by a different amount. The differential axial deformation between two adjacent cross sections will induce axial stresses, giving rise to a warping restraint torsion.

5.4 LATERAL BUCKLING OF BEAMS

When a beam is bent about its axis of greatest flexural rigidity, out-of-plane bending and twisting will occur when the applied load reaches its critical value, unless the beam is provided with a sufficient lateral support. For a geometrically perfect beam, this critical load corresponds to the point of bifurcation of equilibrium when in-plane bending deformation of the member ceases to be stable and out-of-plane bending and twisting deformations become the stable configuration of the member. Here, as in the case of a column, to find the critical load of the beam one must first establish the equilibrium conditions of the beam in a slightly deformed configuration. The critical or lateral buckling load is then obtained as the lowest eigenvalue satisfying the characteristic equation of the differential equations. The following examples will illustrate the procedure for determining this critical load. The assumptions used in the following examples are the following:

1. The beam is geometrically perfect.
2. The applied loads act solely in the plane of the weak axis (or in the plane of the web in the case of an I-beam).
3. The deflection of the member is small.
4. The geometry of the cross section does not change during buckling.

Example 5.1. Simply Supported Rectangular Beam Under Pure Bending
Figure 5.10 shows a simply supported beam of narrow rectangular cross section subjected to a pair of equal and opposite end moments acting in the y-z plane. The simple support condition used in the context of lateral instability of beams means that the ends of the beam are free to rotate about the two principal axes, the x and y axes, but rotation of the end cross section about the z axis is prevented. Under the action of the applied moment, the beam will bend in the y-z plane. This type of bending is known as the *in-plane bending*. As the moments are increased, a stability condition will be reached at which the in-plane deformation of the beam ceases to be stable and a slightly deflected form that corresponds to the out-of-plane bending and twisting of the beam becomes possible. The beam is stable and favored with this new configuration. The lowest load at which this condition occurs is the critical load for the beam.

SOLUTION: To determine this critical load, it is necessary to establish the equilibrium equations governing this slightly deformed stable configuration of the beam. Referring to Fig. 5.10, in which the slightly deformed configuration of the beam is shown, we note that for any cross section three different displacement components are needed to define the deflected position of the cross section. They are the in-plane displace-

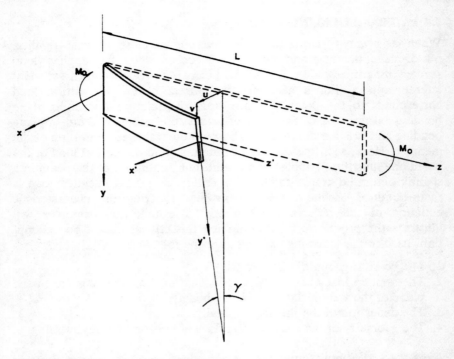

FIGURE 5.10 Lateral buckling of a narrow rectangular strip under uniform moment

ment v, the out-of-plane displacement u, and the rotation of the cross section γ. To facilitate the analysis, we establish two sets of coordinate axes. The x-y-z axes are fixed coordinate axes that are fixed to the original or undeformed position of the member; the x'-y'-z' axes are local coordinate axes that are fixed with the cross section that moves with the deflected position of the member. The x' and y' axes coincide with the principal axes of the cross section. The z' axis is always tangent to the center lines of the deflected position of the member. The procedure to establish the governing differential equations is very similar to that presented in Chapter 2 for columns. First, we establish the expressions for the moments induced by the external loads at an arbitrary section z. These externally induced moments at section z are called external moments. The differential equations are then obtained by equating these external moments to the corresponding internal resisting moments of the cross section. The only difference between the solution process presented in this chapter and that discussed in Chapter 2 is that, in the earlier

5.4 Lateral Buckling of Beams

chapter, only in-plane bending of the member is considered, whereas here both in-plane and out-of-plane bending, as well as twisting, will be considered. Thus, the lateral torsional buckling problem will be more complex than the in-plane buckling problem, as the former is a three-dimensional problem, whereas the latter is a two-dimensional problem.

For the beam shown in Fig. 5.10, the components of external moments acting on a cross section with a distance z from the origin with respect to the x-y-z coordinates are; $(M_x)_{ext} = M_0$, $(M_y)_{ext} = (M_z)_{ext} = 0$. Figure 5.11a-c shows the moments and moment components in the three mutually perpendicular planes. In this chapter, for convenience, we use a

FIGURE 5.11 Components of moments

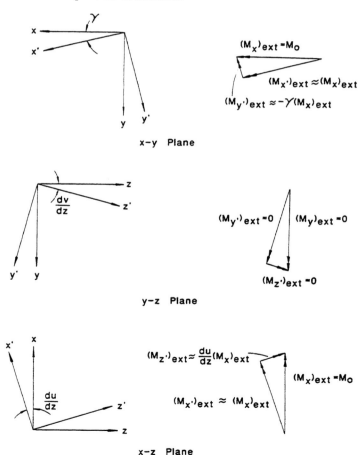

right-handed screw rule to represent the moment vector. For example, imagine now a right-hand screw whose axis is colinear with the x axis. If this screw were turned in the direction of M_0 at the positive face $z = L$, as shown in Fig. 5.10, it would tend to advance in the direction of the positive x axis. Because of this, the applied moment M_0 is positive and represented by the positive moment $(M_x)_{\text{ext}}$ in Fig. 5.11 in the positive x direction. This sign convention for moment is called the *right-handed screw rule*. Using this sign convention for moment, the moments acting on the cross section in the slightly deflected position with respect to the x'-y'-z' coordinate system can now be obtained directly from Fig. 5.11.

$$(M_{x'})_{\text{ext}} \approx (M_x)_{\text{ext}} = M_0 \tag{5.4.1}$$

$$(M_{y'})_{\text{ext}} \approx -\gamma (M_x)_{\text{ext}} = -\gamma M_0 \tag{5.4.2}$$

$$(M_{z'})_{\text{ext}} = \frac{du}{dz}(M_x)_{\text{ext}} = \frac{du}{dz} M_0 \tag{5.4.3}$$

The corresponding internal resisting moments are

$$(M_{x'})_{\text{int}} = -EI_x \frac{d^2v}{dz^2} \tag{5.4.4}$$

$$(M_{y'})_{\text{int}} = EI_y \frac{d^2u}{dz^2} \tag{5.4.5}$$

$$(M_{z'})_{\text{int}} = GJ \frac{d\gamma}{dz} \tag{5.4.6}$$

In writing Eq. (5.4.4) and (5.4.5), it is tacitly assumed that the angle of rotation γ is sufficiently small so that the curvatures and moment of inertia in the y'-z' and x'-z' planes may be represented by their corresponding values in the y-z and x-z planes, respectively. The minus sign in Eq. (5.4.4) indicates that a negative curvature in the y'-z' plane will give a position moment using the right-handed screw rule. Equation (5.4.6) follows from Eq. (5.3.11) with $C_w = 0$. For a narrow rectangular section, warping of the cross section is negligible and so warping restraint torsion can be neglected. Equating the corresponding external and internal moments, we have

$$EI_x \frac{d^2v}{dz^2} + M_0 = 0 \tag{5.4.7}$$

$$EI_y \frac{d^2u}{dz^2} + \gamma M_0 = 0 \tag{5.4.8}$$

$$GJ \frac{d\gamma}{dz} - \frac{du}{dz} M_0 = 0 \tag{5.4.9}$$

5.4 Lateral Buckling of Beams

An inspection of these equations shows that the first equation contains only the variable v and is independent of the other two equations. In fact, this equation describes the in-plane bending behavior of the member that occurs before lateral instability. It is not important for the out-of-plane lateral torsional buckling behavior in a small displacement buckling analysis. The buckling behavior of the beam is described by the last two equations, which are coupled, as they both contain u and γ as variables. If we differentiate Eq. (5.4.9) once with respect to z and substitute the result into Eq. (5.4.8), the two equations can be combined to give

$$\frac{EI_y GJ}{M_0}\frac{d^2\gamma}{dz^2} + \gamma M_0 = 0 \tag{5.4.10}$$

Upon rearranging, and denoting $k^2 = M_0^2/EI_y GJ$, we have the differential equation

$$\frac{d^2\gamma}{dz^2} + k^2\gamma = 0 \tag{5.4.11}$$

which has the same form as Eq. (2.2.12) and hence the general solution is

$$\gamma = A \sin kz + B \cos kz \tag{5.4.12}$$

Since rotations of the end cross sections are prevented, the boundary conditions

$$\gamma(0) = 0 \quad \text{and} \quad \gamma(L) = 0 \tag{5.4.13}$$

apply. Using the first boundary condition in Eq. (5.4.12), we have $B = 0$, and using the second boundary condition, we have

$$A \sin kL = 0$$

If $A = 0$, Eq. (5.4.12) becomes a trivial solution. Thus, for a nontrivial solution, we must have

$$\sin kL = 0 \quad \text{or} \quad kL = n\pi \tag{5.4.14}$$

Since $k^2 = M_0^2/EI_y GJ$, we can obtain

$$M_0 = \frac{n\pi}{L}\sqrt{EI_y GJ} \tag{5.4.15}$$

The critical moment is the lowest value of M_0 that will cause a lateral torsional buckling. It can be obtained by setting $n = 1$ in Eq. (5.4.15), so

$$M_{ocr} = \frac{\pi}{L}\sqrt{EI_y GJ} \tag{5.4.16}$$

It is important to note that the critical moment is a function of both the lateral bending stiffness EI_y and the torsional stiffness GJ. Thus, the coupling effect of out-of-plane deformation and twisting is manifested in the result.

Equation (5.4.16), although derived for a narrow rectangular section, is also valid for box sections composed of narrow rectangular shapes. Like narrow rectangular cross sections, warping for box sections are negligible and so warping constant C_w can be set to zero. However, for box sections, M_{cr} will be much higher due to a significant increase in values of I_y and J.

Example 5.2. Simply Supported I-Section Under Pure Bending
A simply supported I-beam subjected to a pair of equal and opposite end moments applied about the x axis is shown in Fig. 5.12.

SOLUTION: Here, as in the preceding example, two sets of coordinate axes, x-y-z and x'-y'-z', are used to facilitate the analysis. Since no change has been made in the external loadings or support conditions, Eqs. (5.4.1) to (5.4.3) are still applicable here. As to the internal resisting moments, the two equations describing the in-plane [Eq. (5.4.4)] and out-of-plane [Eq. (5.4.5)] bending behavior of the member are also applicable here. In fact, the only equation that needs to be modified is Eq. (5.4.6). For an I-section, in addition to St. Venant torsion, there is a warping restraint torsion; hence, the total torsional resistant offered by the I-section is

$$(M_{z'})_{\text{int}} = GJ\frac{d\gamma}{dz} - EC_w\frac{d^3\gamma}{dz^3} \qquad (5.4.17)$$

FIGURE 5.12 Lateral buckling of an I-section under uniform moment

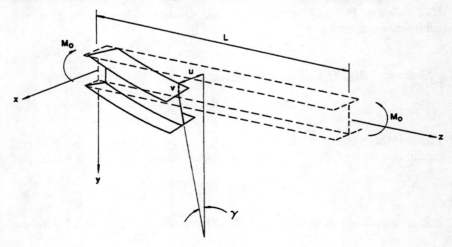

5.4 Lateral Buckling of Beams

By equating the corresponding external and internal moments, the governing differential equations for the I-section under pure bending are obtained

$$EI_x \frac{d^2v}{dz^2} + M_0 = 0 \qquad (5.4.18)$$

$$EI_y \frac{d^2u}{dz^2} + \gamma M_0 = 0 \qquad (5.4.19)$$

$$GJ \frac{d\gamma}{dz} - EC_w \frac{d^3\gamma}{dz^3} - \frac{du}{dz} M_0 = 0 \qquad (5.4.20)$$

Again, the first equation is of no interest to us, as it describes the in-plane behavior of the beam before the lateral buckling. The differential equation describing the behavior of the beam at lateral torsion buckling is obtained by combining Eqs. (5.4.19) and (5.4.20)

$$EC_w \frac{d^4\gamma}{dz^4} - GJ \frac{d^2\gamma}{dz^2} - \frac{M_0^2}{EI_y} \gamma = 0 \qquad (5.4.21)$$

Denoting

$$a = \frac{GJ}{2EC_w} \quad \text{and} \quad b = \frac{M_0^2}{EI_y EC_w} \qquad (5.4.22)$$

Equation (5.4.21) can be written as

$$\frac{d^4\gamma}{dz^4} - 2a \frac{d^2\gamma}{dz^2} - b\gamma = 0 \qquad (5.4.23)$$

Equation (5.4.23) is a fourth-order linear differential equation with constant coefficients, the general solution is

$$\gamma = A \sin mz + B \cos mz + Ce^{nz} + De^{-nz} \qquad (5.4.24)$$

in which m and n are positive, real quantities defined by

$$m = \sqrt{-a + \sqrt{(a^2 + b)}}, \quad n = \sqrt{a + \sqrt{(a^2 + b)}} \qquad (5.4.25)$$

The arbitrary constants A, B, C, and D can be determined from the conditions at the ends of the beam. Since rotation of the cross section at the supports is prevented, we must have

$$\gamma(0) = 0, \quad \gamma(L) = 0 \qquad (5.4.26)$$

The other two boundary conditions can be obtained as follows: Since warping is unrestrained at the ends of the beam, no moments will be developed in the flanges. By setting M_f in Eq. (5.3.1) equal zero and by differentiating Eq. (5.3.5) twice, it can easily be shown that the following conditions must be satisfied.

$$\left.\frac{d^2\gamma}{dz^2}\right|_{z=0} = 0, \quad \left.\frac{d^2\gamma}{dz^2}\right|_{z=L} = 0 \qquad (5.4.27)$$

From the first conditions of Eqs. (5.4.26) and (5.4.27), we obtain

$$B = 0, \quad C = -D \qquad (5.4.28)$$

and from the second conditions of Eqs. (5.4.26) and (5.4.27), we obtain the simultaneous equations

$$A \sin mL - 2D \sinh nL = 0$$
$$Am^2 \sin mL + 2Dn^2 \sinh nL = 0 \qquad (5.4.29)$$

For a nontrivial solution, the determinant of the above equations must vanish

$$(\sin mL)(\sinh nL)(2m^2 + 2n^2) = 0 \qquad (5.4.30)$$

Since m and n are both positive nonzero quantities, and $\sinh nL$ is zero only at $nL = 0$, it follows that for a nontrivial solution we must have

$$\sin mL = 0 \qquad (5.4.31)$$

The smallest value of m satisfying Eq. (5.4.31) is

$$m = \frac{\pi}{L} \qquad (5.4.32)$$

Using Eq. (5.4.25), we have

$$-a + \sqrt{(a^2 + b)} = \left(\frac{\pi}{L}\right)^2 \qquad (5.4.33)$$

and in view of Eq. (5.4.22), we have

$$M_{ocr} = \frac{\pi}{L}\sqrt{EI_y GJ}\sqrt{(1 + W^2)} \qquad (5.4.34)$$

where

$$W = \frac{\pi}{L}\sqrt{\frac{EC_w}{GJ}} \qquad (5.4.35)$$

It should be noted that the critical moment depends not only on the quantities EI_y and GJ, but also on EC_w. In fact, the second square root in Eq. (5.4.34) represents the contribution of warping to the torsional resistance of the beam. For a rectangular or box section, C_w is negligible, so the second square root becomes one and Eq. (5.4.34) reduces to Eq. (5.4.16). For an I-section, M_{cr} will increase if the distance between the two flanges increases. This observation will become evident if one refers to Eq. (5.3.8), the warping constant C_w is proportional to the square of the depth h of the section. Thus, *if I_f remains unchanged*, an increase in h will increase C_w and, so, M_{cr}.

In developing Eq. (5.4.34), it has been assumed that the in-plane deflection has no effect on the lateral torsional buckling behavior of the beam. This assumption is justified when the flexural rigidity EI_x is much larger than the flexural rigidity EI_y so that the in-plane deflection will be negligible compared with that of the out-of-plane deflection. If both rigidities are of the same order of magnitude, the effect of bending in the vertical y-z plane may be important and should be considered in calculating M_{cr}. An approximate solution that includes the effect of in-plane deflection is given by Kirby and Nethercot[3] as

$$M_{ocr} = \frac{\pi}{L} \sqrt{\frac{EI_y GJ}{I_r}} \sqrt{(1+W^2)} \qquad (5.4.36)$$

where

$$I_r = 1 - (I_y/I_x) \qquad (5.4.37)$$

Note that if $I_y = I_x$, I_r becomes zero. Then, From Eq. (5.4.36) it can be seen that M_{cr} becomes infinity. If $I_y > I_x$, I_r becomes negative and M_{cr} becomes imaginary. So, for the cases when I_y equals or exceeds I_x, no solution exists. Thus, one can conclude that the lateral torsional buckling of beams is possible only if the cross section possesses different bending stiffnesses in the two principal planes and the applied loads act in the plane of the weak axis. As a result, lateral torsional buckling will never occur in circular cross sections or square box sections in which all the component plates have the same thicknesses.

5.5 BEAMS WITH OTHER LOADING CONDITIONS

In the preceding section, the critical moment for a beam under equal and opposite end moments has been derived. The moment is constant for the entire length of the beam, and so the resulting differential equation describing the equilibrium conditions of the beam at its slightly deformed state is linear with constant coefficients. Beams in practical situations will, of course, be subjected to a wide variety of loadings, thus producing non-uniform moment along the length of the beam. If the moment in the beam is not constant throughout, the resulting governing differential equation will have variable coefficients. For such cases, closed-form solutions are not available and recourse must be had to numerical and approximate procedures to obtain the critical loads. Some classical numerical solutions for the critical loads of members with non-uniform moment have been presented in the book by Timoshenko and Gere,[2] as well as in the papers by Massonnet,[4] Horne,[5] and Salvadori,[6] among others.

In this section, we will present a simple but effective method to take into account the effect of non-uniform moment on the critical lateral buckling loads of beams. Here, as in the beam-column case, the approach is based on the *equivalent moment concept* and the accuracy of the approach has been found to be quite sufficient for most practical cases.

5.5.1 Unequal End Moments

If a beam is subjected to end moments that are unequal in magnitude (Fig. 5.13), the moment in the beam will be a function of z. Consequently, the resulting governing differential equation will have variable coefficients. Therefore, a numerical procedure, involving the use of series or special functions, is necessary to obtain solutions. The procedure is, evidently, quite cumbersome. Fortunately, for the purpose of design, it has been demonstrated by Salvadori[6] that the effect of moment gradient on the critical moment can easily be accounted for by the use of an *equivalent moment factor* C_b. The critical moment for the beam in Fig. 5.13 can be obtained from

$$M_{cr} = C_b M_{ocr} \tag{5.5.1}$$

FIGURE 5.13 Beam subjected to unequal bending moments

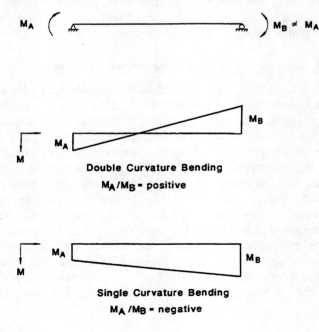

5.5 Beams with Other Loading Conditions

where
$$M_{ocr} = \text{Eq. (5.4.34)}$$

$$C_b = 1.75 + 1.05\left(\frac{M_A}{M_B}\right) + 0.3\left(\frac{M_A}{M_B}\right)^2 \leq 2.3 \qquad (5.5.2)$$

in which (M_A/M_B) is the ratio of the numerically smaller to larger end moments. Its value is positive when the beam bends in double curvature and is negative when the beam bends in single curvature.

A comparison of the theoretical value of M_{cr} with that evaluated from Eqs. (5.5.1) for various values of M_A/M_B is shown in Fig. 5.14. One can see that Eq. (5.5.1) gives a conservative and quite accurate representation of the actual critical moment. It should be noted that the concept of equivalent uniform moment used here for beams is very similar to that used earlier in Chapter 3 for beam-columns. The physical meaning of C_b here is that it represents the amount of increase in the critical uniform moment M_{ocr}, which causes lateral instability, as would the given unequal end moments also cause such instability. Since the moment ratio M_A/M_B is always between -1 and 1, it follows from Eq. (5.5.2) that C_b is always greater than unity. This means that the critical moment M_{cr} for unequal end moments will always be larger than the critical moments M_{ocr} for equal and opposite end moments. Thus, the equal and opposite end moments loading case represents the most severe loading condition for the beam.

FIGURE 5.14 Comparison of theoretical results with Eq. (5.5.1)

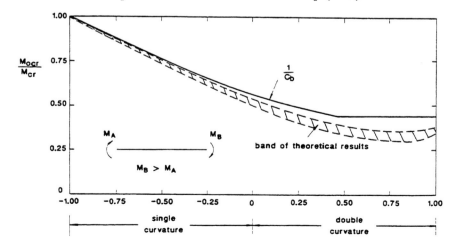

5.5.2 Central Concentrated Load

If a simply supported beam is loaded at midspan by a concentrated force, the moment diagram is bilinear, as shown in Fig. 5.15. Here, as in the case of unequal end moments, the differential equation will contain a variable coefficient.

As an illustration, consider a simply supported I-beam subjected to a concentrated force P applied at the shear center of the middle cross section (Fig. 5.16). To derive the governing differential equation, we need to relate the externally induced moments acting on the beam at its slightly deformed (buckled) configuration to its internal resistance. The procedure is facilitated by using the two coordinate systems: a fixed coordinate (x-y-z) system and a local coordinate (x'-y'-z') system as shown in Fig. 5.16. As the beam buckles laterally, vertical reactions $P/2$ and torsional reactions $Pu_m/2$, where u_m is the lateral out-of-plane displacement of the shear center of the middle cross section, will develop at the supports. By considering a cross section at a distance z from the origin, the various components of external moments acting on that cross section with respect to the x-y-z coordinate are, using the right-handed screw rule for the moment vector,

$$(M_x)_{ext} = \frac{P}{2}\left(\frac{L}{2} - z\right) \tag{5.5.3}$$

$$(M_y)_{ext} = 0 \tag{5.5.4}$$

$$(M_z)_{ext} = -\frac{P}{2}(u_m - u) \tag{5.5.5}$$

Referring to Fig. 5.17, we see that the components of external moments acting on the cross section of the deformed beam with respect

FIGURE 5.15 Simply supported beam loaded at midspan

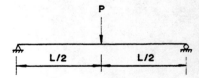

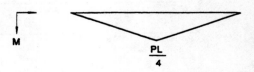

In-plane Moment Diagram

5.5 Beams with Other Loading Conditions

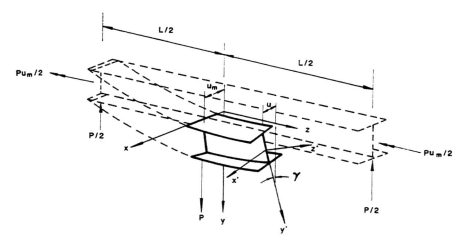

FIGURE 5.16 Lateral buckling of a simply supported i-beam loaded at midspan

to the x'-y'-z' coordinate are

$$(M_{x'})_{ext} \approx (M_x)_{ext} - \frac{du}{dz}(M_z)_{ext} = \frac{P}{2}\left(\frac{L}{2} - z\right) + \frac{du}{dz}\frac{P}{2}(u_m - u) \quad (5.5.6)$$

$$(M_{y'})_{ext} \approx -\gamma(M_x)_{ext} - \frac{dv}{dz}(M_z)_{ext}$$

$$= -\gamma\frac{P}{2}\left(\frac{L}{2} - z\right) + \frac{dv}{dz}\frac{P}{2}(u_m - u) \quad (5.5.7)$$

$$(M_{z'})_{ext} \approx (M_z)_{ext} + \frac{du}{dz}(M_x)_{ext}$$

$$= -\frac{P}{2}(u_m - u) + \frac{du}{dz}\frac{P}{2}\left(\frac{L}{2} - z\right) \quad (5.5.8)$$

The minus sign for the terms dv/dz and du/dz in the figure accounts for the fact that the slopes dv/dz and du/dz are negative with a positive z (see Fig. 5.16).

The internal resisting moments are

$$(M_{x'})_{int} = -EI_x\frac{d^2v}{dz^2} \quad (5.5.9)$$

$$(M_{y'})_{int} = EI_y\frac{d^2u}{dz^2} \quad (5.5.10)$$

$$(M_{z'})_{int} = GJ\frac{d\gamma}{dz} - EC_w\frac{d^3\gamma}{dz^3} \quad (5.5.11)$$

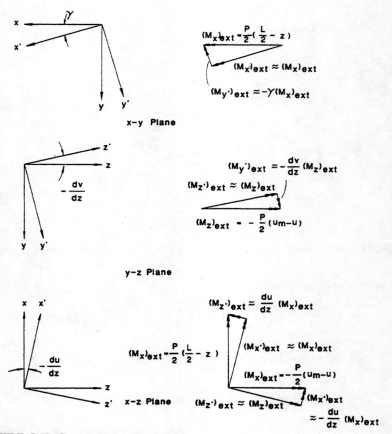

FIGURE 5.17 Components of moments

The minus sign in Eq. (5.5.9) indicates that the positive moment $(M_{x'})_{int}$ produces a negative curvature d^2v/dz^2, based on the right-handed screw rule for moment.

By equating the corresponding external and internal moments and neglecting the higher order terms, we obtain the following equilibrium equations

$$EI_x \frac{d^2v}{dz^2} + \frac{P}{2}\left(\frac{L}{2} - z\right) = 0 \qquad (5.5.12)$$

$$EI_y \frac{d^2u}{dz^2} + \gamma \frac{P}{2}\left(\frac{L}{2} - z\right) = 0 \qquad (5.5.13)$$

$$GJ \frac{d\gamma}{dz} - EC_w \frac{d^3\gamma}{dz^3} + \frac{P}{2}(u_m - u) - \frac{du}{dz} \frac{P}{2}\left(\frac{L}{2} - z\right) = 0 \qquad (5.5.14)$$

5.5 Beams with Other Loading Conditions

Note that the second term in Eq. (5.5.6) and (5.5.7) is neglected in writing Eq. (5.5.12) and (5.5.13), because the quantities (du/dz), (dv/dz), and $(u_m - u)$ are all small. The reader should recognize that Eq. (5.5.12), which describes the in-plane bending behavior of the beam, is uncoupled with the other two equations. Therefore, it is not important in the present buckling analysis. The lateral torsional buckling behavior of the beam is described by Eq. (5.5.13) and Eq. (5.5.14). By eliminating u from Eqs. (5.5.13) and (5.5.14) and noting that $du_m/dz = 0$, we can write a simple differential equation as

$$EC_w \frac{d^4\gamma}{dz^4} - GJ \frac{d^2\gamma}{dz^2} + \frac{1}{EI_y}\left[\frac{P}{2}\left(\frac{L}{2}-z\right)\right]^2 \gamma = 0 \qquad (5.5.15)$$

This differential equation has a variable coefficient in its third term. The solution for this differential equation can be obtained[2] by the method of infinite series. The results are plotted as solid lines in Fig. 5.18. The curves correspond to the cases when the load acts on the upper flange, at the shear center, and on the lower flange of the cross section, respectively.

The case where the load acts on the upper flange is the most detrimental, because of the increase in the torque arm as the beam buckles laterally. On the other hand, the least detrimental case is when the load acts on the lower flange; this is because of the decrease in the torque arm as the beam buckles laterally. These observations can be explained with reference to Fig. 5.1. If the load acts on the upper flange, Eq. (5.5.5) will become

$$(M_z)_{\text{ext}} = -\frac{P}{2}\left(u_m + \frac{\gamma_m h}{2} - u\right) \qquad (5.5.16)$$

whereas, if the load acts on the lower flange, Eq. (5.5.5) will become

$$(M_z)_{\text{ext}} = -\frac{P}{2}\left(u_m - \frac{\gamma_m h}{2} - u\right) \qquad (5.5.17)$$

where u_m and γ_m are the out-of-plane lateral displacement and twist of the cross section at the beam's midspan, respectively.

The term $\gamma_m h/2$ (or $-\gamma_m h/2$) represents the amount of increase (or decrease) in the torque arm of the applied load causing an increase (or decrease) in the external moment $(M_z)_{\text{ext}}$. Evidently, if $(M_z)_{\text{ext}}$ is larger, P_{cr} will be smaller and vice versa. For the purpose of design, it is convenient to approximate the theoretical values of P_{cr} by Eq. (5.5.1).

$$M_{cr} = \frac{P_{cr} L}{4} = C_b M_{ocr} \qquad (5.5.18)$$

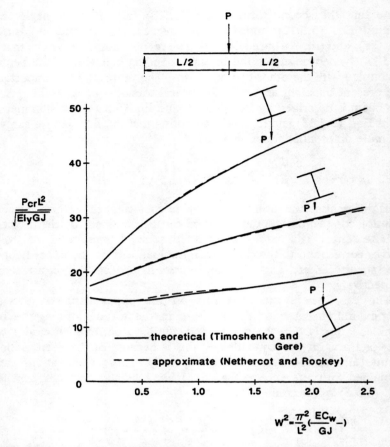

FIGURE 5.18 Comparison of theoretical and approximate solutions

with

$$C_b = \begin{cases} AB & \text{for load at bottom flange} \\ A & \text{for load at shear center} \\ A/B & \text{for load at top flange} \end{cases} \quad (5.5.19)$$

The values A and B are given by Nethercot and Rockey[7] as

$$A = 1.35 \quad (5.5.20)$$

$$B = 1 + 0.649W - 0.180W^2 \quad (5.5.21)$$

in which $W = (\pi/L)\sqrt{(EC_w/GJ)}$.

The approximate solutions for P_{cr} using Eqs. (5.5.18) to (5.5.21) are plotted as dotted lines on the same figure. It can be seen that the approximate solutions give an excellent representation of the theoretical exact solutions.

5.5.3 Other Loading Conditions

The effect of the distribution of load along the unbraced length of a simply supported beam on its elastic lateral torsional buckling strength has been investigated numerically by a number of researchers. The results are discussed in various books,[2,8,9] and papers.[10-15] For simplicity, approximate solutions in the form of Eq. (5.5.1) are often used to obtain the critical loads. The approximate solutions for some commonly encountered loading cases with the loads applied at the shear center of the cross section are summarized in Table 5.2a. By using the expressions for M_{cr} in the third column and the value of C_b in the fourth column, together with M_{ocr} given in Eq. (5.4.34), the corresponding approximate values for the critical loads can easily be computed from Eq. (5.5.1).

For loadings whose moment diagrams do not resemble any of those given in Table 5.2a, an empirical formula given by Kirby and Nethercot[3] for C_b can be used:

$$C_b = \frac{12}{3(M_1/M_{max}) + 4(M_2/M_{max}) + 3(M_3/M_{max}) + 2} \quad (5.5.22)$$

where M_1, M_2, and M_3 are the moments at the quarter point, midpoint, and three-quarter point of the beam, respectively, and M_{max} is the maximum moment of the beam, as shown in Table 5.2b (Problem 5.4).

If the location of the load is not at the shear center, the values of the critical loads will be different. For the two load cases shown in Table 5.3, Nethercot and Rockey[7] and Nethercot[16] have proposed expressions for C_b to be used in Eq. (5.5.1) to give approximate values for the critical loads. Figure 5.19 shows a comparison of the theoretical critical load obtained by Timoshenko and Gere[2] for the case of uniformly distributed load with the approximate solutions proposed by Nethercot and Rockey[7] (Table 5.3). Good agreement between the two solutions is observed.

5.6 BEAMS WITH OTHER SUPPORT CONDITIONS

The discussion heretofore pertains only to beams that are torsionally simply supported. That is, the ends of the beams are free to rotate and warp about the weak axis, but are restrained against rotation about the centroidal axis. A change in support conditions will undoubtedly have a pronounced effect on the resistance of the beam to lateral torsional buckling. In this section we shall examine the effects of support

Table 5.2 Values of C_b for Different Loading Cases (All Loads are Applied at Shear Center of the Cross Section)

(a)

Loadings	Bending Moment Diagrams	M_{cr}	C_b
M, L, M (equal end moments, single curvature)	rectangle	M_{cr}	1.00
M, L (one end moment)	triangle	M_{cr}	1.75
M, L, M (equal end moments, reverse curvature)	two triangles	M_{cr}	2.30
P at midspan, $L/2$, $L/2$	triangle	$\dfrac{P_{cr} L}{4}$	1.35
w uniform over L	parabola	$\dfrac{w_{cr} L^2}{8}$	1.13
P, P at $L/4$, $L/2$, $L/4$	trapezoid	$\dfrac{P_{cr} L}{4}$	1.04
P at $L/4$, $3L/4$	triangle	$\dfrac{3 P_{cr} L}{16}$	1.44

(b)

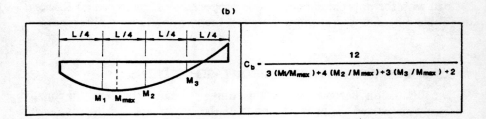

$$C_b = \dfrac{12}{3 (M_1/M_{max}) + 4 (M_2/M_{max}) + 3 (M_3/M_{max}) + 2}$$

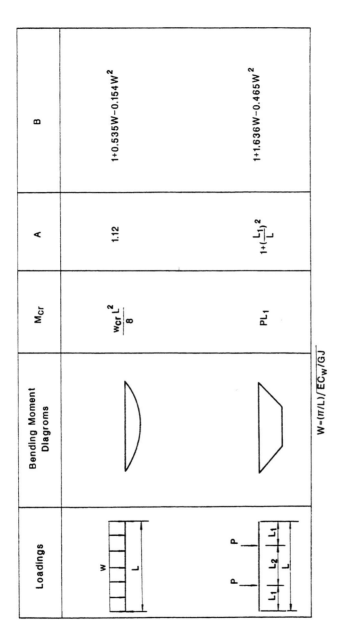

Loadings	Bending Moment Diagrams	M_{cr}	A	B
w, L	(parabolic)	$\dfrac{w_{cr}L^2}{8}$	1.12	$1 + 0.535W - 0.154W^2$
P, P, L_1, L_2, L_1	(trapezoidal)	PL_1	$1 + \left(\dfrac{L_1}{L}\right)^2$	$1 + 1.636W - 0.465W^2$

$W = (\pi/L)\sqrt{EC_w/GJ}$

$$C_b = \begin{cases} AB & \text{for load at bottom flange} \\ A & \text{for load at shear center} \\ A/B & \text{for load at top flange} \end{cases}$$

Table 5.3 Values of C_b for Distributed and Concentrated Loads (Loads Applied at Bottom Flange, Shear Center, or Top Flange)

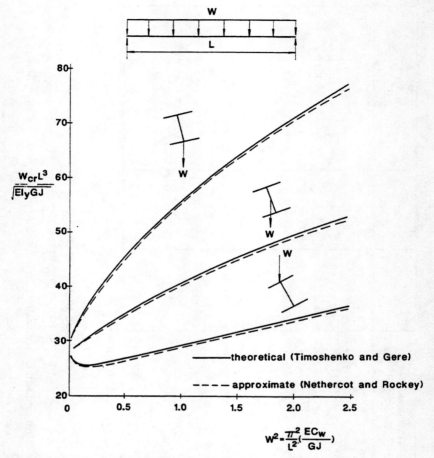

FIGURE 5.19 Comparison of theoretical and approximate solutions

conditions on the critical loads and lateral buckling behavior of beams, and, in particular, the use of the effective length concept to account for the support conditions.

5.6.1 Cantilever Beams

The critical lateral buckling loads for cantilevers are different from that of simply supported beams because of the obvious difference in boundary conditions at the supports. The elastic buckling load for a cantilever under a uniform moment caused by an end moment M_0 applied at the free end can be obtained directly from the solution of the simply

5.6 Beams with Other Support Conditions

supported beam by imagining the beam to be consisted of two cantilevers of equal length joined together at the fixed ends. Thus, the critical moment for the cantilever beam can be obtained from Eq. (5.4.34) by replacing L by $2L$

$$M_{cr} = \left(\frac{\pi}{2L}\right)(\sqrt{EI_y GJ})\sqrt{\left(1 + \frac{\pi^2 EC_w}{(2L)^2 GJ}\right)} \qquad (5.6.1)$$

For other loading conditions, recourse must be made to numerical procedures to obtain solutions.[17,18] Figures 5.20 and 5.21 show the results for two load cases: cantilever beam with a concentrated load at the free end and cantilever beam with a uniform distributed load. For both of these load cases, the figures present the critical loads corresponding to

FIGURE 5.20 Critical loads of a cantilever subjected to a concentrated force at the free end

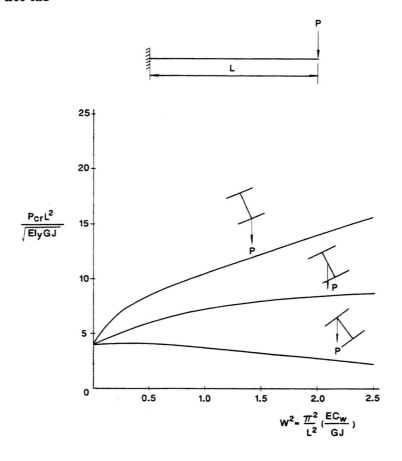

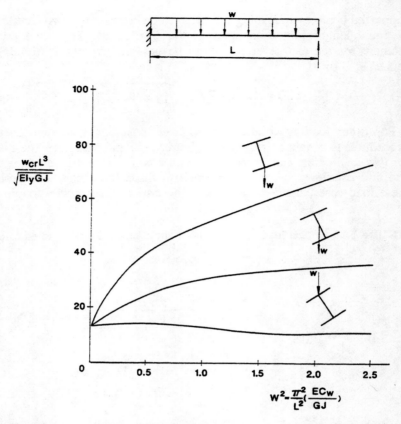

FIGURE 5.21 Critical loads of a cantilever subjected to uniform distributed load

loading on bottom flange, at the shear center, and on the top flange. These plots are applicable to cantilever beams for which the root or the fixed end is completely fixed against lateral displacement and warping while the tip or the free end is completely free. For other support conditions, Nethercot[16] has shown that the equation

$$M_{cr} = \left(\frac{\pi}{KL}\right)\sqrt{EI_y GJ}\sqrt{\left(1 + \frac{\pi^2 EC_w}{(KL)^2 GJ}\right)} \qquad (5.6.2)$$

gives a conservative estimate of M_{cr} for most applications. In Eq. (5.6.2), K is the *effective length* factor of the beam. The values of K for various restraint conditions at the root and at the tip of the cantilever are given in Table 5.4. The table is applicable to both end load and uniformly distributed load cases. Equation (5.6.2) is also applicable to other support conditions so long as K is properly defined, as shown in the following sections.

5.6 Beams with Other Support Conditions

Table 5.4 Effective Length Factors for Cantilevers with Various End Conditions (Adapted from Ref. 16)

Restraint Conditions		Effective Length	
At root	At tip	Top flange loading	All other cases
(built-in)	(free)	1.4L	0.8L
	(lateral restraint)	1.4L	0.7L
	(torsional restraint)	0.6L	0.6L
(lateral restraint at top)	(free)	2.5L	1.0L
	(lateral restraint)	2.5L	0.9L
	(torsional restraint)	1.5L	0.8L
(torsional restraint)	(free)	7.5L	3.0L
	(lateral restraint)	7.5L	2.7L
	(torsional restraint)	4.5L	2.4L

5.6.2 Fixed-Ended Beams

If the ends of the beams are fixed against lateral displacement and warping but free to rotate about the strong axis (Fig. 5.22), the boundary conditions at the ends for lateral bending are

$$u|_{z=0} = u|_{z=L} = \frac{du}{dz}\bigg|_{z=0} = \frac{du}{dz}\bigg|_{z=L} = 0 \tag{5.6.3}$$

and for twisting and warping

$$\gamma|_{z=0} = \gamma|_{z=L} = \frac{d\gamma}{dz}\bigg|_{z=0} = \frac{d\gamma}{dz}\bigg|_{z=L} = 0 \tag{5.6.4}$$

If the beam is under a uniform moment, the differential equation [Eq. (5.4.23)] and the general solution [Eq. (5.4.24)] for the simply supported beam under a uniform moment, which were presented in Section 5.4, are still applicable. By using the boundary conditions in Eq. (5.6.4), it can be

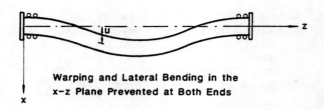

Warping and Lateral Bending in the x–z Plane Prevented at Both Ends

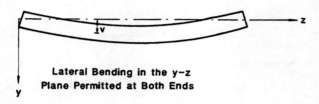

Lateral Bending in the y–z Plane Permitted at Both Ends

FIGURE 5.22 Fixed-end beams

shown that the characteristic equation is

$$\left(\frac{m^2 - n^2}{2mn}\right) \sin mL \sinh nL + \cos mL \cosh nL - 1 = 0 \qquad (5.6.5)$$

where m, n are defined in Eq. (5.4.25).

Unlike the characteristic equation for the simply supported beam, for which a solution is readily obtained [Eq. (5.4.31)], the solution for Eq. (5.6.5) can only be obtained by trial and error. However, an easier way to obtain the critical moment for this case is to realize that the inflection points occur at a distance $L/4$ from the ends. Thus, the critical value for the moment can be obtained by simply replacing L in Eq. (5.4.34) by $L/2$, giving

$$M_{cr} = \frac{\pi}{L/2} \sqrt{EI_y GJ} \sqrt{\left(1 + \frac{\pi^2 EC_w}{(L/2)^2 GJ}\right)} \qquad (5.6.6)$$

Comparing Eq. (5.6.6) with Eq. (5.6.2) gives $K = 1/2$ for this fixed-ended beam case under a uniform moment. Figure 5.23 shows a plot of Eq. (5.6.6). Also shown in the figure is the critical moment for the corresponding simply supported beam. It can be seen that the critical moment for a fixed-ended beam is considerably higher than that of a simply supported beam.

For other types of loadings that produce non-uniform distribution of moments along the length of the beam, the differential equations will

5.6 Beams with Other Support Conditions

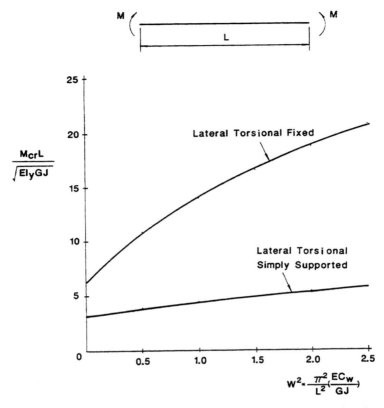

FIGURE 5.23 Comparison of critical moments for a lateral torsional simply supported and lateral torsional fixed beam

have variable coefficients and recourse to numerical procedures is inevitable. If the fixed-ended beam is subjected to a concentrated load at midspan or if it is subjected to a uniformly distributed load along the entire unbraced length, Nethercot and Rockey[7] have presented the following equation for the critical moments

$$M_{cr} = C_{bs}M_{ocr} \tag{5.6.7}$$

where

$$M_{cr} = \begin{cases} \dfrac{P_{cr}L}{4} & \text{for the concentrated load case} \\ \dfrac{w_{cr}L^2}{8} & \text{for the uniformly distributed load case} \end{cases} \tag{5.6.8}$$

Table 5.5 Expressions of A and B for a Fixed-End Beam $[W = (\pi/L)\sqrt{EC_w/GJ}]$

Load Case	A	B
uniform load w over length L	$1.643+1.771W-0.405W^2$	$1+0.625W-0.339W^2$
point load P at midspan ($L/2$, $L/2$)	$1.916+1.851W-0.424W^2$	$1+0.923W-0.466W^2$

and

$$C_{bs} = \begin{cases} AB & \text{for bottom flange loading} \\ A & \text{for shear center loading} \\ A/B & \text{for upper flange loading} \end{cases} \quad (5.6.9)$$

The expressions for A and B are given in Table 5.5 and M_{ocr} is given by Eq. (5.4.34). It should be mentioned that C_{bs} used in Eq. (5.6.7) is different from C_b used earlier in Eq. (5.5.1). The term C_b only accounts for the effect of moment gradient on the critical lateral buckling loads, whereas the term C_{bs} accounts for both the effect of moment gradient and end conditions of the beam.

The effective length KL for the beams shown in Table 5.5 can be obtained by equating Eqs. (5.6.7) with (5.6.2) and solve for the effective length factor K

$$K = \frac{\pi\sqrt{EI_y GJ}}{\sqrt{2}L(C_{bs}M_{ocr})}\left\{1 + \sqrt{\left(1 + \frac{4(C_{bs}M_{ocr})^2}{EI_y GJ}\frac{EC_w}{GJ}\right)}\right\}^{1/2} \quad (5.6.10)$$

Note that the effective length factor depends on a number of parameters. These include the unbraced length L, the material properties E and G, the cross-section geometry C_w and J, the types of loadings, and the location of the load with respect to the shear center of the cross section. Since Eq. (5.6.10) is rather cumbersome to use for practical purpose, it is common to use a K-factor of unity in design.

If the restraint conditions are different with respect to lateral bending and twisting, the effective length factor of the beam that corresponds to lateral bending K_b and the effective length factor that corresponds to twisting K_t will be different. In such cases, for beams under a *uniform*

5.6 Beams with Other Support Conditions

Table 5.6 Effective Length Factors for Beams Under Uniform Moment with Various Boundary Conditions (Adapted from Ref. 19)

Boundary Conditions		K_b	K_t
$z = 0$	$z = L$		
$u = u'' = \gamma = \gamma'' = 0$	$u = u'' = \gamma = \gamma'' = 0$	1.000	1.000
$u = u'' = \gamma = \gamma'' = 0$	$u = u'' = \gamma = \gamma' = 0$	0.904	0.693
$u = u'' = \gamma = \gamma'' = 0$	$u = u' = \gamma = \gamma'' = 0$	0.626	1.000
$u = u'' = \gamma = \gamma'' = 0$	$u = u' = \gamma = \gamma' = 0$	0.693	0.693
$u = u'' = \gamma = \gamma' = 0$	$u = u'' = \gamma = \gamma' = 0$	0.883	0.492
$u = u' = \gamma = \gamma'' = 0$	$u = u' = \gamma = \gamma' = 0$	0.431	0.693
$u = u' = \gamma = \gamma' = 0$	$u = u' = \gamma = \gamma' = 0$	0.492	0.492
$u = u' = \gamma = \gamma'' = 0$	$u = u' = \gamma = \gamma'' = 0$	0.434	1.000
$u = u' = \gamma = \gamma'' = 0$	$u = u'' = \gamma = \gamma' = 0$	0.606	0.492

moment, the critical moment is expressed as

$$M_{ocr} = \left(\frac{\pi}{K_b L}\right)\sqrt{EI_y GJ}\sqrt{\left(1 + \frac{\pi^2 EC_w}{(K_t L)^2 GJ}\right)} \quad (5.6.11)$$

Values for K_b and K_t for many boundary conditions with respect to lateral bending and twisting respectively are given by Vlasov[19] and are shown in Table 5.6. In the table, a prime represents differentiation with respect to z. For simplicity, the values for K_b and K_t can be taken as the following:

1. 1.00 if both ends are simply supported
2. 0.70 if one end is simply supported and the other end is fixed
3. 0.50 if both ends are fixed

However, if the degree of end fixity is questionable, a conservative measure is to use one for the effective length factors.

5.6.3 Other End Conditions

If the ends of the beams are prevented from warping but are unrestrained in bending about the weak axis (Fig. 5.24), the out-of-plane conditions for lateral bending will be

$$u|_{z=0} = u|_{z=L} = \frac{d^2 u}{dz^2}\bigg|_{z=0} = \frac{d^2 u}{dz^2}\bigg|_{z=L} = 0 \quad (5.6.12)$$

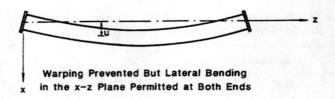
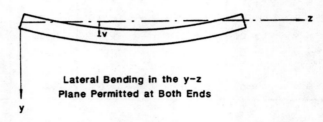

FIGURE 5.24 Warping prevented—lateral bending–permitted beam

and for twisting and warping

$$\gamma|_{z=0} = \gamma|_{z=L} = \frac{d\gamma}{dz}\bigg|_{z=0} = \frac{d\gamma}{dz}\bigg|_{z=L} = 0 \qquad (5.6.13)$$

The critical loads for this case are given by Nethercot and Rockey[7] in the form of Eqs. (5.6.7) to (5.6.9) for the two load cases given in Table 5.7. The values for A and B to be used in Eq. (5.6.9) are also given in the same table.

Table 5.7 Expressions of A and B for a Warping Prevented but Lateral Bending Permitted Beam $[W = (\pi/L)\sqrt{EC_w/GJ}]$

Load Case	A	B
w, L (uniformly distributed)	$1.2+0.402W+0.416W^2$	$1+0.571W-0.225W^2$
P at L/2	$1.43+0.463W+0.485W^2$	$1+0.619W-0.317W^2$

5.6 Beams with Other Support Conditions

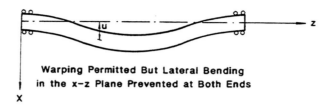

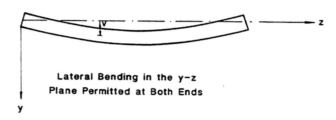

FIGURE 5.25 Warping permitted—lateral bending–prevented beam

If the ends of the beams are prevented from bending about the weak axis but are free to warp (Fig. 5.25), the out-of-plane boundary conditions for lateral bending will be

$$u|_{z=0} = u|_{z=L} = \frac{du}{dz}\bigg|_{z=0} = \frac{du}{dz}\bigg|_{z=L} = 0 \qquad (5.6.14)$$

and for twisting and warping

$$\gamma|_{z=0} = \gamma|_{z=L} = \frac{d^2\gamma}{dz^2}\bigg|_{z=0} = \frac{d^2\gamma}{dz^2}\bigg|_{z=L} = 0 \qquad (5.6.15)$$

Equations (5.6.7) to (5.6.9) are also applicable.[7] Table 5.8 gives the values of A and B in accordance with Eq. (5.6.9) for the two load cases shown.

An alternative approach to calculate the elastic critical loads for beams of doubly symmetric sections with out-of-plane bending and torsional simply supported as well as fixed-ended conditions are presented by Clark and Hill.[20] In their approach, the critical moment is expressed in the simple form

$$M_{cr} = \frac{C_4}{L}\sqrt{EI_y GJ} \qquad (5.6.16)$$

Table 5.8 Expressions of A and B for a Warping Permitted but Lateral Bending Prevented Beam [$W = (\pi/L)\sqrt{EC_w/GJ}$]

Load Case	A	B
w, L (uniform load)	$1.9 + 0.006W - 0.120W^2$	$1 + 0.806W - 0.100W^2$
P at L/2	$2.0 + 0.304W - 0.074W^2$	$1 + 1.047W - 0.207W^2$

where

$$C_4 = C_1 \frac{\pi}{K}\left\{\sqrt{\left[1 + \left(\frac{\pi}{KL}\right)^2\left(\frac{EC_w}{GJ}\right)(C_2^2 + 1)\right]} \pm C_2\left(\frac{\pi}{KL}\right)\sqrt{\frac{EC_w}{GJ}}\right\} \quad (5.6.17)$$

The coefficient C_1 accounts for the type of load and support conditions and the coefficient C_2 accounts for the location of the load vertically with respect to the shear center of the cross section. The plus sign in Eq. (5.6.17) is used if the load is applied at the bottom flange and the minus is used if the load is applied on the top flange. C_2 is zero if the load is applied at the shear center or if the load is an end moment loading.

Values for C_1 and C_2 for various load cases are tabulated in Table 5.9. Note that for the end moment load cases, C_1 is equal to C_b and is obtained from Eq. (5.5.2).

5.6.4 Continuous Beam

In the previous sections it has been assumed that the beams were supported laterally only at the ends. When a simply supported I-beam subjected to a uniform moment has an additional lateral support at its midspan, the lateral buckling mode will be a complete sine wave. The critical buckling moment can therefore be obtained directly from Eq. (5.4.31) by using the second lowest value of m in the solution, that is,

$$m = \frac{2\pi}{L} \quad (5.6.18)$$

giving

$$M_{ocr} = \frac{\pi}{(L/2)}\sqrt{EI_y GJ}\sqrt{\left(1 + \frac{\pi^2 EC_w}{(L/2)^2 GJ}\right)} \quad (5.6.19)$$

5.6 Beams with Other Support Conditions

Table 5.9 Values of C_1 and C_2 in Eq. (5.6.16) (Adapted from Ref. 20)

Case No.	Loading	Bending Moment Diagram	Condition of Restraint Against Rotation about Vertical Axis at Ends	Equiv. Length Factor, K	Value of Coefficients C_1^*	C_2
Beams Restrained Against Lateral Displacement at Both Ends of Span						
1.	$M(\rightleftarrows)M$, L	M ...	Simple support / Fixed	1.0 / 0.5	1.0 / 1.0	0 / 0
2.	$M(\rightleftarrows)M/2$	M ... M/2	Simple support / Fixed	1.0 / 0.5	1.3 / 1.3	0 / 0
3.	$M(\rightleftarrows)$	M ...	Simple support / Fixed	1.0 / 0.5	1.8 / 1.8	0 / 0
4.	$M(\rightleftarrows)M/2$	M ... M/2	Simple support / Fixed	1.0 / 0.5	2.4 / 2.3	0 / 0
5.	$M(\rightleftarrows)M$	M ... M	Simple support / Fixed	1.0 / 0.5	2.6 / 2.3	0 / 0
6.	w (udl)	$\frac{wL}{8}$ ...	Simple support / Fixed	1.0 / 0.5	1.1 / 1.0	0.45 / 0.29
7.	w (udl)	$\frac{wL}{12}$ ... $M_{cr} = wL/24$	Simple support / Fixed	1.0 / 0.5	1.3 / 0.9	1.55 / 0.82
8.	P at mid	$\frac{PL}{4}$ ...	Simple support / Fixed	1.0 / 0.5	1.4 / 1.1	0.55 / 0.42
9.	P	$\frac{PL}{8}$ ... PL/8	Simple support / Fixed	1.0 / 0.5	1.7 / 1.0	1.42 / 0.84
10.	P/2, P/2 at L/4	$\frac{PL}{8}$ ...	Simple support	1.0	1.0	0.42
Cantilever Beams						
11.	P	PL ...	Warping restrained at supported end	1.0	1.3	0.64
12.	w	$\frac{wL}{2}$ ...	Warping restrained at supported end	1.0	2.1	

*Approximate minimum values for C_1 are given. Range of values are given in Ref. 5.20

For other types of loadings, such as a concentrated load at midspan or a uniformly distributed load over the entire span of the beam, numerical solutions for the critical loads are available.[2] However, for simplicity, Eqs. (5.6.7) to (5.6.9) can be used again.[7] Table 5.10 shows the applicable values of A and B for these two loading cases.

For a continuous beam with more than two spans, partial end restraint will develop between adjacent spans. The determination of the lateral

Table 5.10 Expressions of A and B for a Two-Span Continuous Beam with Lateral Support at Center and Restraint Equal at Both Ends $[W = (\pi/L)\sqrt{EC_w/GJ}]$

Load Case	A	B
w, L/2, L/2	$2.093 + 3.117W - 0.947W^2$	$1.073 + 0.044W$
P, L/2, L/2	$2.95 + 4.070W - 1.143W^2$	1

Top View

buckling load for this type of continuous beam is beyond the scope of this book. In general, the lateral buckling load depends on the relative stiffness of the segments, the type and relative magnitude of the loads on the beam, and the type of bracing or constraint used for the intermediate supports. The readers are referred to the papers by Salvadori[21] and Hartmann[22] and the books by Trahair[23] and by Chen and Atsuta[24] for a detailed discussion of the subject. For design purpose, it is common to evaluate the critical load for each span separately by assuming the ends of the span are simply supported. The lowest value of the critical loads obtained for each individual span will furnish a conservative estimate of the critical load for the continuous beam.

5.7 INITIALLY CROOKED BEAMS

In deriving the differential equation for the lateral buckling of beam under a uniform moment in Section 5.4, one assumption we used is that the beam is geometrically perfect. Beams in reality are seldom perfect, the presence of initial curvatures and twists will cause them to bend and twist at the beginning of loading. In this section, we shall examine the lateral buckling behavior of such beams.

Consider a simply supported beam of narrow rectangular cross section subjected to an equal and opposite end moment M_0. The beam has an initial out-of-straightness and twist given by

$$u_0 = \delta_0 \sin \frac{\pi z}{L} \qquad (5.7.1)$$

$$\gamma_0 = \theta_0 \sin \frac{\pi z}{L} \qquad (5.7.2)$$

5.7 Initially Crooked Beams

where δ_0 and θ_0 are the initial out-of-straightness and twist at the midspan of the beam, the out-of-plane bending and torsional equations [Eqs. (5.4.8) and (5.4.9)] become

$$EI_y \frac{d^2u}{dz^2} + (\gamma + \gamma_0)M_0 = 0 \tag{5.7.3}$$

$$GJ \frac{d\gamma}{dz} - \left(\frac{du}{dz} + \frac{du_0}{dz}\right)M_0 = 0 \tag{5.7.4}$$

Differentiating Eq. (5.7.4) once and substituting the result into Eq. (5.7.3), we obtain

$$EI_y GJ \frac{d^2\gamma}{dz^2} + \gamma M_0^2 = EI_y M_0 \frac{d^2 u_0}{dz^2} - \gamma_0 M_0^2 \tag{5.7.5}$$

In view of Eqs. (5.7.1) and (5.7.2), we can write Eq. (5.7.5) as

$$EI_y GJ \frac{d^2\gamma}{dz^2} + \gamma M_0^2 = -EI_y M_0 \delta_0 \left(\frac{\pi}{L}\right)^2 \sin \frac{\pi z}{L}$$

$$- \theta_0 M_0^2 \sin \frac{\pi z}{L} \tag{5.7.6}$$

If δ_0 and θ_0 are related by

$$\frac{\delta_0}{\theta_0} = \frac{M_{ocr}}{\pi^2 EI_y / L^2} \tag{5.7.7}$$

where M_{ocr} is given by Eq. (5.4.16), Eq. (5.7.6) can be written as

$$EI_y GJ \frac{d^2\gamma}{dz^2} + \gamma M_0^2 = -\theta_0 M_0^2 \left(\frac{M_{ocr}}{M_0} + 1\right) \sin \frac{\pi z}{L} \tag{5.7.8}$$

or

$$\frac{d^2\gamma}{dz^2} + k^2 \gamma = -k^2 \theta_0 \left(\frac{M_{ocr}}{M_0} + 1\right) \sin \frac{\pi z}{L} \tag{5.7.9}$$

in which $k^2 = M_0^2 / EI_y GJ$.

The general solution of Eq. (5.7.9) consists of an homogeneous solution and a particular solution. The homogeneous solution is given by Eq. (5.4.12) and the particular solution is given by

$$\gamma_p = C \sin \frac{\pi z}{L} + D \cos \frac{\pi z}{L} \tag{5.7.10}$$

Substituting Eq. (5.7.10) into Eq. (5.7.9) and combining terms gives

$$\left\{ C\left[k^2 - \left(\frac{\pi}{L}\right)^2\right] + k^2 \theta_0 \left(\frac{M_{ocr}}{M_0} + 1\right) \right\} \sin \frac{\pi z}{L}$$

$$+ \left\{ D\left[k^2 - \left(\frac{\pi}{L}\right)^2\right] \right\} \cos \frac{\pi z}{L} = 0 \tag{5.7.11}$$

Equation (5.7.11) can be satisfied for all values of z only if both the coefficients of the sine and cosine terms vanish. That is

$$C = \frac{-k^2 \theta_0 \left(\dfrac{M_{ocr}}{M_0} + 1 \right)}{k^2 - \left(\dfrac{\pi}{L} \right)^2} \quad (5.7.12)$$

and either

$$D = 0 \quad (5.7.13)$$

or

$$k^2 = \left(\frac{\pi}{L} \right)^2 \quad (5.7.14)$$

However, if Eq. (5.7.14) is valid, then the solution reduces to Eq. (5.4.16); which is not the case to be investigated here. So, Eq. (5.7.13) must be valid. Thus the particular solution reduces to

$$\gamma_p = \left[\frac{-k^2 \theta_0 \left(\dfrac{M_{ocr}}{M_0} + 1 \right)}{k^2 - \left(\dfrac{\pi}{L} \right)^2} \right] \sin \frac{\pi z}{L} \quad (5.7.15)$$

Realizing that $k^2 = M_0^2 / EI_y GJ$ and using $M_{ocr} = (\pi/L)\sqrt{EI_y GJ}$, Eq. (5.7.15) can be simplified to

$$\gamma_p = \left[\frac{\theta_0 M_0 / M_{ocr}}{1 - (M_0 / M_{ocr})} \right] \sin \frac{\pi z}{L} \quad (5.7.16)$$

Combining the homogeneous solution given in Eq. (5.4.12) and the particular solution given in Eq. (5.7.16), the general solution to the differential equation (5.7.9) can be written as

$$\gamma = A \sin kz + B \cos kz + \left[\frac{\theta_0 M_0 / M_{ocr}}{1 - (M_0 / M_{ocr})} \right] \sin \frac{\pi z}{L} \quad (5.7.17)$$

Using the first boundary condition of Eq. (5.4.13), we obtain

$$B = 0$$

and the second boundary condition of Eq. (5.4.13) gives

$$A \sin kL = 0$$

Thus, either A or $\sin kL$ must vanish. However, if we let $\sin kL = 0$, we are limiting the solution to Eq. (5.4.16); therefore, we must have $A = 0$. Hence, with $A = B = 0$, Eq. (5.7.17) becomes

$$\gamma = \left[\frac{\theta_0 M_0 / M_{ocr}}{1 - (M_0 / M_{ocr})} \right] \sin \frac{\pi z}{L} \quad (5.7.18)$$

Backsubstituting Eq. (5.7.18) into Eq. (5.7.3) or Eq. (5.7.4), and making use of Eqs. (5.7.1), (5.7.2), and (5.7.7), it can be shown that

$$u = \left[\frac{\delta_0 M_0/M_{ocr}}{1-(M_0/M_{ocr})}\right]\sin\frac{\pi z}{L} \quad (5.7.19)$$

The total twist and out-of-plane deflection are obtained by adding Eqs. (5.7.18) and (5.7.19) to Eqs. (5.7.2) and (5.7.1), respectively,

$$\gamma_{total} = \gamma_0 + \gamma = \left[\frac{1}{1-(M_0/M_{ocr})}\right]\theta_0 \sin\frac{\pi z}{L} = A_F \gamma_0 \quad (5.7.20)$$

and

$$u_{total} = u_0 + u = \left[\frac{1}{1-(M_0/M_{ocr})}\right]\delta_0 \sin\frac{\pi z}{L} = A_F u_0 \quad (5.7.21)$$

where

$$A_F = \frac{1}{1-(M_0/M_{ocr})} \quad (5.7.22)$$

is the *amplification factor*.

The reader should note the similarity of Eq. (5.7.22) and Eq. (2.6.20). An expression similar to that of Eq. (5.7.22) can be derived for a geometrically imperfect I-beam under a uniform moment, provided, first, that the initial out-of-straightness and twist are given by Eqs. (5.7.1) and (5.7.2), respectively, and, second, that they are related by Eq. (5.7.7).

Figure 5.26 shows a plot of the load-deformation relationship of an initially crooked beam. Unlike a geometrically perfect beam, in which bifurcation of equilibrium takes place at the critical load, a geometrically imperfect beam bends out-of-plane and twists as soon as the load is applied. As the deformation increases, the load approaches the critical load asymptotically if fully elastic behavior is assumed.

5.8 INELASTIC BEAMS

The solutions for the critical loads presented in the preceding sections are valid only for the cases in which yielding of material does not take place anywhere in the beam. In other words, a fully elastic behavior is assumed for the beam. This assumption is reasonable for beams of high slenderness ratio (L/r_y). However, for beams of intermediate slenderness ratios, yielding will occur in some fibers of the beam before the attainment of the critical load. Since portions of the beam are inelastic when buckling commences, only the elastic portion of the cross section will remain effective in providing resistance to lateral buckling. As a result, the critical load will be reduced.

For simply supported I-beams subjected to equal and opposite end moments, if there are no residual stresses the distribution of yielding

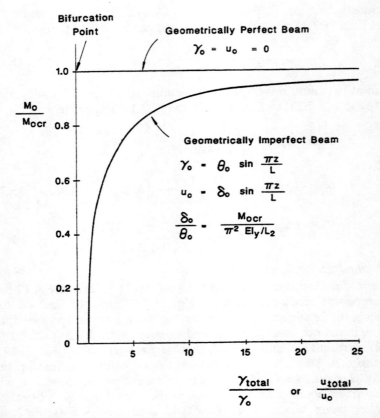

FIGURE 5.26 Load-deformation relationship of a geometrically imperfect beam

across the section is symmetric about the horizontal principal axis and is constant along the entire length of the beam. The inelastic critical moment can be obtained from a simple modification of Eq. (5.4.34), as done in reference 25:

$$M_{ocr} = \frac{\pi}{L}\sqrt{(EI_y)_e(GJ)_e}\sqrt{\left(1 + \frac{\pi^2(EC_w)_e}{L^2(GJ)_e}\right)} \quad (5.8.1)$$

where $(EI_y)_e$, $(GJ)_e$, and $(EC_w)_e$ are the *effective* bending rigidity, torsional rigidity, and warping rigidity, respectively. Their values can be estimated by using the tangent modulus concept discussed in Section 2.7.1.

A nondimensional plot of the critical moment M_{ocr}/M_y where M_y is the yield moment versus the slenderness ratio L/r_y is shown in Fig. 5.27 for a typical stress-relieved, hot-rolled I-beam without residual stresses (curve

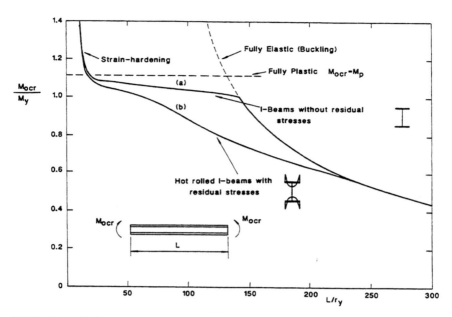

FIGURE 5.27 Beam strength curves (equal and opposite moments)

a). In the inelastic range ($M_{ocr}/M_y > 1$), the variation of M_{ocr}/M_y with L/r_y is almost linear. Also shown in the figure is the M_{ocr}/M_y versus L/r_y plot for a typical hot-rolled I-beam with residual stresses (curve b). It can be seen that the presence of residual stresses greatly reduces the lateral torsional buckling strength of the beam in the inelastic range. Note that the moment at which first yield occurs is significantly below M_y. This is because for such sections, compressive residual stresses as high as $0.3F_y$, where F_y is the yield stress, may occur at the flange tips, thus yielding can be initiated readily at the flange tips, and, as the applied moment increases, yielding will spread to other parts of the cross section, further reducing the resistance capacity of the beam to lateral buckling.

It should be mentioned that Eq. (5.8.1) is not applicable for beams with residual stresses, since this equation is only valid for doubly symmetric sections. If residual stresses are considered in the analysis, the distribution of yielding across the section will *not* be symmetric about the horizontal principal axis. The critical moment equation valid for *monosymmetric* section must be derived and used to replace that of Eq. (5.8.1). The discussion of lateral buckling of monosymmetric sections is beyond the scope of this book. The discussions of this type of section can be found in reference 26.

To obtain the inelastic buckling loads for beams with a more general loading case other than that of equal and opposite end moments, recourse must be had to numerical methods.[24,27,28] The reason for this is that for beams under general loadings, the in-plane bending moment will vary along the beam, and so the distribution of yielding will also vary from cross section to cross section. Nethercot and Trahair[29] have presented numerical solutions for some hot-rolled I-beams with a number of different loading arrangements. Some of the results are shown in Fig. 5.28, in which the nondimensional moment M_B/M_p (M_p = plastic moment capacity of the section) is plotted as a function of the modified slenderness $\sqrt{M_y/M_{ocr}}$. Note that the most severe loading case is the one with equal and opposite end moments that bends the beam in a single curvature and the least severe case is the one with equal end moment that bends the beam in a double curvature. For the former loading case, yielding occurs along the entire length of the beam, whereas for the latter loading case, yielding is usually confined to small portions at or near the supports of the beam.

For very *stocky* beams, lateral instability usually will not occur. As a result, the failure mode will be that of the formation of a *collapse mechanism* when sufficient *plastic hinges* (locations where the moment equals the plastic moment M_p of the section) have formed. For hot-rolled I-sections, the ratio of plastic moment to yield moment, i.e., M_p/M_y, is approximately 1.12. The plastic moment capacity of a typical hot-rolled I-section is shown as a dashed line in Fig. 5.27. Although, in reality, a fully plastic beam can withstand a moment above M_p because of the

FIGURE 5.28 Beam strength curves (unequal moments)

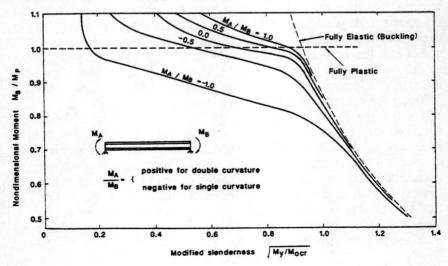

strain hardening of the material, this strain-hardening effect is usually neglected in design consideration.

5.9 DESIGN CURVES FOR STEEL BEAMS

It is quite evident from the preceding discussions that the lateral instability behavior of real beams is complex. Some of the important parameters are (1) the lateral unsupported length of the beam, (2) the cross-sectional geometry, (3) the material behavior (elastic versus inelastic), (4) the magnitude, type, and location of the applied loads, and (5) the type of lateral supports, as well as the end conditions of the beam, etc. Because of the complexity of this problem, certain simplifying assumptions are inevitable in order to make use of the analytical results for design purposes. In this section, we shall briefly discuss some aspects of the design rules as specified by the AISC Specifications.[30,31] Three different design approaches, already familiar to the reader, will be presented here: they are ASD, PD, and LRFD.

5.9.1 ALLOWABLE STRESS DESIGN

In ASD, a design is said to be satisfactory if the maximum stress developed in the member falls below some specified value that is usually obtained as a fraction of the yield stress or ultimate stress.

Local Buckling

COMPACT SECTION

For design purpose, it is customary to distinguish sections that are *compact* from those that are *noncompact*. A compact section is one that can develop the full *plastic moment capacity* M_p and sustain a large hinge rotation without local buckling. A section is considered to be compact if the following conditions as contained in the AISC Specification are satisfied:

1. The flanges are continuously connected to the web or webs.
2. The width-thickness ratio of *unstiffened projecting elements* of the compression flange is less than or equal to $65/\sqrt{F_y}$, where F_y is the yield stress of the material in ksi. Thus, for an I-section, this can be represented mathematically as

$$\frac{b_f}{2t_f} \leq \frac{65}{\sqrt{F_y}} \qquad (5.9.1)$$

   in which b_f equals the width of the flange and t_f equals the thickness of the flange.
3. The width-thickness ratio of the *stiffened elements* of the compression flange is less than or equal to $190/\sqrt{F_y}$.

4. The depth-thickness ratio of the web or webs must satisfy the criterion given by the following applicable equation. For $f_a/F_y \leq 0.16$,

$$\frac{d}{t_w} \leq \frac{640}{\sqrt{F_y}}\left(1 - 3.74\frac{f_a}{F_y}\right) \tag{5.9.2}$$

For $f_a/F_y > 0.16$,

$$\frac{d}{t_w} \leq \frac{257}{\sqrt{F_y}} \tag{5.9.3}$$

In Eqs. (5.9.2) and (5.9.3) [AISC Eqs. (1.5-4a) and (1.5-4b)],
$d =$ the depth of the section
$t_w =$ the thickness of the web
$f_a =$ the applied axial stress in the section (in ksi)
$F_y =$ the yield stress of the material (in ksi)

Sections that do not satisfy the requirements stated above are referred to as noncompact sections. For noncompact sections, local buckling may occur before the attainment of M_p.

Lateral Buckling

LATERAL UNBRACED LENGTH
In addition to the compactness of the section, the lateral unbraced length of the member will also play a predominant role in determining the strength of the beam. Thus, the applicable allowable bending stress in beams as contained in the AISC Specification is determined by the compactness of the section and the lateral unbraced length of the member. Defining

$$L_c = \text{smaller of} \begin{cases} \dfrac{76b_f}{\sqrt{F_y}} & (5.9.4a) \\ \dfrac{20000}{(d/A_f)F_y} & (5.9.4b) \end{cases}$$

$$L_u = \text{larger of} \begin{cases} \dfrac{20000 C_b}{(d/A_f)F_y} & (5.9.5a) \\ r_T\sqrt{\dfrac{102000 C_b}{F_y}} & (5.9.5b) \end{cases}$$

where
$b_f =$ width of flange (in inches)
$d =$ depth of section (in inches)
$A_f =$ area of compression flange (in square inches)
$r_T =$ radius of gyration of a section comprising the compression flange plus 1/3 of the compression web area, taken about an axis in the plane of the web (in inches)
$F_y =$ yield stress of the material (in ksi)

5.9 Design Curves for Steel Beams

The allowable bending stress F_b for the member will be discussed as follows:

ALLOWABLE STRESS F_b

The *allowable bending stress* F_b in the beam with lateral unbraced length L_b is as follows.

Compact Section with $L_b \leq L_c$. If the unbraced length L_b is less than or equal to L_c, the beam is considered to be adequately braced, and so lateral instability will not be a factor. For such cases, the full plastic moment of the section can be developed. Thus, the allowable bending stresses for doubly symmetric I-section (except hybrid girders and members of A514 steel) bent about the strong axis are

$$F_b = 0.66 F_y \qquad (5.9.6a)$$

and, for doubly symmetric I-section (except members of A514 steel) bent about the weak axis,

$$F_b = 0.75 F_y \qquad (5.9.6b)$$

For strong axis bending, the value 0.66 corresponds to a safety factor of 1:0.66 or 1.5 against yield. The use of 1.5 here as the safety factor rather than the usual safety factor of 1.67 is attributed to the fact that for compact section that is adequately braced, the full plastic moment of the section can be developed. The ability to develop plastic moment means that the section has a reserve strength over that of first yield (that is, the yielding of the extreme fibers). This reserve strength is measured by the ratio M_p/M_y, where M_p is the full plastic moment and M_y is the moment at first yield (Fig. 5.29). For hot-rolled, wide-flange sections, the ratio M_p/M_y is about 1.12. Thus, if we divide the safety factor against first yield of 1.67 by 1.12, we get a safety factor of 1.5 against full yield. For weak axis bending, an even lower safety factor of $1/0.75 = 1.33$ is used because the reserve strength, that is, the ratio M_p/M_y, is about 1.5. For weak axis bending, this value is much higher than that of the strong axis bending for all hot-rolled, wide-flange shapes.

Semicompact Section with $L_b \leq L_c$. Semicompact sections are sections for which the web or webs satisfy the compactness requirement, namely Eq. (5.9.2) or Eq. (5.9.3), but for which the flange has a $b_f/2t_f$ value larger than $65/\sqrt{F_y}$ but less than $95/\sqrt{F_y}$. For such sections, the allowable bending stresses are reduced from the allowable value of $0.66F_y$ for strong axis bending and $0.75F_y$ for weak axis bending, respectively, as follows: for strong axis bending:

$$F_b = F_y \left[0.79 - 0.002 \left(\frac{b_f}{2t_f} \right) \sqrt{F_y} \right] \qquad (5.9.7a)$$

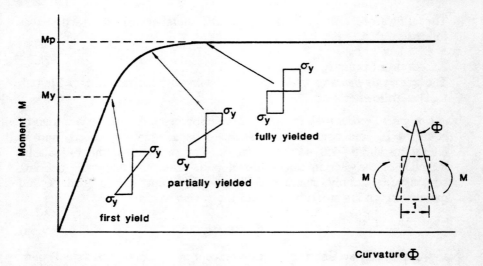

FIGURE 5.29 Yield moment and plastic moment

for weak axis bending:

$$F_b = F_y\left[1.075 - 0.005\left(\frac{b_f}{2t_f}\right)\sqrt{F_y}\right] \quad (5.9.7b)$$

Note that Eq. (5.9.7a) and Eq. (5.9.7b) are merely straight lines interpolating the allowable bending stresses between $0.66F_y$ for compact section and $0.60F_y$ for noncompact section bent about the strong axis, and between $0.75F_y$ for compact section and $0.60F_y$ for noncompact section bent about the weak axis, respectively.

Noncompact Section with $L_b \leq L_c$.

$$F_b = 0.60F_y \quad (5.9.8)$$

Compact or Noncompact Section with $L_c < L_b \leq L_u$.

$$F_b = 0.60F_y \quad (5.9.9)$$

Note that in Eqs. (5.9.8) and (5.9.9) the number 0.60 corresponds to the usual safety factor of $1/0.60 = 1.67$ used in ASD to bring the load to service load range.

5.9 Design Curves for Steel Beams

Compact or Noncompact Section with $L_b > L_u$ Bent About the Strong Axis.

For

$$r_T\sqrt{\frac{102{,}000C_b}{F_y}} \leq L_b \leq r_T\sqrt{\frac{510{,}000C_b}{F_y}}$$

$$F_b = \text{larger of} \begin{cases} \left[\dfrac{2}{3} - \dfrac{F_y(L_b/r_T)^2}{1530 \times 10^3 C_b}\right]F_y & (5.9.10a) \\ \dfrac{12{,}000C_b}{L_b d/A_f} & (5.9.10b) \end{cases}$$

but must not exceed $0.60F_y$, and for

$$L_b \geq r_T\sqrt{\frac{510{,}000C_b}{F_y}}$$

then

$$F_b = \text{larger of} \begin{cases} \dfrac{170{,}000C_b}{(L_b/r_T)^2} & (5.9.11a) \\ \dfrac{12{,}000C_b}{L_b d/A_f} & (5.9.11b) \end{cases}$$

but must not exceed $0.60F_y$. A schematic plot of F_b as a function of L_b is shown in Fig. 5.30. In the above equations, the units are kips and inches.

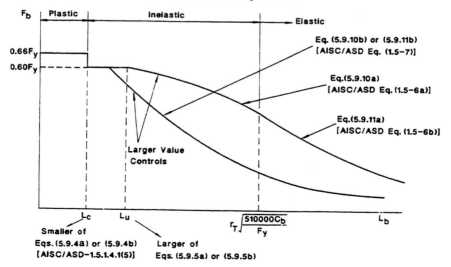

FIGURE 5.30 Schematic plot of F_b versus L_b in ASD

Equations (5.9.10a) and (5.9.11a) are Eqs. (1.5-6a) and (1.5-6b) in the AISC Specification, respectively, and Eqs. (5.9.10b) and (5.9.11b) correspond to Eq. (1.5–1.7) in the AISC Specification. In these equations, C_b is a parameter to account for the moment gradient in the member as defined in Eq. (5.5.2) for unequal end moments. Although values for C_b for other load conditions are available (Table 5.2a), a conservative value of unity is used for design purposes, when the bending moment at any point within an unbraced length is larger than that at both ends of the unbraced length. In such a case, the reduction in stiffness due to the yielding material usually exceeds the benefits of the less severe pattern of nonuniform moment distribution and the use of $C_b = 1.0$ is therefore more appropriate. C_b is also taken as unity for frames braced against joint translation and for cantilever beams.

Equations (5.9.10a) and (5.9.11a) were developed by assuming that the torsional resistance of the beam is negligible compared to warping resistance. The former equation represents lateral buckling of the compression beam flange in the inelastic range, while the latter equation represents lateral buckling of the compression beam flange in the elastic range. The unbraced length L_b that marks the point of demarcation of inelastic and elastic buckling is given by

$$L_b = r_T \sqrt{\frac{510{,}000 C_b}{F_y}}$$

It is obtained by equating Eq. (5.9.10a) to Eq. (5.9.11a).

Equation (5.9.10b) was developed by assuming that the warping resistance of the beam is negligible compared to the torsional resistance.

BASIS OF AISC/ASD RULES

Equations (5.9.10a,b) and (5.9.11a,b) have their origin in Eq. (5.4.34). Their development is briefly discussed as follows. By replacing L by L_b and expanding, Eq. (5.4.34) can be written as

$$M_{ocr} = \sqrt{\left(\frac{\pi^2 E I_y GJ}{L_b^2}\right) + \left(\frac{\pi^4 E I_y E C_w}{L_b^4}\right)} \qquad (5.9.12)$$

The critical stress can be obtained by dividing Eq. (5.9.12) by S_x (the section modulus of the section about the strong axis).

$$F_{ocr} = \sqrt{\left(\frac{\pi^2 E I_y GJ}{S_x^2 L_b^2}\right) + \left(\frac{\pi^4 E I_y E C_w}{S_x^2 L_b^4}\right)} \qquad (5.9.13)$$

The first term of Eq. (5.9.13) represents the torsional strength of the section, whereas the second term represents the warping strength of the

5.9 Design Curves for Steel Beams

section. If we take

$$G = \frac{E}{2(1+v)} = \frac{E}{2.6}$$

$$I_y = Ar_y^2, \quad C_w = I_y h^2/4$$

$$S_x = 2Ar_x^2/d, \quad h \approx 0.95d$$

$$r_x \approx 0.41d, \quad J \approx 0.28At_f^2$$

where

- v = Poisson's ratio ($= 0.3$ for steel)
- r_x, r_y = radius of gyration about the x and y axis, respectively
- d = overall depth of beam
- t_f = flange thickness
- h = distance between centroids of flanges

and substitute it into Eq. (5.9.13), we can write

$$F_{ocr} = \sqrt{\left(\frac{3E}{L_b d/r_y t_f}\right)^2 + \left[\frac{14E}{(L_b/r_y)^2}\right]^2} \tag{5.9.14}$$

For simplicity, one of the square terms in Eq. (5.9.14) can be neglected. For example, if the section is a shallow, thick-flanged section, the torsional strength will predominate. By retaining the first term (which represents torsional strength) and disregarding the second term (which represents warping strength), the critical stress is approximated by

$$F_{ocr} = \frac{3E}{L_b d/r_y t_f} \tag{5.9.15}$$

On the other hand, if the section is a deep section with relatively thin flanges and web, the warping strength will predominate. By retaining the second term and disregarding the first term, the critical stress is approximated by

$$F_{ocr} = \frac{14E}{(L_b/r_y)^2} \tag{5.9.16}$$

Note that both Eqs. (5.9.15) and (5.9.16) represent a quite conservative estimate of the true critical stress. Consequently, for design purposes, they can be used in place of the more complicated form of Eq. (5.9.14). To reduce the overconservatism, it is the larger value of these two approximate equations that controls the design.

CONSIDERATION OF TORSIONAL STRENGTH ONLY

If only torsional strength is considered, the critical stress is approximated by Eq. (5.9.15). By taking $r_y \approx 0.22b_f$ and using $A_f = b_f t_f$, Eq. (5.9.15) can be written as

$$F_{ocr} = \frac{0.66E}{L_b d/A_f} \tag{5.9.17}$$

To account for the moment gradient, the parameter C_b is used; thus,

$$F_{cr} = C_b F_{ocr} = \frac{0.66 E C_b}{L_b d/A_f} \tag{5.9.18}$$

Finally, substitution of $E = 29,000$ ksi into Eq. (5.9.18) and dividing the result by a safety factor of 1.67 gives

$$F_b \approx \frac{12,000 C_b}{L_b d/A_f} \quad \text{(in ksi)} \tag{5.9.19}$$

Equation (5.9.19) is Eq. (5.9.10b) and Eq. (5.9.11b) [AISC Eq. (1.5-7)].

CONSIDERATION OF WARPING STRENGTH ONLY

If only warping strength is considered, the critical stress is approximated by Eq. (5.9.16). However, it should be mentioned that in developing the warping restraint strength of the flange, it is assumed that there is no interaction between the flange and the web. Nevertheless, in reality, the web that is attached to the flange will always provide some restraint to the compression flange. To account for this effect, the AISC introduces a parameter r_T defined as "the radius of gyration of a section comprising the compression flange plus one-third of the compression web area, taken about an axis in the plane of the web." Thus, by substituting $r_T \approx 1.2 r_y$ into Eq. (5.9.16), we have

$$F_{ocr} = \frac{\pi^2 E}{(L_b/r_T)^2} \tag{5.9.20}$$

Equation (5.9.20) has the same form as the Euler column-buckling stress equation. Thus, the warping strength of the flange can be regarded as the column-buckling strength of the compression flange. By introducing the parameter C_b to account for the moment gradient, we can write Eq. (5.9.20) as

$$F_{cr} = C_b F_{ocr} = \frac{\pi^2 E C_b}{(L_b/r_T)^2} \tag{5.9.21}$$

Taking $E = 29,000$ ksi and dividing the result by a safety factor of 1.67, we obtain

$$F_b = \frac{170,000 C_b}{(L_b/r_T)^2} \tag{5.9.22}$$

Equation (5.9.22) is Eq. (5.9.11a) [AISC Eq. (1.5-6b)].

5.9 Design Curves for Steel Beams

Equation (5.9.22) is valid so long as all the fibers in the column compression flange remain elastic. This, in turn, can be realized so long as the unbraced length of the beam is large. If it is small, inelastic buckling will occur and the SSRC (or CRC) parabola will again be used to present this inelastic buckling strength:

$$F_{cr} = \left[1 - \frac{F_y}{4\pi^2 E}\left(\frac{L_b}{r_T}\right)^2\right]F_y \qquad (5.9.23)$$

Using the term C_b to account for moment gradient and dividing the equation by a safety factor of 1.50, we obtain

$$F_b = \left[\frac{2}{3} - \frac{F_y(L_b/r_T)^2}{1720 \times 10^3 C_b}\right]F_y \qquad (5.9.24)$$

The number 1720×10^3 in the second term is replaced by 1530×10^3 to account for the fact that as L_b/r_T gets larger, the reserved strength due to plastification (progressive yielding from extreme fibers to fibers at the neutral axis) may not be realized fully because of a possible occurrence of local or lateral torsional instability before the attainment of the full plastic moment M_p of the section. Thus,

$$F_b = \left[\frac{2}{3} - \frac{F_y(L_b/r_T)^2}{1530 \times 10^3 C_b}\right]F_y \qquad (5.9.25)$$

Equation (5.9.25) is Eq. (5.9.10a) [AISC Eq. (1.5-6a)].

In summary, Eq. (5.9.10b) or (5.9.11b) [AISC Eq. (1.5-7)] was developed by considering the torsional strength of the section only. Equations (5.9.10a) and (5.9.11a) [AISC Eqs. (1.5-6a) and (1.5-6b)] were developed by considering the warping strength of the section only Eq. (5.9.10a) is applicable to inelastic buckling of the compression flange of the beam and Eq. (5.9.11a) is applicable to elastic buckling of the compression flange.

It should be noted that Eq. (5.9.10a,b) and Eq. (5.9.11a,b) are applicable only to members bent about their strong axes. For members bent about their weak axes, lateral torsional buckling will not occur. As a result, only the compactness of the section is important in determining the allowable bending stress. If the section is compact $F_b = 0.75 F_y$; if the section is noncompact, $F_b = 0.60 F_y$; if the section is semicompact, Eq. (5.9.7b) is used to interpolate the allowable bending stress.

5.9.2 Plastic Design

Since the purpose of PD is to develop the maximum plastic strength of the structure, local buckling or lateral torsional instability of members must be prevented from occurring before the attainment of the plastic strength. To assure this, the AISC Specification provides the limits on the width-to-thickness ratios of flanges and webs as well as the limits on the lateral unbraced length of the beam.

Local Buckling

LIMITS FOR WIDTH-TO-THICKNESS RATIO FOR FLANGES

F_y, ksi	$b_f/2t_f$
36	8.5
42	8.0
45	7.4
50	7.0
55	6.6
60	6.3
65	6.0

LIMITS FOR WIDTH-TO-THICKNESS RATIO FOR WEBS
For $P/P_y \leq 0.27$

$$\frac{d}{t_w} = \frac{412}{\sqrt{F_y}}\left(1 - 1.4\frac{P}{P_y}\right) \tag{5.9.26}$$

and for $P/P_y > 0.27$

$$\frac{d}{t_w} = \frac{257}{\sqrt{F_y}} \tag{5.9.27}$$

where F_y is in ksi. Equations (5.9.26) and (5.9.27) correspond to AISC Equations (2.7-1a) and (2.7-1b), respectively.

Lateral Buckling

LIMITS FOR LATERAL UNBRACED LENGTH

For $-0.5 < \frac{M}{M_p} < 1.0$

$$\frac{L_b}{r_y} \leq \frac{1375}{F_y} + 25 \tag{5.9.28}$$

For $-1.0 < \frac{M}{M_p} \leq -0.5$

$$\frac{L_b}{r_y} = \frac{1375}{F_y} \tag{5.9.29}$$

5.9 Design Curves for Steel Beams

where

M/M_p = the ratio of the end moments at the ends of the unbraced segment, positive when the segment is bent in reverse curvature and negative when bent in single curvature

r_y = the radius of gyration of the member about its weak axis in inches

F_y = the yield stress of the material (in ksi)

Equations (5.9.28) and (5.9.29) are AISC Eqs. (2.9-1a) and (2.9-1b), respectively. They are applicable for members bent about their strong axis, but not applicable for members bent about their weak axis, nor are they applicable in the region of the last plastic hinge formation in the development of a failure mechanism. This is because for members bent about their weak axis, lateral instability is not a factor, and so the length limitation should not be applied. For the region of last hinge formation, ductility (large rotation capacity) of the hinge is not required. Therefore, in this region, and in the regions not adjacent to a plastic hinge, the maximum unbraced length and allowable stress are determined from the elastic equations (5.9.10a,b) and (5.9.11a,b) instead.

Strength Reduction

If the unbraced length of the member exceeds the limits set forth in Eqs. (5.9.28) or (5.9.29), a reduced ultimate bending strength of a member subjected to moment alone is given in the AISC Specification as

$$M_m = \left[1.07 - \frac{(L_b/r_y)\sqrt{F_y}}{3160}\right]M_p \leq M_p \qquad (5.9.30)$$

in which F_y is in ksi and M_p is the plastic moment capacity of the cross section in inch-kips. Equation (5.9.30) [AISC Eq. (2.4-4)] was developed based on a semiempirical approach. It is plotted in Fig. 5.31 as a solid line. The dashed-dotted line in the figure represents a more exact analysis by Galambos,[27] considering the effect of residual stress. It can be seen that Eq. (5.9.30) gives a reasonably good approximation to the more exact analysis.

5.9.3 Load and Resistance Factor Design

LRFD is based on the concept of limit states. For a beam, there are three possible types of failure limits: (1) plastic yielding, (2) lateral instability, and (3) local buckling. In the following, we shall limit our attention to the

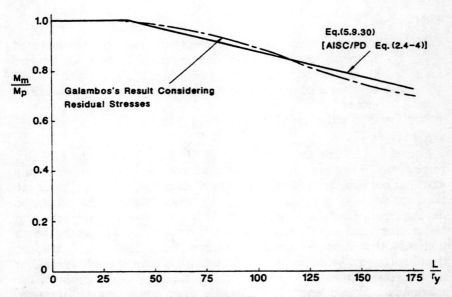

FIGURE 5.31 Comparison of Eq. (5.9.30) with numerical results of Galambos

first two failure modes or limits, namely, yielding and lateral instability. The subject of local buckling is treated elsewhere.[2]

To avoid local buckling, one must use compact sections. The limitations for the width-to-thickness ratios for sections that can be regarded as *compact* under the AISC/LRFD Specification[31] are the following:

Flange:
$$\frac{b_f}{2t_f} \leq \frac{65}{\sqrt{F_y}} \tag{5.9.31}$$

Web:
$$\frac{h_c}{t_w} \leq \frac{640}{\sqrt{F_y}} \tag{5.9.32}$$

where

b_f = width of flange
t_f = thickness of flange
h_c = depth of web clear of fillets
t_w = thickness of web
F_y = yield stress of the material in ksi

Defining
$$L_p = \frac{300 r_y}{\sqrt{F_y}} \tag{5.9.33}$$

5.9 Design Curves for Steel Beams

$$L_r = \frac{r_y X_1}{(F_y - F_r)} \sqrt{(1 + \sqrt{(1 + X_2(F_y - F_r)^2)})} \quad (5.9.34)$$

$$L_b = \text{unbraced length of the member} \quad (5.9.35)$$

where

r_y = radius of gyration about the weak axis in inches
F_y = yield stress in ksi

$$X_1 = \frac{\pi}{S_x} \sqrt{\frac{EAGJ}{2}} \quad (5.9.36)$$

$$X_2 = \frac{4C_w}{I_y} \left(\frac{S_x}{GJ}\right)^2 \quad (5.9.37)$$

S_x = section modulus about strong axis (in in.^{3})
E = elastic modulus (in ksi)
G = shear modulus (in ksi)
A = cross section area (in in.^{2})
J = torsional constant, (in in.^{4})
C_w = warping constant (in in.^{6})
I_y = moment of inertia about weak axis (in in.^{4})
F_r = compressive residual stress in flange; taken as 10 ksi for rolled shapes and 16.5 ksi for welded shapes

If $L_b \leq L_p$, the beam is considered to have adequate lateral support. If $L_p < L_b \leq L_r$, the beam is considered to be laterally unsupported and *inelastic lateral buckling* may occur. If $L_b > L_r$, *the beam is considered to be laterally unsupported and elastic lateral buckling* may occur. The nominal moment M_n, specified in the AISC/LRFD Specification[31] and developed on the basis of a study reported in reference 32 for doubly symmetric I-sections for various ranges of L_b, is summarized as follows, and illustrated in Fig. 5.32.

COMPACT SECTION WITH $L_b \leq L_p$

$$M_n = M_p \quad (5.9.38)$$

for section bent about strong and weak axes.

COMPACT SECTION WITH $L_p < L_b \leq L_r$ BENT ABOUT STRONG AXIS

$$M_n = C_b \left[M_p - (M_p - M_r)\left(\frac{L_b - L_p}{L_r - L_p}\right) \right] \leq M_p \quad (5.9.39)$$

where C_b = Eq. (5.5.2) with $M_1 = M_A$ and $M_2 = M_B$. It is taken as unity for unbraced cantilevers and for members where the moment within a significant portion of the unbraced segment is greater than or equal to the

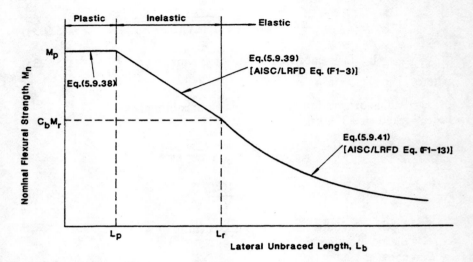

FIGURE 5.32 Schematic plot of M_n versus L_b in LRFD

larger of the segment end moments and

$$M_r = S_x(F_y - F_r) \tag{5.9.40}$$

in which

S_x = section modulus about the strong axis in in.[3]
F_y = yield stress in ksi

Equation (5.9.40) can be written in the form $M_r = M_y - F_r S_x$. Here, as in ASD, a smaller value of moment than M_y is used as the *transition moment* M_r from inelastic to elastic behavior to account for the presence of residual stress in the section. In LRFD, the maximum compressive residual stress in rolled shapes is assumed to be $F_r = 10$ ksi and in welded shapes, it is 16.5 ksi, as given in Eq. (5.9.40). As a result, the moment $F_r S_x$ is subtracted from M_y.

Equation (5.9.39) represents a straight line interpolation from $M_n = M_p$ for $L_b = L_p$ to $M_n = M_r$ for $L_b = L_r$. The term C_b can be introduced to convert the case of unequal end moments with $|M_1| < |M_2| = M_n$ to the case of equal end moments with $M_n = C_b M_{ocr}$.

COMPACT SECTION WITH $L_b > L_r$ BENT ABOUT STRONG AXIS

$$M_n = C_b M_{ocr} = C_b \frac{\pi}{L_b} \sqrt{\left(EI_y GJ + \frac{\pi^2}{L_b^2} EI_y EC_w\right)} \leq C_b M_r \tag{5.9.41}$$

5.9 Design Curves for Steel Beams

Note that Eq. (5.9.41) is identical to the exact elastic buckling solution given by Eq. (5.5.1). In other words, the theoretical lateral torsional buckling moment of an elastic beam under equal end moments is used directly in LRFD.

Equations (5.9.39), (5.9.34), and (5.9.41) are contained in the AISC/

FIGURE 5.33 Comparison of beam curves for ASD, PD, and LRFD

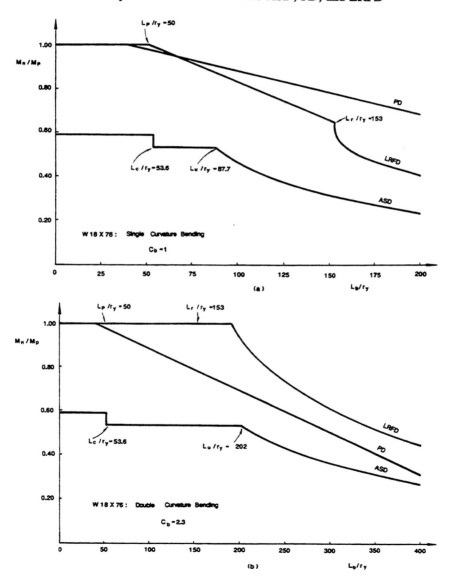

LRFD Specification[31] as Eqs. (F1-3), (F1-6), and (F1-13), respectively. Figure 5.32 shows schematically the variation of the nominal moment capacity M_n with the unbraced length L_b. The LRFD beam curve is generally more liberal and much simpler than that of the ASD beam curve.

The equations given above are applicable only to compact nonhybrid I-sections in which yielding or lateral torsional instability is the limit state. The applicable expressions for other sections, including symmetric box sections, solid rectangular bars, hybrid sections, and tee and double-angle sections, are given elsewhere.[31]

Figure 5.33a,b shows a comparison of the variation of the nondimensionalized nominal moment capacity M_n/M_p as a function of the slenderness ratio L_b/r_y of a beam using the three design approaches (ASD, PD, and LRFD) discussed in the preceding sections. The beam is a W18 × 76 section with a yield stress $F_y = 36$ ksi and a compressive residual stress in the flange $F_r = 10$ ksi. The Young's modulus E is taken as 29,000 ksi and the shear modulus G is taken as $0.385E$. Figure 5.33a corresponds to the case in which $M_A/M_B = -1$ (that is, $C_b = 1.0$) and Fig. 5.33b corresponds to the case in which $M_A/M_B = 1$ (that is, $C_b = 2.3$). Note that the nominal moment capacity of the beam is larger for double-curvature bending than for the single-curvature bending.

5.10 OTHER DESIGN APPROACHES

Certainly, by now we see that designing for lateral torsional buckling is a complex and challenging problem. This is because the resistance of the beam against lateral torsional buckling is dependent on many factors, such as the geometry and end conditions of the beam, the amount and nature of bracing, the type and manner of the loadings, etc. Nevertheless, there are two undebatable facts. A beam with a low slenderness ratio should be able to develop its full plastic moment M_p, and a beam with a large slenderness ratio should behave in a fully elastic manner, so that the elastic critical moment M_{cr} should represent the nominal moment capacity of the beam. The main problem is that the behavior of a beam with intermediate slenderness ratios is neither fully plastic nor fully elastic, and so neither the plastic moment M_p nor the elastic critical moment M_{cr} can be used to represent the nominal moment capacity M_n of the beam. The nominal moment capacity for a beam with an intermediate slenderness ratio should fall between M_p and M_{cr} and its value should provide a smooth transition from M_p to M_{cr}. For most specifications, the choice of this transition moment is rather empirical. For example, in the ASD method, a parabola [Eq. (5.9.25)] is used to represent this transition range; and in the LRFD method, a straight line [Eq. (5.9.39)] is used for this transition range. In the discussion that follows, we will

5.10 Other Design Approaches

present two other empirical methods to represent beam strength in the transition range.

5.10.1 Structural Stability Research Council Approach

In the Structural Stability Research Council (SSRC) approach, it is assumed that the stability behavior of beams is analogous to the stability behavior of columns, and so the lateral-torsional buckling strength of a beam is represented by the SSRC multiple-column curves. Using the Rondal–Maquoi form [Eq. (2.11.17)] with P/P_y replaced by M_n/M_p and λ_c replaced by λ_b, where $\lambda_b = \sqrt{M_p/M_{cr}}$, we have

$$\frac{M_n}{M_p} = \frac{(1+\eta+\lambda_b^2) - \sqrt{(1+\eta+\lambda_b^2)^2 - 4\lambda_b^4]}}{2\lambda_b^2} \quad (5.10.1)$$

where η is defined the same as for columns in Eq. (2.11.18).

5.10.2 European Committee on Constructional Steelwork Approach

In the European Convention of Constructional Steelworks (ECCS) approach, the nominal moment capacity for the transition range is represented by

$$\frac{M_n}{M_p} = \left(\frac{1}{1+\lambda_b^{2n}}\right)^{1/n} \quad (5.10.2)$$

where n is a coefficient that varies from 2.5 to 1.5, depending on the criteria set by the various national code-writing bodies in Western Europe and λ_b is the beam slenderness parameter defined as $\lambda_b = \sqrt{M_p/M_{cr}}$.

Example 5.3

Calculate the nominal moment capacity of a W16 × 36 section bent about its strong axis using the following:

1. AISC/ASD approach
2. AISC/LRFD approach
3. SSRC approach
4. ECCS approach

The beam is simply supported and subjected to a uniform moment. Use $E = 29,000$ ksi, $G/E = 0.385$, $F_y = 36$ ksi, $L_b = 150$ in., $F_r = 10$ ksi.

SOLUTION: W16 × 36 section properties: $A = 10.6$ in.^{2}, $A_f = 3$ in.^{2}, $d = 15.86$ in., $b_f = 6.985$ in., $r_T = 1.79$ in., $S_x = 56.5$ in.^{3}, $I_y = 24.5$ in.^{4}, $r_y = $

1.52 in., $J = 0.545$ in.^{4}, $Z_x = 64.0$ in.^{3}, $C_w = 1450$ in.^{6}

$$C_b = 1.0 \quad \text{(constant moment)}$$

$$M_p = Z_x F_y = 2304 \text{ in-kips}$$

$$M_{cr} = C_b M_{ocr} = M_{ocr} = \frac{\pi}{L_b} \sqrt{\left[EI_y GJ + \left(\frac{\pi^2}{L_b^2}\right) EI_y EC_w\right]}$$

$$= 2765 \text{ in-kips}$$

1. *AISC/ASD*:

$$L_c = \text{smaller of} \begin{cases} \dfrac{76 b_f}{\sqrt{F_y}} = 88.5 \text{ in.} \\ \dfrac{20{,}000}{(d/A_f) F_y} = 105 \text{ in.} \end{cases}$$

$$= 88.5 \text{ in.}$$

$$L_u = \text{larger of} \begin{cases} \dfrac{20{,}000 C_b}{(d/A_f) F_y} = 105 \text{ in.} \\ r_T \sqrt{\dfrac{102{,}000 C_b}{F_y}} = 95.3 \text{ in.} \end{cases}$$

$$= 105 \text{ in.}$$

$$r_T \sqrt{\frac{102{,}000 C_b}{F_y}} = 95.3 \text{ in.}$$

$$r_T \sqrt{\frac{510{,}000 C_b}{F_y}} = 213 \text{ in.}$$

Since $L_b > L_u$ and

$$r_T \sqrt{\frac{102{,}000 C_b}{F_y}} < L_b < r_T \sqrt{\frac{510{,}000 C_b}{F_y}}$$

therefore use Eq. (5.9.10), that is,

$$F_b = \text{larger of} \begin{cases} \left[\dfrac{2}{3} - \dfrac{F_y (L_b / r_T)^2}{1530 \times 10^3 C_b}\right] F_y = 18.05 \text{ ksi} \\ \dfrac{12{,}000 C_b}{L_b (d/A_f)} = 15.1 \text{ ksi} \end{cases}$$

$$= 18.05 \text{ ksi}$$

and $M_n = S_x F_b = 1020$ in-kips

2. AISC/LRFD

$$L_p = \frac{300r_y}{\sqrt{F_y}} = 76 \text{ in.}$$

$$L_r = \frac{r_y X_1}{F_y - F_r} \sqrt{[1 + \sqrt{[1 + X_2(F_y - F_r)^2]}]}$$

$$= 219 \text{ in.}$$

Since $L_p < L_b < L_r$, therefore, by using Eq. (5.9.39) with $M_r = S_x(F_y - F_r)$, we get

$$M_n = C_b \left[M_p - (M_p - M_r) \left(\frac{L_b - L_p}{L_r - L_p} \right) \right]$$

$$= 1872 \text{ in-kips}$$

3. SSRC

$$\lambda_b = \sqrt{\frac{M_p}{M_{cr}}} = \sqrt{\frac{2304}{2765}} = 0.913$$

Using SSRC curve 2, that is, where $\alpha = 0.293$ in Eq. (2.11.18) for η, we have, from Eq. (5.10.1),

$$M_n = \frac{(1 + \eta + \lambda_b^2) - \sqrt{[(1 + \eta + \lambda_b^2)^2 - 4\lambda_b^4]}}{2\lambda_b^2} M_p$$

$$= 1177 \text{ in-kips}$$

4. ECCS

$$\lambda_b = \sqrt{\frac{M_p}{M_{cr}}} = 0.913$$

Using $n = 2.5$ in Eq. (5.10.2) gives

$$M_n = \left(\frac{1}{1 + \lambda_b^{2n}} \right)^{1/n} M_p$$

$$= 1893 \text{ in-kips}$$

5.11 SUMMARY

If an I-beam is loaded in the plane of its weak axis and is not adequately braced laterally, the beam will bend out-of-plane and twist when the load has reached a critical value. This phenomenon is known as the *lateral torsional buckling*. For a geometrically perfect beam, lateral torsional instability is a typical bifurcation problem. We can use the concept of neutral equilibrium to obtain the critical load. In this approach, a slightly deformed state of the beam corresponding to its buckled position is first drawn and equilibrium equations are then written with respect to this deformed configuration. The eigenvalue solution to the characteristic equation of the resulting equilibrium equation gives the *critical load* of the beam. Depending on the type and nature of the loading condition, the resulting linear differential equations may have constant or variable coefficients. For the later case, it is often necessary to resort to numerical techniques to obtain solutions.

The critical loads for beams subjected to loads that induce moment gradients can be obtained approximately by introducing the parameter C_b, as in Eq. (5.5.1). This parameter was developed on the basis of *equivalent moment concept*. Thus, the product $C_b M_{ocr}$ in Eq. (5.5.1) represents the magnitude of a pair of equal and opposite end moments that would cause lateral instability in the beam, the same as real loadings which would also cause such instability. The value of C_b is always greater than unity, which indicates that the most severe loading case is the one in which the member is subjected to a pair of equal and opposite end moment. In view of this, it is the usual practice in design to let C_b equal unity for members where the moment within a significant portion of the unbraced segment is greater than or equal to the larger of the segment end moments. This situation usually arises for unbraced cantilevers and for members in unbraced frames for which single curvature bending occurs.

In addition to the effect of moment gradient, the support conditions of the beam also play an important role in affecting its lateral instability behavior. These effects can be accounted for by the use of the C_{bs} factor, as shown in Eq. (5.6.7). Alternatively, one can use the *effective length factor K*, expressed in Eq. (5.6.10). However, because of its complexity in application, and because of the sometimes questionable support conditions for the real beams, the effective length concept is not currently used in design. For design purpose, the actual unbraced length (that is, $K = 1$) is usually used.

An alternative approach recommended in the third edition of the SSRC Guide[26] to account for the effects of moment gradient and support conditions is also presented in this chapter. The applicable equations are given in Eqs. (5.6.16) and (5.6.17).

If the beam is not geometrically perfect, out-of-plane deformation and twist will occur as soon as the load is applied, making the problem a *load-deflection*, rather than an eigenvalue problem. For a perfectly elastic behavior, an *amplification factor* analogous to that of an imperfect column can be developed. By using this amplification factor, the total rotational and translational deformations of the imperfect beam can easily be obtained as a function of the corresponding initial deformations [Eqs. (5.7.20) and (5.7.21)].

The assumption of a perfectly elastic behavior is justified only if the slenderness ratio of the beam is large. If this is not the case, material yielding will occur before buckling, with the result that only the elastic core of the cross section is effective in resisting the buckling deformation. The study of the inelastic buckling behavior of I-beams is beyond the scope of this book; it is given in detail in reference 24. Nevertheless, the readers should realize that material yielding has a detrimental effect on the buckling strength of the beam. Consequently, the buckling load of an inelastic beam usually falls significantly below the elastic buckling load. For design purposes, the inelastic buckling strength of beams with intermediate slenderness ratio is represented, therefore, by a simple curve-fitting technique. For example, the ASD uses the parabola approximation [Eq. (5.9.10a)], whereas the PD and LRFD uses a straight line approximation [Eqs. (5.9.30) and (5.9.39), respectively].

For beams with low slenderness ratio, or in cases where bending is about the weak axis, lateral instability is not a factor. For such cases, the full plastic moment M_p of the beam can be developed and so the limit state is *yielding*. The point of demarcation between failure by *full yielding* of the cross section and *inelastic buckling* for beams bent about their strong axes is given by the length L_c expressed in Eq. (5.9.4a,b) for ASD, the conditions expressed in Eqs. (5.9.28) and (5.9.29) for PD and L_p expressed in Eq. (5.9.33) for LRFD.

For beams with large slenderness ratio, the limit state is *elastic buckling*. Consequently, the elastic bifurcation solution [Eq. (5.4.34)] can be used to represent the beam strength. For example, the ASD uses a conservative approach by ignoring either the torsional stiffness or the warping stiffness contribution of Eq. (5.4.34) to develop Eqs. (5.9.11a) and Eq. (5.9.11b), respectively, to represent the beam strength in the elastic range. The LRFD on the other hand, adopts the exact equation (5.4.34) to represent the elastic beam strength.

PROBLEMS

5.1 Derive the differential equation governing the lateral torsional buckling behavior of a simply supported I-beam subjected to unequal end moments as shown in Fig. P5.1.

FIGURE P5.1

5.2 Derive the differential equation governing the lateral torsional buckling behavior of a tip-loaded cantilever beam of rectangular cross section if the load is applied at the following:

a. at the centroid of the cross section
b. at the top of the cross section
c. at the bottom of the cross section

5.3 Plot the critical moment M_{ocr} as a function of the length L for a rectangular beam subjected to a pair of equal and opposite end moments (Fig. P5.3a)

FIGURE P5.3

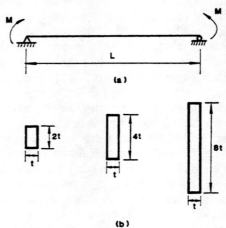

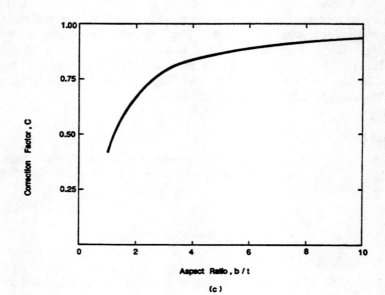

Problems

with the cross sections shown in Fig. P5.3b if

$$t = 1 \text{ in.} \quad (25.4 \text{ mm})$$
$$E = 29,000 \text{ ksi} \quad (2 \times 10^5 \text{ MN/m}^2)$$
$$G = 12,000 \text{ ksi} \quad (8.3 \times 10^4 \text{ MN/m}^2)$$
$$J = C\tfrac{1}{3}bt^3$$

where C is a correction factor that is a function of b/t as shown in Fig. P5.3c.
What observations can you make?

5.4 Show that Eq. (5.5.22) can be used to approximate the values for C_b given in the fourth column of Table 5.2a (Page 334).

5.5 Plot the beam curves for a simply supported W8 × 31 section using

a. LRFD approach
b. SSRC approach
c. ECCS approach

Use M_n/M_p as the ordinate and $\lambda_b = \sqrt{M_p/M_{cr}}$ as the abscissa.
The material properties are:

$$E = 29,000 \text{ ksi}, \quad G/E = 0.385, \quad F_y = 36 \text{ ksi}$$

Using $C_b = 1.0$ (single-curvature bending) and $C_b = 2.3$ (double-curvature bending).

5.6 Compare the three design methods (ASD, PD, and LRFD) by plotting the nondimensional $(M_{cr}/M_p) - (L/r_y)$ curves for a W24 × 55 beam. Plot points of the unbraced length limits for the elastic, inelastic, yield, and plastic ranges. Use $E = 29,000$ ksi, $G/E = 0.385$, $F_y = 43.5$ ksi, and $F_r = 12.6$ ksi.

(i) $M_1/M_2 = -1$, $\quad C_b = 1.0$
(ii) $M_1/M_2 = -0.5$, $\quad C_b = 1.3$

5.7 Find the maximum end moment capacity of a simply supported beam with W24 × 55 section, where

$$L_b = 150 \text{ in.}, \quad C_b = 1.0, \quad F_y = 36 \text{ ksi}, \quad F_r = 10 \text{ ksi}, \quad E = 29,000 \text{ ksi},$$
$$G/E = 0.385, \quad I_y = 29.1 \text{ in.}^4, \quad S_x = 114 \text{ in.}^4,$$
$$Z_x = 134 \text{ in.}^3, \quad r_y = 1.34 \text{ in.}, \quad J = 1.18 \text{ in.}^4, \quad C_w = 3870 \text{ in.}^6,$$

using the LRFD method.

5.8 Explain why the actual value of C_b factor in LRFD specification is limited to $C_b = 1.0$ when the maximum moment occurs between the ends.

5.9 A 12 ft. long steel cantilever has the cross section shown in Prob. P5.10.

Determine the elastic buckling value of the concentrated load P that acts at the free end of the top flange using the relevant solution curves given in this chapter.

5.10 A simply supported steel I-beam whose properties are given below has a central concentrated load applied in such a way that lateral deflection and twist are prevented at midspan ($u_m = \gamma_m = 0$). If the beam span is 35 ft, then

a. determine the elastic buckling load M_{cr}
b. determine the nominal moment capacity M_n specified in the LRFD Specification

where

$A = 18.35$ in.^{2}, $I_x = 1330$ in.^{4}, $r_x = 8.54$ in., $S_x = 126.4$ in.^{3}, $Z_x = 144.1$ in.^{3},
$I_y = 57.5$ in.^{4}, $r_y = 1.77$ in., $r_T = 1.93$ in., $C_w = 5968$ in.^{6}, $J = 1.97$ in.^{4},
$E = 30,000$ ksi, $G = 12,000$ ksi, $F_y = 45$ ksi

(see Fig. P5.10).

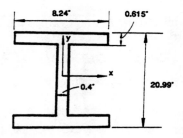

FIGURE P5.10

5.11 Derive the differential equation governing the lateral torsional buckling behavior of a simply supported I-beam subjected to a uniformly distributed load as shown in Fig. P5.11.

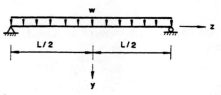

FIGURE P5.11

REFERENCES

1. Timoshenko, S. P. *Strength of Materials: Part II.* Third edition. Van Nostrand, Princeton, NJ, 1956, pp. 240–246.
2. Timoshenko, S. P., and Gere, J. M. *Theory of Elastic Stability.* Second edition. Engineering Societies Monographs, McGraw-Hill, NY, 1961, p. 223.

3. Kirby, P. A., and Nethercot, D. A. Design for Structural Stability. Constrado Monographs, Granada Publishing, Suffolk, UK, 1979.
4. Massonnet, C. Buckling of thin-walled bars with open cross section. Hommage de la Faculté des Sc. Appl. University of Liège, à l'A.I.Lg. G. Thone, editor (in French), 135–146, 1947.
5. Horne, M. R., The flexural-torsional buckling of members of symmetric I-section under combined thrust and unequal terminal moments. Quarterly Journal of Mechanics and Applied Mathematics. Vol. 7, Part 4, 1954, pp. 410–426.
6. Salvadori, M. G. Lateral buckling of I-beams. ASCE Transaction. Vol. 120, 1955, 1165–1177.
7. Nethercot, D. A., and Rockey, K. C. A unified approach to the elastic lateral buckling of beams. The Structural Engineer. Vol. 49, No. 7, July 1971, 321–330.
8. Bleich, F. Buckling Strength of Metal Structures. Engineering Societies Monographs, McGraw-Hill, NY, 1952.
9. Galambos, T. V. Structural Members and Frames. Prentice-Hall, Englewood Cliffs, NJ, 1968.
10. Brown, P. T., and Trahair, N. S. Finite integral solution of differential equations. Civil Engineering Transactions. Institute of Engineers, Australia, CE10, No. 2, Oct. 1968, 193–196.
11. Barsoum, R. S., and Gallagher, R. H. Finite element analysis of torsional and torsional-flexural stability problems. International Journal of Numerical Methods in Engineering, Vol. 2, 1970, 335–352.
12. Powell, G., and Klingner, R. Elastic lateral buckling of steel beams. Journal of the Structural Division. ASCE, Vol. 96, No. ST9, 1970, 1919–1932.
13. Nethercot, D. A., and Rockey, K. C. Finite element solutions for the buckling of columns and beams. International Journal of Mechanical Sciences, Vol. 13, 1971, 945–949.
14. Lee, G. C. A Survey of Literature on the Lateral Instability of Beams. Welding Research Council Bulletin, No. 60, August 1960.
15. Clark, J. W., and Hill, H. N. Lateral buckling of beams. Journal of the Structural Division. ASCE, Vol. 86, No. ST7, July, 1960, 175–196.
16. Nethercot, D. A. Elastic Lateral Buckling of Beams, Chapter 1 in the book: Beams and Beam-Columns—Stability and Strength. R. Narayanan, editor. Elsevier Applied Science Publishers, London and New York, 1983.
17. Anderson, J. M., and Trahair, N. S. Stability of monosymmetric beams and cantilevers. Journal of the Structural Division. ASCE, Vol. 98, No. ST1, Jan. 1972, 269–286.
18. Nethercot, D. A. The effective lengths of cantilevers as governed by lateral buckling. The Structural Engineer. Vol. 51, No. 5, May, 1973, 161–168.
19. Vlasov, V. Z. Thin-Walled Elastic Beams. Y. Schechtman, Translator. Moscow, 1959. Israel Program for Scientific Translation, Jerusalem, 1961.
20. Clark, J. W., and Hill, H. N. Lateral buckling of beams and girders. Transactions. ASCE, Vol. 127, No. ST7, Part II, July 1962, 180–201.
21. Salvadori, M. G. Lateral buckling of beams of rectangular cross section under bending and shear. Proceedings of the First U.S. Congress of Applied Mechanics, ASME, New York, 1951, 403.

22. Hartmann, A. J. Elastic lateral buckling of continuous beams. Journal of the Structural Division. ASCE, Vol. 93, No. ST4, August 1967, 11–26.
23. Trahair, N. S. The Behavior and Design of Steel Structures. Chapman & Hall, London, 1977.
24. Chen, W. F., and Atsuta, T. Theory of Beam-Columns. Vol. 2: Space Behavior and Design. McGraw-Hill, NY, 1977.
25. Flint, A. R. Stability and strength of stocky beams. Journal of Mechanics and Physics of Solids. Vol. 1, No. 90, 1953.
26. Johnston, B. G., editor. SSRC Guide to Stability Design Criteria for Metal Structures. Third edition. John Wiley, NY, 1976.
27. Galambos, T. V. Inelastic lateral buckling of beams. Journal of the Structural Division. ASCE, Vol. 89, No. ST5, October 1963, 217.
28. Lay, M. G., and Galambos, T. V. The inelastic behavior of beams under moment gradient. Journal of the Structural Division. ASCE, Vol. 93, No. ST1, February 1967, 381–399.
29. Nethercot, D. A., and Trahair, N. S. Inelastic lateral buckling of determinate beams. Journal of the Structural Division. ASCE, Vol. 102, No. ST4, April 1976, 701–717.
30. Specification for the Design, Fabrication and Erection of Structural Steel for Buildings. AISC, Chicago, November 1978.
31. Load and Resistance Factor Design Specification for Structural Steel Buildings. AISC, Chicago, November, 1986.
32. Yura, J. A., Galambos, T. V. and Ravindra, M. K. The bending resistance of steel beams. Journal of the Structural Division. ASCE, Vol. 104, No. ST9, September, 1978, 1355–1369.

Chapter 6

ENERGY AND NUMERICAL METHODS

6.1 INTRODUCTION

In the preceding chapters, the critical loads of columns (Chaper 2) and beams (Chapter 5) have been determined by the eigenvalue (or bifurcation) approach. In an eigenvalue analysis, the members are assumed to be geometrically perfect. As a result of this assumption, the lateral deflections of a centrally loaded column or a beam subjected to in-plane forces will not occur until the applied load reaches a critical value. At this critical load, a small disturbance on the member will produce a large lateral deflection.

To obtain the critical load, we used the method of neutral equilibrium in which a linear differential equation is first written down for the member in a slightly deformed state. The solution to the characteristic equation derived from this governing differential equation then gives the critical load of the member.

To account for the inelasticity in the eigenvalue analysis, the concept of *effective modulus* (tangent modulus and reduced modulus) was used: here, the elastic modulus is simply replaced by an effective modulus. Since the effect of inelasticity is to reduce the stiffness of the member, the *plastic buckling load* in an inelastic analysis is always smaller than that of the elastic buckling load.

As we said, eigenvalue analysis assumes that a member is geometrically perfect. If the member is geometrically *imperfect,* lateral deflection will begin as soon as the load is applied. As a result, the method of neutral equilibrium, which is for determining the critical load for a perfect member, cannot be applied for the case of an imperfect member. Instead, a more complex analysis, known as the *load-deflection (or stability*) *approach*, must be used. In this approach, the complete load-deflection response of the member is traced from the start of loading up to the critical load. Such an analysis for the relatively simple case of *elastic frameworks* has been described in Section 4.5, Chapter 4. Rigorous load-deflection analysis for inelastic, imperfect members is rather complicated. As a consequence, recourse must be had to numerical means. We shall discuss two such numerical methods in the later part of this chapter.

In this chapter, we will present two approximate methods of stability analyses. They are the *energy method* and the *numerical method*. In the energy method, an approximate value for the critical load is determined by examining the variation and balance of energies before and after buckling in a structural system. It is valid only for structural systems that are elastic and conservative. For an inelastic system, recourse must be had to numerical methods to obtain solutions. We will present two such methods in this chapter. They are the *Newmark method* and the *numerical integration method*. Since iterations are involved in these methods, it is most convenient and efficient to implement the solution procedures in a computer.

6.2 PRINCIPLE OF VIRTUAL WORK

In this section, we will explain the *principle of virtual work* as a precursor to the discussion of the energy method.

In applying the principle of virtual work to structural analysis, we first apply a *virtual displacement* to a structural system and then write down the corresponding work done by the system on this virtual (or imaginary) displacement. The work done by the forces in the system on the virtual displacement is called *virtual work*. The *principle of virtual work* states that a system is in its equilibrium state if the virtual work done by all the forces acting on it is zero for *every* virtual displacement.

The validity of this principle can be shown easily by considering a particle in a system subjected to forces $F_1, F_2, \ldots, F_n$ as shown in Fig. 6.1. If we denote F_R as the resultant of the forces acting on this particle, the virutal work done by the resultant force F_R acting through the virtual displacement δr is

$$\delta W = F_R \delta r_R \tag{6.2.1}$$

where δr_R is the component of virtual displacement in the direction of F_R.

However, if the system of forces $F_1, F_2, \ldots, F_n$ is in equilibrium, their resultant force F_R must be zero. Consequently, the work done by the resultant force is also zero. Hence, if the system of forces acting on the particle is in equilibrium, Eq. (6.2.1) vanishes

$$\delta W = F_R \delta r_R = 0 \tag{6.2.2}$$

6.2 Principle of Virtual Work

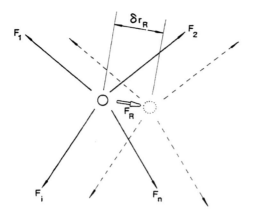

FIGURE 6.1 Particle subjected to a system of forces

Equation (6.2.2) is the mathematical statement of the principle of virtual work. It is important to observe that the virtual displacement δr_R need not be infinitesimal; it can be *finite* so long as the system of forces remains constant in magnitude and direction during the virtual displacement. Thus, the principle of virtual work is applicable regardless of whether the virtual displacements are small or finite. Furthermore, it should be noted that the virtual displacement is purely imaginary. It does not need to bear any relationship to the actual displacement experienced by the system under the action of the applied forces. The only requirement is that these virtual displacements are *kinematically admissible*, that is, they are conformable with the kinematic constraints of the structural system.

The requirement that the virtual displacements be kinematically admissible is introduced as a matter of convenience rather than as an absolute necessity. By using virtual displacements that are kinematically admissible, the number of arbitrary displacements, which are needed in establishing the equilibrium condition of the system, is greatly reduced.[1] For example, consider a rigid mass confined in a frictionless tube acted on by two forces F_1 and F_2 (Fig. 6.2a). We shall use the principle of virtual work to determine whether the mass is in equilibrium under the applied forces. Let us introduce an arbitrary displacement to the mass. If we ignore the geometric constraints and introduce an arbitrary displacement δx to the mass in the x-direction (Fig. 6.2b) (the symbol δ denotes a variation), the virtual work done by the mass is

$$\delta W = R_x \delta x \qquad (6.2.3)$$

in which R_x is the x-component reaction force acting on the mass by the tube.

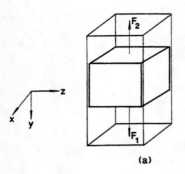
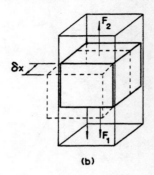

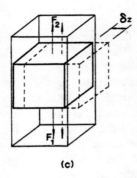
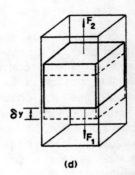

FIGURE 6.2 Rigid mass in a frictionless tube

Similarly, if we introduce an arbitrary displacement δz to the mass in the z-direction (Fig. 6.2c), the virtual work done by the mass is

$$\delta W = R_z \delta_z \tag{6.2.4}$$

in which R_z is the z-component reaction force acting on the mass by the tube.

Now, if we introduce an arbitrary displacement δy to the mass in the y-direction (Fig. 6.2d), the virtual work done by the mass is

$$\delta W = F_1 \delta y - F_2 \delta y \tag{6.2.5}$$

The minus sign appearing in the second term is due to the fact that the direction of F_2 and δy are opposite to each other.

If the mass is in equilibrium, then the virtual work done by the forces upon any arbitrary displacement must be zero. In other words, if the

6.2 Principle of Virtual Work

mass is in equilibrium, then Eq. (6.2.3) to Eq. (6.2.5) must vanish.

$$\delta W = R_x \delta x = 0 \qquad (6.2.6)$$

$$\delta W = R_z \delta z = 0 \qquad (6.2.7)$$

$$\delta W = F_1 \delta y - F_2 \delta y = 0 \qquad (6.2.8)$$

Since none of the arbitrary displacements (δx, δz, δy) are zero, from Eq. (6.2.6) we must have $R_x = 0$ and from Eq. (6.2.7), $R_z = 0$. Finally, from Eq. (6.2.8), we must have $F_1 = F_2$. However, the fact that $R_x = R_z = 0$ is quite obvious by inspection of Fig. 6.2a. Therefore, the only useful information we can obtain concerning the equilibrium conditions of the rigid mass is from Eq. (6.2.8), which states that if the mass is in equilibrium, the applied forces F_1 and F_2 must be equal in magnitude, opposite in direction and applied on the same straight line. Note that since Eq. (6.2.8) was obtained by introducing an arbitrary displacement δy that does not violate the condition of geometric constraints of the mass, it follows that we need only consider the kinematically admissible displacement (i.e., virtual displacement) in order to establish the equilibrium condition of the mass.

In the light of this example, we can conclude that a necessary and sufficient condition for the mass in equilibrium is when

$$\delta W = (F_1 - F_2) \delta y = F_y \delta y = 0 \qquad (6.2.9)$$

In Eq. (6.2.9), $F_y = (F_1 - F_2)$ is the resultant of the applied forces in the y-direction and δy is the virtual displacement in the y-direction. Note that the values of the forces used in calculating the virtual work are fixed at their full value during the virtual displacement. This is because the virtual displacement is purely imaginary. Since the displacement is not real, a change in values of the forces, which may occur during a real displacement of the system, will not occur for a virtual displacement.

Another important point is that if we use only arbitrary displacements that do not violate the kinematic constraints of the system, we need to consider only the work done by the applied forces in establishing the equilibrium condition of the system. This is because the reactions that are associated with rigid constraints will do no work and hence need not be considered in calculating the virtual work of the system.

The rigid mass example discussed above has only one degree of freedom because it is confined to move only vertically. As a result, the only virtual displacement that is kinematically permissible is δy.

For a rigid system that has n degrees of freedom, Eq. (6.2.9) becomes

$$\delta W = \sum_{i=1}^{n} F_i \delta q_i = 0 \qquad (6.2.10)$$

in which q_i are the *generalized displacements* of the system and δq_i are independent variations of the generalized displacements. F_i are the *generalized forces* acting in the direction of q_i. The word *generalized* is used here to emphasize that the displacements can be either translational or rotational and that the forces can be concentrated forces or moments. If the forces are distributed, Eq. (6.2.10) still applies, provided that the summation sign is replaced by an integral sign.

For systems that are not rigid, deformation will occur when forces are applied to the systems. As a result, in addition to doing *external work*, the systems will do *internal work* when virtual displacements are introduced. Thus, the virtual work equation for the system that is not rigid, viz. a deformable system, takes the form

$$\delta W = \delta W_{ext} + \delta W_{int} = 0 \qquad (6.2.11)$$

where

δW_{ext} = external virtual work
δW_{int} = internal virtual work

Equation (6.2.11) states that a deformable system is in equilibrium if the total virtual work done by the external forces and the virtual work done by the internal forces equals zero for any virtual displacement.

FIGURE 6.3 Deformation system

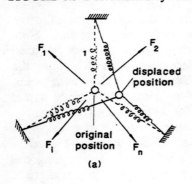
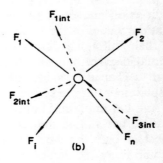

6.2 Principle of Virtual Work

To show the validity of Eq. (6.2.11), let us consider a system that consists of a particle attached to three springs (Fig. 6.3a) subjected to a system of external forces $F_{1\,ext}, F_{2\,ext}, \ldots, F_{n\,ext}$. Under the action of this set of external forces, the particle will displace. This displacement will induce internal forces $F_{1\,int}, F_{2\,int}, F_{3\,int}$ in the springs (Fig. 6.3b). Denoting $F_{R\,ext}$ and $F_{R\,int}$ as the resultants of the set of external forces and the set of internal forces, respectively, and F_R as the resultant of all the forces (external and internal) acting on the particle (Fig. 6.3c), if a virtual displacement δr_R is introduced to the particle in its equilibrium position in the direction of F_R (Fig. 6.3d), the total virtual work done on the particle will be

$$\delta W = \delta W_{ext} + \delta W_{int}$$
$$= F_{R\,ext}\delta r_R - F_{R\,int}\delta r_R$$
$$= F_R \delta r_R \qquad (6.2.12)$$

where $F_R = F_{R\,ext} - F_{R\,int}$.

If the system of external and internal forces acting on the particle is in equilibrium, F_R is zero. As a result, Eq. (6.2.12) vanishes and thus verifies Eq. (6.2.11).

As an example to demonstrate the use of Eq. (6.2.11), consider a spring-mass assemblage (Fig. 6.4a). Because of the weight of the mass, the spring will stretch by an amount Δ. As a result of the elongation, an internal force will develop in the spring. Let us denote this internal spring force by F_{spring}. The external force acting on the system is the weight of the mass W_{mass}. If the system is in equilibrium, then both the spring and the mass must be in equilibrium. In Fig. 6.4b a free body of the mass is shown. It is obvious from the diagram that if the mass is to be in equilibrium, the spring force F_{spring} must be balanced by the weight of the

FIGURE 6.4 Spring-mass system

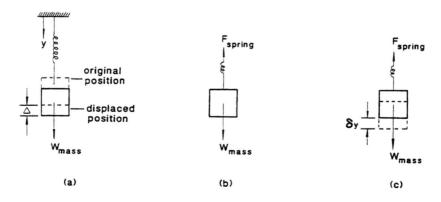

(a)   (b)   (c)

mass W_{mass}. In other words, we must have

$$F_{spring} = W_{mass} \qquad (6.2.13)$$

Now, let us establish this equilibrium condition using the principle of virtual work. Suppose we introduce a virtual displacement δy from the equilibrium position of the system (Fig. 6.4c). The virtual work done by the external force on the mass is

$$\delta W_{ext} = W_{mass} \delta y \qquad (6.2.14)$$

and the virtual work done by the internal force on the mass is

$$\delta W_{int} = -F_{spring} \delta y \qquad (6.2.15)$$

The internal virtual work is negative because the direction of internal force F_{spring} is opposite to that of the virtual displacement δy.

Using Eq. (6.2.11), we have

$$\delta W_{ext} + \delta W_{int} = W_{mass} \delta y - F_{spring} \delta y$$
$$= (W_{mass} - F_{spring}) \delta y = 0 \qquad (6.2.16)$$

Since δy is arbitrary and so not necessarily equal to zero, the only way that Eq. (6.2.16) can be satisfied is when

$$W_{mass} - F_{spring} = 0 \qquad (6.2.17)$$

or

$$W_{mass} = F_{spring} \qquad (6.2.18)$$

Equation (6.2.18) expresses the equilibrium condition of the system. Note that the same equilibrium condition can be established by using the conventional free-body approach [Eq. (6.2.13)]. In fact, for this simple example, it is much easier and more direct to use the free-body approach than the virtual work equation to establish the equilibrium condition of the system. However, for more complicated systems the virtual work equation works better.

If we restrict ourselves to elastic systems subjected to conservative forces, an important principle in mechanics known as the *principle of stationary total potential energy* can be developed from the virtual work principle. This is described in what follows.

6.3 PRINCIPLE OF STATIONARY TOTAL POTENTIAL ENERGY

When external forces are applied to an assemblage of deformable bodies in a structural system, the system will deform. If the system returns to its original configuration when the applied forces are gradually removed, the system is said to exhibit *elastic* behavior. An elastic system can be linear or nonlinear. For a linear elastic system, the stress and strain relationship

6.3 Principle of Stationary Total Potential Energy

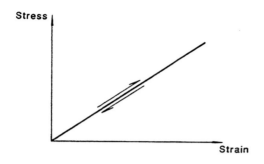

(a) Linear Elastic System

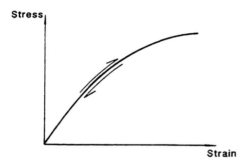

(b) Nonlinear Elastic System

FIGURE 6.5 Linear and nonlinear elastic system

of the material is related linearly (Fig. 6.5a). For a *nonlinear* elastic system, the stress and strain relationship is related by some nonlinear functions (Fig. 6.5b). In either case, loading and unloading will follow the same path as indicated in Fig. 6.5. In other words, for elastic systems, there is a unique one-to-one relationship between stress and strain.

For an elastic system, the work done by the external forces on the system is stored as *strain energy* in the system. Thus, if a virtual displacement is introduced to the system, the external virtual work done by the external forces on the system δW_{ext} is stored as virtual strain energy δU in the system

$$\delta W_{ext} = \delta U \qquad (6.3.1)$$

From the principle of virtual work [Eq. (6.2.11)], we can write

$$\delta W_{ext} = -\delta W_{int} \qquad (6.3.2)$$

By comparing Eqs. (6.3.1) and (6.3.2), the variation of internal virtual

work can be related to the variation of virtual strain energy in the system by
$$\delta W_{int} = -\delta U \qquad (6.3.3)$$

If the external forces acting on the system are conservative (that is, if the work done by these forces is path independent and fully recoverable in a loading/unloading cycle), the increase in the external work done on the system will correspond to a decrease in the potential energy of the system. As a result, we can also write
$$\delta W_{ext} = -\delta V \qquad (6.3.4)$$
where V is the *potential energy* of the system.

Upon substituting Eqs. (6.3.3) and (6.3.4) into Eq. (6.2.11), we obtain
$$-\delta V - \delta U = 0 \qquad (6.3.5)$$
or
$$\delta(U + V) = 0 \qquad (6.3.6)$$

The sum of the strain energy and potential energy of the system is the *total potential energy* of the system. Using the symbol Π to denote the total potential energy of a system, Eq. (6.3.6) can be written as
$$\delta \Pi = 0 \qquad (6.3.7)$$

Equation (6.3.7) is the mathematical statement of the *principle of stationary total potential energy*. In words, this principle states that if an elastic system acted on by conservative forces is in equilibrium, the total potential energy of the system must be *stationary*.

A stationary value may correspond to a minimum or maximum value of the total potential energy function. A minimum value indicates that the equilibrium is *stable* and a maximum value indicates that the equilibrium is *unstable*. To investigate whether a system is in a stable or unstable equilibrium condition, while at the same time ensuring that the first variation of the total potential energy vanishes (that is, $\delta \Pi = 0$), one must also evaluate the second variation of the total potential energy ($\delta^2 \Pi$) of the system. If $\delta^2 \Pi > 0$, the system is in a stable equilibrium condition, and if $\delta^2 \Pi < 0$, the system is in an unstable equilibrium condition.

For a continuous system (that is, a system with an infinite number of degrees of freedom), the use of the principle of stationary total potential energy to establish the equilibrium conditions of the system requires us to resort to a special branch of calculus known as the *calculus of variations*. This is described in the following section, where we shall attempt to use the principle of stationary total potential energy together with the calculus of variation to establish the equilibrium conditions describing the in-plane buckling behavior of a hinged-hinged column and the lateral buckling behavior of a simply supported beam of rectangular cross section under pure bending.

6.4 CALCULUS OF VARIATIONS

The *calculus of variations* is concerned with the evaluation of stationary values or extremals (maximum or minimum quantities) of *functionals*. Functionals are definite integrals of function(s). The function(s) in these integrals are unknown(s) and the calculus of variations is used to determine the conditions for which these function(s) will assume a stationary value.

The calculus of variations differs from ordinary calculus in that the former deals with functionals while the latter deals with functions. Functionals are functions of functions, whereas functions are just functions of variables. In applying the calculus of variations to extremize a functional, one does not obtain the function. Instead, one obtains conditions that the function(s) must satisfy to ensure that the functional will assume a stationary value. In contrast, in applying the ordinary calculus to extremize a function, one obtains the variable(s) for which the function will assume an extremum value.

Detailed discussion of the theory of the calculus of variations is beyond the scope of this book. However, we shall use two examples to demonstrate the application of calculus of variation in conjunction with the principle of stationary total potential energy in establishing the necessary conditions for the solutions of the following:

1. The in-plane buckling problem of a hinged-hinged column.
2. The lateral torsional buckling problem of a rectangular beam under pure bending.

6.4.1 In-Plane Buckling of a Hinged-Hinged Column

Figure 6.6 shows a perfect column in a slightly deformed configuration at the instance of buckling. Here, we shall use the calculus of variation in conjunction with the principle of stationary total potential energy to determine the conditions that must be satisfied by the column if it is to be in equilibrium in this slightly deformed configuration.

In applying the principle of stationary total potential energy [Eq. (6.3.6) or (6.3.7)], we need to evaluate the strain energy U and the potential energy V of the column. They are derived separately in what follows.

Strain Energy

In general, the strain energy function of a three-dimensional elastic solid element (linear or nonlinear) can be expressed as

$$U = \int dU = \int_V \int_{\varepsilon_0}^{\varepsilon_t} \sigma_{ij}\, d\varepsilon_{ij}\, dV \qquad (6.4.1)$$

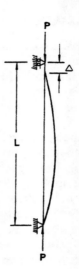

FIGURE 6.6 Hinged-hinged column

where, for brevity, we have used

$\sigma_{ij}\, d\varepsilon_{ij} = \sigma_x\, d\varepsilon_x + \sigma_y\, d\varepsilon_y + \tau_{xy}\, d\gamma_{xy} + \cdots$
σ_{ij} = stress tensor (Fig. 6.7)
ε_{ij} = strain tensor
V = volume of the element
ε_0 = initial value of the strain tensor
ε_f = final value of the strain tensor

If the system is linear elastic, Eq. (6.4.1) can be simplified to

$$U = \frac{1}{2}\int_V \sigma_{ij}\varepsilon_{ij}\, dV \qquad (6.4.2)$$

For the column shown in Fig. 6.6, the strain energy consists of two components; that is due to the axial shortening effect U_a and the bending curvature effect U_b. They are described as follows:

Strain Energy Due to Axial Shortening Effect, U_a. As a result of the

FIGURE 6.7 Stress tensor

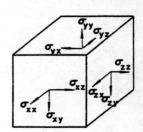

6.4 Calculus of Variations

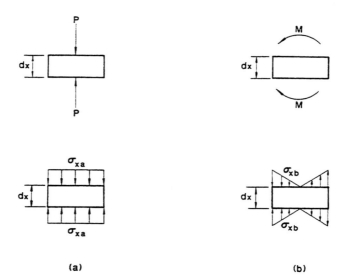

FIGURE 6.8 Axial and bending stresses in an elastic column of buckling

axial force P, axial stress σ_{xa} will be present in the cross section (Fig. 6.8a); this stress can be expressed as

$$\sigma_{xa} = \frac{P}{A} \qquad (6.4.3)$$

where

P = axial force
A = cross-sectional area

The axial strain associated with this stress is

$$\varepsilon_{xa} = \frac{\sigma_{xa}}{E} = \frac{P}{EA} \qquad (6.4.4)$$

Since the other components of stress, except the axial stresses due to axial force and to bending moment σ_{xb}, are zero, we can substitute σ_{xa} for σ_{ij} and ε_{xa} for ε_{ij} in Eq. (6.4.2) to obtain U_a as

$$U_a = \frac{1}{2} \int \left(\frac{P}{A}\right)\left(\frac{P}{EA}\right) dV \qquad (6.4.5)$$

Substitute dV by $dA\, dx$, where A is the area of the cross section, and we have

$$U_a = \frac{1}{2}\int_0^L \int_A \frac{1}{E}\left(\frac{P}{A}\right)^2 dA\, dx$$

$$= \frac{1}{2}\int_0^L \frac{P^2}{EA} dx \qquad (6.4.6)$$

Equation (6.4.6) can be expressed in terms of the axial strain $\varepsilon_{xa} = du/dx$, where u is the axial displacement by recognizing that

$$P = EA\varepsilon_{xa} = EA\frac{du}{dx} \qquad (6.4.7)$$

therefore

$$U_a = \frac{1}{2}\int_0^L EA\left(\frac{du}{dx}\right)^2 dx \qquad (6.4.8)$$

Strain Energy Due to Bending, U_b. Figure 6.8b shows the stress distribution over the cross section of the column due to a bending moment M. From the basic mechanics of materials, it can be shown that the bending stress σ_{xb} is given by

$$\sigma_{xb} = \frac{My}{I} \qquad (6.4.9)$$

where
 M = bending moment
 y = distance from neutral axis to the point where σ_{xb} acts
 I = moment of inertia of the cross section

The corresponding bending strain is

$$\varepsilon_{xb} = \frac{\sigma_{xb}}{E} = \frac{My}{EI} \qquad (6.4.10)$$

Since simple bending is basically a one-dimensional problem, the strain energy of the column due to bending can be obtained by substituting σ_{ij} by σ_{xb} and ε_{ij} by ε_{xb} into Eq. (6.4.2)

$$U_b = \frac{1}{2}\int_V \left(\frac{My}{I}\right)\left(\frac{My}{EI}\right) dV = \int_V \frac{1}{E}\left(\frac{My}{I}\right)^2 dV \qquad (6.4.11)$$

If we replace dV by $dA\, dx$, where A is the area of the cross section, we can write Eq. (6.4.11) as

$$U_b = \frac{1}{2}\int_0^L \int_A \frac{1}{E}\left(\frac{My}{I}\right)^2 dA\, dx \qquad (6.4.12)$$

or

$$U_b = \frac{1}{2}\int_0^L \frac{1}{E}\left(\frac{M}{I}\right)^2 \left(\int_A y^2\, dA\right) dx \qquad (6.4.13)$$

6.4 Calculus of Variations

Recognizing that $\int_A y^2 \, dA = I$, the moment of inertia of the cross section, we can write

$$U_b = \frac{1}{2} \int_0^L \frac{M^2}{EI} dx \qquad (6.4.14)$$

An alternate form for U_b in terms of the curvature d^2v/dx^2 instead of the moment M can be obtained if we substitute $M = -EI\left(\frac{d^2v}{dx^2}\right)$ into Eq. (6.4.14)

$$U_b = \frac{1}{2} \int_0^L EI\left(\frac{d^2v}{dx^2}\right)^2 dx \qquad (6.4.15)$$

where v is the lateral deflection of the column.

By combining Eq. (6.4.8) with Eq. (6.4.15), the total strain energy of the column in Fig. 6.6 is

$$U = U_a + U_b$$
$$= \frac{1}{2} \int_0^L EA\left(\frac{du}{dx}\right)^2 dx + \frac{1}{2} \int_0^L EI\left(\frac{d^2v}{dx^2}\right)^2 dx \qquad (6.4.16)$$

Potential Energy

Since the work done by the external forces can be represented by a decrease in potential energy of the system,

$$W_{ext} = -V \qquad (6.4.17)$$

it follows that

$$V = -W_{ext} \qquad (6.4.18)$$

In other words, we can calculate the potential energy of the system by evaluating the work done on the system by the external forces.

For the centrally loaded column shown in Fig. 6.6, the external work done on the system is

$$W_{ext} = P\Delta \qquad (6.4.19)$$

where Δ is the shortening of the column.

It is important to observe that the factor one-half does not appear in Eq. (6.4.19) because at buckling the force P is acting at its full constant value when the column passes from the straight to the slightly bent configuration. As a result, the factor one-half, which should appear if the force is increased linearly from zero to its full value during the deformation process, does not appear here.

The shortening of the column Δ in Eq. (6.4.19) consists of two parts: an axial shortening Δ_a and a bending shortening Δ_b. The axial shortening

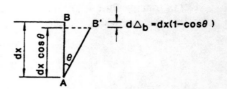

FIGURE 6.9 Bending shortening of an infinitesimal element

can be obtained by integrating the axial strain $\varepsilon_{xa} = du/dx$ over the length of the column, that is

$$\Delta_a = \int_0^L \varepsilon_{xa}\, dx = \int_0^L \frac{du}{dx}\, dx \tag{6.4.20}$$

The bending shortening, on the other hand, requires more computations. Consider Fig. 6.9, in which an infinitesimal element of length dx is shown. Before bending occurs, the element assumes the position AB. After bending, the element assumes the position AB'. The differential shortening due to this change in the element is

$$d\Delta_b = dx(1 - \cos\theta) \tag{6.4.21}$$

where θ is the angle between the undeformed and deformed elements.

The trigonometric function $\cos\theta$ can be expressed in series form as

$$\cos\theta = 1 - \tfrac{1}{2}\theta^2 + \tfrac{1}{24}\theta^4 - \cdots \tag{6.4.22}$$

Using only the first two terms in the series and substituting them into Eq. (6.4.21), we have

$$d\Delta_b = \tfrac{1}{2}\theta^2\, dx \tag{6.4.23}$$

For small bending deformation, θ can be represented by dv/dx. Thus, Eq. (6.4.23) can be written as

$$d\Delta_b = \frac{1}{2}\left(\frac{dv}{dx}\right)^2 dx \tag{6.4.24}$$

The total shortening of the column due to bending can be obtained by integrating Eq. (6.4.24) over the length of the column:

$$\Delta_b = \int_0^L d\Delta_b = \frac{1}{2}\int_0^L \left(\frac{dv}{dx}\right)^2 dx \tag{6.4.25}$$

Finally, in view of Eqs. (6.4.18) and (6.4.19), the potential energy of the column in a slightly bent configuration is

$$\begin{aligned} V &= -P\Delta \\ &= -P(\Delta_a + \Delta_b) \\ &= -P\int_0^L \left(\frac{du}{dx}\right) dx - \frac{P}{2}\int_0^L \left(\frac{dv}{dx}\right)^2 dx \end{aligned} \tag{6.4.26}$$

Total Potential Energy of the Column

By combining Eqs. (6.4.16) and (6.4.26), the total potential energy of the column can be expressed as

$$\Pi = U + V$$
$$= \frac{1}{2}\int_0^L EA\left(\frac{du}{dx}\right)^2 dx + \frac{1}{2}\int_0^L EI\left(\frac{d^2v}{dx^2}\right)^2 dx$$
$$- P\int_0^L \left(\frac{du}{dx}\right) dx - \frac{P}{2}\int_0^L \left(\frac{dv}{dx}\right)^2 dx \qquad (6.4.27)$$

For equilibrium, the first variation of the total potential energy of the column must vanish

$$\delta\Pi = \delta(U + V) = \delta U + \delta V = 0 \qquad (6.4.28)$$

since

$$\delta U = \int_0^L EA\left(\frac{du}{dx}\right)\delta\left(\frac{du}{dx}\right) dx + \int_0^L EI\left(\frac{d^2v}{dx^2}\right)\delta\left(\frac{d^2v}{dx^2}\right) dx$$
$$= EA\int_0^L \left(\frac{du}{dx}\right)\left(\frac{d\,\delta u}{dx}\right) dx + EI\int_0^L \left(\frac{d^2v}{dx^2}\right)\left(\frac{d^2\,\delta v}{dx^2}\right) dx \qquad (6.4.29)$$

and

$$\delta V = -P\int_0^L \delta\left(\frac{du}{dx}\right) dx - P\int_0^L \left(\frac{dv}{dx}\right)\delta\left(\frac{dv}{dx}\right) dx$$
$$= -P\int_0^L \frac{d\,\delta u}{dx} dx - P\int_0^L \left(\frac{dv}{dx}\right)\left(\frac{d\,\delta v}{dx}\right) dx \qquad (6.4.30)$$

therefore, Eq. (6.4.28) can be written as

$$\delta\Pi = EA\int_0^L \left(\frac{du}{dx}\right)\left(\frac{d\,\delta u}{dx}\right) dx + EI\int_0^L \left(\frac{d^2v}{dx^2}\right)\left(\frac{d^2\,\delta v}{dx^2}\right) dx$$
$$- P\int_0^L \frac{d\,\delta u}{dx} dx - P\int_0^L \left(\frac{dv}{dx}\right)\left(\frac{d\,\delta v}{dx}\right) dx = 0 \qquad (6.4.31)$$

By examining Eqs. (6.4.29) and (6.4.30), the reader should recognize the similarity in operation between the variational operator δ and the differential operator d. Furthermore, it should be noted that the two operators are commutative.

To transform the terms involving derivatives of δu or δv so that a common factor of δu or δv will appear in the terms of Eq. (6.4.31), we need to make use of integration by parts. For the first term

$$EA\int_0^L \left(\frac{du}{dx}\right)\left(\frac{d\,\delta u}{dx}\right) dx$$

letting

$$p = \frac{du}{dx} \quad \text{and} \quad dq = \frac{d\,\delta u}{dx} dx \qquad (6.4.32)$$

and knowing

$$\int_0^L p \, dq = pq \Big|_0^L - \int_0^L q \, dp \qquad (6.4.33)$$

we have

$$EA \int_0^L \left(\frac{du}{dx}\right)\left(\frac{d\,\delta u}{dx}\right) dx = EA \frac{du}{dx} \delta u \Big|_0^L - EA \int_0^L \frac{d^2 u}{dx^2} \delta u \, dx \qquad (6.4.34)$$

Similarly, for the second term

$$EI \int_0^L \left(\frac{d^2 v}{dx^2}\right)\left(\frac{d^2 \delta v}{dx^2}\right) dx$$

if we let

$$p = \frac{d^2 v}{dx^2} \quad \text{and} \quad dq = \frac{d^2 \delta v}{dx^2} dx \qquad (6.4.35)$$

and using Eq. (6.4.33), we have

$$EI \int_0^L \left(\frac{d^2 v}{dx^2}\right)\left(\frac{d^2 \delta v}{dx^2}\right) dx = EI \left(\frac{d^2 v}{dx^2}\right)\left(\frac{d\,\delta v}{dx}\right) \Big|_0^L$$
$$- EI \int_0^L \left(\frac{d^3 v}{dx^3}\right)\left(\frac{d\,\delta v}{dx}\right) dx \qquad (6.4.36)$$

The derivative of δv in the last term of Eq. (6.4.36) can be eliminated by using integration by parts again. Letting

$$p = \frac{d^3 v}{dx^3} \quad \text{and} \quad dq = \left(\frac{d\,\delta v}{dx}\right) dx \qquad (6.4.37)$$

we can obtain

$$-EI \int_0^L \left(\frac{d^3 v}{dx^3}\right)\left(\frac{d\,\delta v}{dx}\right) dx = -EI \left(\frac{d^3 v}{dx^3} \delta v\right) \Big|_0^L$$
$$+ EI \int_0^L \frac{d^4 v}{dx^4} \delta v \, dx \qquad (6.4.38)$$

Because $\delta v = 0$ at $x = 0$ and $x = L$, and because we are concerned only with virtual displacements that are kinematically admissible, the first term of Eq. (6.4.38) vanishes. With this in mind, and upon substituting Eq.

6.4 Calculus of Variations

(6.4.38) into Eq. (6.4.36), we have

$$EI\int_0^L \left(\frac{d^2v}{dx^2}\right)\left(\frac{d^2\delta v}{dx^2}\right)dx = EI\left(\frac{d^2v}{dx^2}\right)\left(\frac{d\delta v}{dx}\right)\Big|_0^L$$
$$+ EI\int_0^L \frac{d^4v}{dx^4}\delta v\, dx \quad (6.4.39)$$

For the third term of Eq. (6.4.31)

$$-P\int_0^L \frac{d\,\delta u}{dx}\,dx$$

Using integration by parts [Eq. (6.4.33)] with

$$p = 1 \quad \text{and} \quad dq = \frac{d\,\delta u}{dx}dx \quad (6.4.40)$$

we can write

$$-P\int_0^L \frac{d\,\delta u}{dx}dx = -P\,\delta u\,\Big|_0^L \quad (6.4.41)$$

Finally, for the last term of Eq. (6.4.31)

$$-P\int_0^L \left(\frac{dv}{dx}\right)\left(\frac{d\,\delta v}{dx}\right)dx$$

by letting

$$p = \frac{dv}{dx} \quad \text{and} \quad dq = \frac{d\,\delta v}{dx}dx \quad (6.4.42)$$

we can obtain

$$-P\int_0^L \left(\frac{dv}{dx}\right)\left(\frac{d\,\delta v}{dx}\right)dx = -P\left(\frac{dv}{dx}\right)\delta v\,\Big|_0^L$$
$$+ P\int_0^L \frac{d^2v}{dx^2}\delta v\, dx \quad (6.4.43)$$

Realizing that $\delta v = 0$ at $x = 0$ and $x = L$, Eq. (6.4.43) becomes

$$-P\int_0^L \left(\frac{dv}{dx}\right)\left(\frac{d\,\delta v}{dx}\right)dx = P\int_0^L \frac{d^2v}{dx^2}\delta v\, dx \quad (6.4.44)$$

Substituting Eqs. (6.4.34), (6.4.39), (6.4.41), and (6.4.44) into Eq. (6.4.31), upon rearranging we obtain

$$\delta\Pi = \left(EA\frac{du}{dx} - P\right)\delta u\Big|_0^L - \int_0^L EA\frac{d^2u}{dx^2}dx\,\delta u$$
$$+ \left(EI\frac{d^2v}{dx^2}\right)\left(\frac{d\,\delta v}{dx}\right)\Big|_0^L + \int_0^L \left(EI\frac{d^4v}{dx^4} + P\frac{d^2v}{dx^2}\right)dx\,\delta v$$
$$= 0 \quad (6.4.45)$$

or

$$\left(EA\frac{du}{dx}-P\right)\delta u|_{x=L}-\left(EA\frac{du}{dx}-P\right)\delta u|_{x=0}-\int_0^L EA\frac{d^2u}{dx^2}dx\,\delta u$$

$$+\left(EI\frac{d^2v}{dx^2}\right)\left(\delta\frac{dv}{dx}\right)\bigg|_{x=L}-\left(EI\frac{d^2v}{dx^2}\right)\left(\delta\frac{dv}{dx}\right)\bigg|_{x=0}$$

$$+\int_0^L\left(EI\frac{d^4v}{dx^4}+P\frac{d^2v}{dx^2}\right)dx\,\delta v$$

$$=0 \qquad (6.4.46)$$

Since δv and δu are arbitrary and can take on any value for the interval 0 to L, to ensure that Eq. (6.4.46) is satisfied regardless of the value of δv and δu, each and every term in Eq. (6.4.46) must vanish. Therefore, we have, from the first term, either

$$\left(EA\frac{du}{dx}-P\right)\bigg|_{x=L}=0 \qquad (6.4.47)$$

or

$$\delta u|_{x=L}=0 \qquad (6.4.48)$$

and from the second term

$$\left(EA\frac{du}{dx}-P\right)\bigg|_{x=0}=0 \qquad (6.4.49)$$

or

$$\delta u|_{x=0}=0 \qquad (6.4.50)$$

and from the third term

$$\int_0^L EA\frac{d^2u}{dx^2}dx=0$$

or

$$EA\frac{d^2u}{dx^2}=0 \qquad (6.4.51)$$

and from the fourth term

$$EI\frac{d^2v}{dx^2}\bigg|_{x=L}=0 \qquad (6.4.52)$$

or

$$\delta\left(\frac{dv}{dx}\right)\bigg|_{x=L}=0 \qquad (6.4.53)$$

and from the fifth term

$$EI\frac{d^2v}{dx^2}\bigg|_{x=0}=0 \qquad (6.4.54)$$

6.4 Calculus of Variations

or
$$\delta\left(\frac{dv}{dx}\right)\bigg|_{x=0} = 0 \qquad (6.4.55)$$

and from the sixth term
$$\int_0^L \left(EI\frac{d^4v}{dx^4} + P\frac{d^2v}{dx^2}\right) dx = 0$$

or
$$EI\frac{d^4v}{dx^4} + P\frac{d^2v}{dx^2} = 0 \qquad (6.4.56)$$

Close scrutiny of Eq. (6.4.46) shows that the first three terms contain only the axial displacement, u, as the variable, whereas the last three terms contain only the transverse or lateral displacement, v, as the variable. Since they are uncoupled, we can treat them separately. In fact, the first three terms in Eq. (6.4.46), and therefore the conditions expressed in Eqs. (6.4.47) to (6.4.51), pertain only to the axial shortening effect on the column, whereas the last three terms in Eq. (6.4.46), and so the conditions expressed in Eqs. (6.4.52) to (6.4.56), pertain only to the bending curvature effect on the column. When an axial force is applied to a column, it undergoes axial shortening. This behavior is described by Eqs. (6.4.47) to (6.4.51). When the axial force is increased to the point where bifurcation of equilibrium is reached, the column can be in equilibrium either in a straight or in a slightly bent configuration. The behavior of the column in the slightly bent configuration is described by Eqs. (6.4.52) to (6.4.56). In Chapter 2, when the buckling behavior of columns was described, only the bending deformation was considered, the axial deformation was ignored. This is because the axial deformation that occurs before buckling has no bearing on the evaluation of the critical loads of the columns. The critical loads were obtained by an eigenvalue analysis that considered only the bending behavior of the columns at the point of neutral equilibrium where the column can be in equilibrium in both the original straight and slightly bent configurations. Therefore, the equations describing the axial compression of the columns were never written. However, in this section a more general formulation, taking into account both axial and bending deformation, is employed. As a result, we obtain two sets of equations: one depicting the axial compression effect, and the other depicting the bending deformation effect. Since these two sets of equations are independent of each other, we shall examine them separately.

Axial Effect

Equations (6.4.47) and (6.4.49) express the natural (or force) boundary conditions of a column under an axial force P, whereas Eqs. (6.4.48) and

(6.4.50) express the geometric (or kinematic) boundary conditions of the column. For the column shown in Fig. 6.6, it is obvious that the axial displacement at the end $x = L$ is not zero, therefore $\delta u \neq 0$ at $x = L$. As a result, we must have from Eq. (6.4.47)

$$EA \frac{du}{dx}\bigg|_{x=L} = P \tag{6.4.57}$$

At $x = 0$, the end is pinned, therefore we must have

$$\delta u|_{x=0} = 0 \quad \text{or} \quad u|_{x=0} = 0 \tag{6.4.58}$$

In other words, the axial force at the top end of the column is equal to P and the axial displacement at the bottom end of the column is equal to zero.

Equation (6.4.51) is simply the differential equation for an axially loaded column subjected to concentrated force at its ends.

Bending Effect

Equations (6.4.52) and (6.4.54) express the natural (or force) boundary conditions of a column in a slightly bent configuration, whereas Eqs. (6.4.53) and (6.4.55) express the geometric (or kinematic) boundary conditions of the column. For the column shown in Fig. 6.6, it is obvious that the end slopes are not zero, therefore $\delta(dv/dx) \neq 0$ at $x = 0$ and $x = L$. Consequently, from Eqs. (6.4.52) and (6.4.54) we must have

$$EI \frac{d^2 v}{dx^2}\bigg|_{x=L} = 0 \tag{6.4.59}$$

and

$$EI \frac{d^2 v}{dx^2}\bigg|_{x=0} = 0 \tag{6.4.60}$$

Recall that $M = -EI(d^2 v/dx^2)$. Thus, Eqs. (6.4.59) and (6.4.60) indicate that there is no bending moment at the ends of the column.

Equation (6.4.56) is simply the differential equation for a column in a slightly bent configuration.

As have been demonstrated in the above example, by using the principle of stationary potential energy, not only are we able to obtain the differential equations of equilibrium of the member, but we obtain the boundary conditions as well. In using the principle, the only two quantities required are the strain energy function U and the potential energy function V of the system. Here, it is not necessary to draw the free-body diagrams, as would have been required in the conventional approach. Because of this, the energy principle will be more appropriate for problems of more complicated nature.

Note that if we have considered a fixed-fixed column or a fixed-pinned in this example, we will have obtained the same result. The reason is that the expressions for the strain energy U and the potential energy V will not change. Although it is quite obvious that U will not change because it is independent of the boundary conditions of the member, the fact that V, which is dependent upon end conditions and loadings of the members, will not change for a fixed-fixed or a fixed-pinned column is a fact that deserves some explanation. For example, for a fixed-fixed column, the expression for V is

$$V = -P\int_0^L \left(\frac{du}{dx}\right) dx - \frac{P}{2}\int_0^L \left(\frac{dv}{dx}\right)^2 dx - \left(M\frac{dv}{dx}\right)\bigg|_{x=0} - \left(M\frac{dv}{dx}\right)\bigg|_{x=L}$$
(6.4.61)

The first two terms in Eq. (6.4.61) were explained earlier in this section: they are the work done by the axial force P as the force P travels through an axial shortening Δ_a and a bending shortening Δ_b. The third and fourth terms represent the work done as the end moment acts through an end rotation dv/dx at $x = 0$ and $x = L$, respectively. However, for a fixed-fixed column

$$\frac{dv}{dx}\bigg|_{x=0} = \frac{dv}{dx}\bigg|_{x=L} = 0$$

therefore, Eq. (6.4.61) reduces to

$$V = -P\int_0^L \left(\frac{du}{dx}\right) dx - \frac{P}{2}\int_0^L \left(\frac{dv}{dx}\right)^2 dx \qquad (6.4.62)$$

which is the same as that for a pinned-pinned column [Eq. (6.4.26)].

The same argument applies for a fixed-pinned column. At the fixed end, the work done by the end moment equals zero because the end rotation is zero. As a result, the only work done is that resulting from P acting through the axial and bending shortening of the member. Figure 6.10 summarizes the three cases discussed. Since the bending deformation is of primary concern, the conditions shown in the figure correspond only to the effect of bending.

To further demonstrate the use of the principle of stationary potential energy to tackle problems of elastic stability, we will establish the conditions describing the lateral torsional buckling of a rectangular section in the following section.

6.4.2 Lateral Torsional Buckling of a Rectangular Beam Under Pure Bending

Figure 6.11 shows a rectangular beam subjected to pure bending. Here, as in Chapter 5, two sets of axes are established. A fixed global coordinate axes (x, y, z) relative to the undeformed member and a movable local coordinate axes (x', y', z') relative to the deformed

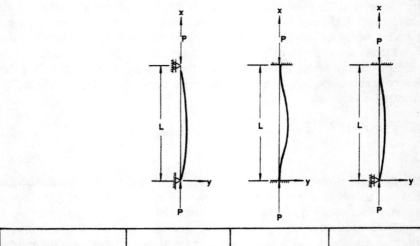

FIGURE 6.10 Boundary conditions of columns with idealized end conditions

member will be used. The assumptions of linear elastic behavior, small deformations, and no distortion in member cross-sectional shape during buckling still apply. To establish the conditions governing the lateral torsional buckling of this member using the principle of stationary total potential energy in conjunction with the calculus of variations, we need to develop expressions for the strain energy U and the potential energy V. They are derived separately in the following.

Strain Energy

The strain energy for this member consists of three components: the strain energy due to bending about the x' axis $U_{bx'}$, the strain energy due to bending about the y' axis $U_{by'}$, and the strain energy due to twisting about the z' axis $U_{tz'}$. These components of strain energy correspond, respectively, to in-plane bending, out-of-plane bending, and twisting of the member at the instant of lateral torsional buckling. All these components of strain energy can be derived from Eq. (6.4.2) by

6.4 Calculus of Variations

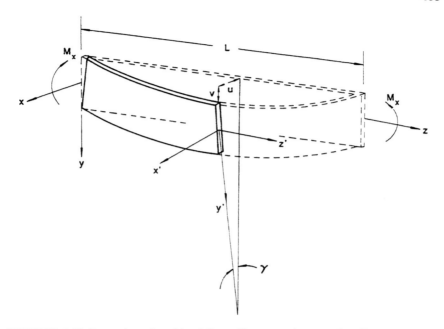

FIGURE 6.11 Lateral torsional buckling of beam under pure bending

identifying the stress and strain components associated with the particular deformations. However, here we will use an alternative approach. We will make use of the fact that for an elastic system the strain energy stored in the system is equal to the negative of the internal work performed in the system, which in turn is equal to the external work done on the system: that is,

$$U = -W_{\text{int}} \qquad (6.4.63)$$

Thus, if we can obtain expressions for W_{ext}, the strain energy can be obtained from Eq. (6.4.63).

From a free body of an infinitesimal segment of the beam at the instant of buckling, like that shown in Fig. 6.12a, we can identify the various components of moments $M_{x'}$, $M_{y'}$, and $M_{z'}$ acting on the cross sections. The shear forces acting on the cross sections are not shown because it is assumed that shear deformations are negligible compared to bending and twisting deformations, so that the strain energy due to shear can be ignored. Consequently, only the action of the bending moments $M_{x'}$, $M_{y'}$, and the twisting moment $M_{z'}$ will contribute to the strain energy of the member.

Strain Energy Due to In-Plane Bending, $U_{bx'}$. The increment of external work performed as a result of in-plane bending is equal to

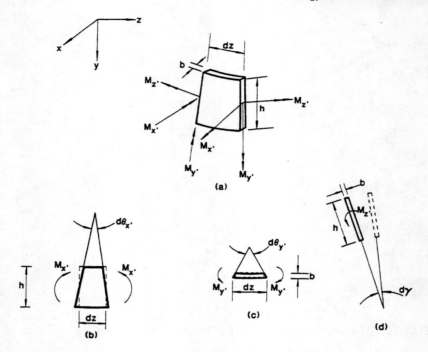

FIGURE 6.12 Deformation of an infinitesimal element

one-half the product of the in-plane bending moment $M_{x'}$ and the change in angle $d\theta_{x'}$ (Fig. 6.12b).

$$dW_{ext} = \tfrac{1}{2} M_{x'} \, d\theta_{x'} \qquad (6.4.64)$$

The factor one-half is introduced because $M_{x'}$ increases from zero to its full value linearly in the load sequence. Since

$$M_{x'} = EI_x \frac{d^2v}{dz^2} \qquad (6.4.65)$$

where

I_x = moment of inertia of the cross section about the x axis

$\dfrac{d^2v}{dz^2}$ = curvature of the member in the y–z plane

and

$$d\theta_{x'} = \frac{d^2v}{dz^2} dz \qquad (6.4.66)$$

6.4 Calculus of Variations

we have

$$dW_{\text{ext}} = \frac{1}{2} EI_x \left(\frac{d^2v}{dz^2}\right)^2 dz \qquad (6.4.67)$$

and so

$$W_{\text{ext}} = \frac{1}{2} \int_0^L EI_x \left(\frac{d^2v}{dz^2}\right)^2 dz \qquad (6.4.68)$$

In view of Eq. (6.4.63), we have

$$U_{bx'} = \frac{1}{2} \int_0^L EI_x \left(\frac{d^2v}{dz^2}\right)^2 dz \qquad (6.4.69)$$

Strain Energy Due to Out-Of-Plane Bending, $U_{by'}$. The increment of external work performed as a result of out-of-plane bending is equal to one-half the product of the out-of-plane bending moment $M_{y'}$ and the change in angle $d\theta_{y'}$ (Fig. 6.12c).

$$dW_{\text{ext}} = \frac{1}{2} M_{y'} \, d\theta_{y'} \qquad (6.4.70)$$

Since

$$M_{y'} = EI_y \frac{d^2u}{dz^2} \qquad (6.4.71)$$

where

I_y = moment of inertia of the cross section about the y axis

$\dfrac{d^2u}{dz^2}$ = curvature of the member about x-z plane

and

$$d\theta_{y'} = \frac{d^2u}{dz^2} dz \qquad (6.4.72)$$

we have

$$dW_{\text{ext}} = \frac{1}{2} EI_y \left(\frac{d^2u}{dz^2}\right)^2 dz \qquad (6.4.73)$$

and so

$$W_{\text{ext}} = \frac{1}{2} \int_0^L EI_y \left(\frac{d^2u}{dz^2}\right)^2 dz \qquad (6.4.74)$$

From Eq. (6.4.63), we can write

$$U_{by'} = \frac{1}{2}\int_0^L EI_y\left(\frac{d^2u}{dz^2}\right)^2 dz \qquad (6.4.75)$$

Strain Energy Due to Twisting, $U_{tz'}$. The increment of external work performed as a result of twisting is equal to one-half the product of the twisting moment $M_{z'}$ and the change in the angle of twist $d\gamma$ (Fig. 6.12d)

$$dW_{ext} = \tfrac{1}{2}M_{z'}\, d\gamma \qquad (6.4.76)$$

Since, from mechanics of material (see Eq. 5.2.5, page 311)

$$M_{z'} = GJ\frac{d\gamma}{dz} \qquad (6.4.77)$$

where

G = shear modulus
J = torsional constant
    = $hb^3/3$ for thin rectangular cross section
$\dfrac{d\gamma}{dz}$ = change in angle of twist per infinitesimal distance dz

and

$$d\gamma = \frac{d\gamma}{dz}dz \qquad (6.4.78)$$

we have

$$dW_{ext} = \frac{1}{2}GJ\left(\frac{d\gamma}{dz}\right)^2 dz \qquad (6.4.79)$$

and so

$$W_{ext} = -\frac{1}{2}\int_0^L GJ\left(\frac{d\gamma}{dz}\right)^2 dz \qquad (6.4.80)$$

Finally, from Eq. (6.4.63), we have

$$U_{tz'} = \frac{1}{2}\int_0^L GJ\left(\frac{d\gamma}{dz}\right)^2 dz \qquad (6.4.81)$$

Combining Eqs. (6.4.69), (6.4.75), and (6.4.81), the total strain energy U stored in the beam is

$$\begin{aligned}U &= U_{bx'} + U_{by'} + U_{tz'} \\ &= \frac{1}{2}\int_0^L EI_x\left(\frac{d^2v}{dz^2}\right)^2 dz + \frac{1}{2}\int_0^L EI_y\left(\frac{d^2u}{dz^2}\right)^2 dz \\ &\quad + \frac{1}{2}\int_0^L GJ\left(\frac{d\gamma}{dz}\right)^2 dz \end{aligned} \qquad (6.4.82)$$

6.4 Calculus of Variations

With the strain energy expression for the beam in Fig. 6.11 developed, the next step is to develop the expression for the potential energy for the beam.

Potential Energy

The potential energy of the beam in its buckled configuration consists of two parts: the potential energy developed as a result of in-plane bending (V_1) and the potential energy developed as a result of out-of-plane bending and twisting (V_2). They are described as follows:

Potential Energy Due to In-Plane Bending, V_1. The potential energy due to in-plane bending of the beam is equal to the negative product of the applied moments M_x and the end rotations (Fig. 6.13a):

$$V_1 = -\left[M_x\left(-\frac{dv}{dz}\right)\bigg|_{z=L} + M_x\left(\frac{dv}{dz}\right)\bigg|_{z=0}\right] = M_x\left(\frac{dv}{dz}\right)\bigg|_0^L \quad (6.4.83)$$

The minus sign for dv/dz in the first term of Eq. (6.4.83) is because the slope is negative at end $z = L$.

FIGURE 6.13 In-plane and out-of-plane deformation of the beam at buckling

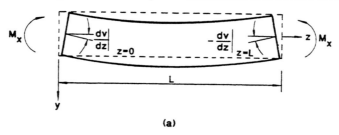

(a)

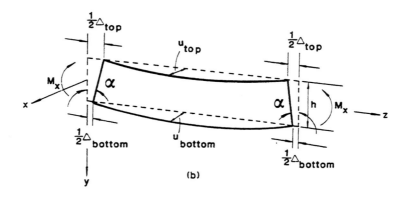

(b)

Potential Energy Due to Out-of-Plane Bending and Twisting, V_2. The potential energy due to out-of-plane bending and twisting is equal to the negative product of the applied moments M_x and the rotation of the end cross sections α as the beam buckles

$$V_2 = -2M_x \alpha \quad (6.4.84)$$

where, as in Fig. 6.13b,

$$\alpha = \frac{1}{2} \frac{\Delta_{\text{top}} - \Delta_{\text{bottom}}}{h} \quad (6.4.85)$$

In Eq. (6.4.85), Δ_{top} and Δ_{bottom} are the bending shortening of the top and bottom fibers, respectively, of the cross section, while h is the height of the cross section.

These bending shortenings can be expressed as

$$\Delta_{\text{top}} = \frac{1}{2} \int_0^L \left(\frac{du_{\text{top}}}{dz}\right)^2 dz \quad (6.4.86)$$

and

$$\Delta_{\text{bottom}} = \frac{1}{2} \int_0^L \left(\frac{du_{\text{bottom}}}{dz}\right)^2 dz \quad (6.4.87)$$

From Fig. 6.14, it can be seen that

$$u_{\text{top}} = u + \frac{\gamma h}{2} \quad (6.4.88)$$

$$u_{\text{bottom}} = u - \frac{\gamma h}{2} \quad (6.4.89)$$

Substituting Eqs. (6.4.88) and (6.4.89) into Eqs. (6.4.86) and (6.4.87), respectively, we have

$$\Delta_{\text{top}} = \frac{1}{2} \int_0^L \left[\frac{d}{dz}\left(u + \frac{\gamma h}{2}\right)\right]^2 dz$$
$$= \frac{1}{2} \int_0^L \left[\left(\frac{du}{dz}\right)^2 + h\left(\frac{du}{dz}\right)\left(\frac{d\gamma}{dz}\right) + \frac{h^2}{4}\left(\frac{d\gamma}{dz}\right)^2\right] dz \quad (6.4.90)$$

$$\Delta_{\text{bottom}} = \frac{1}{2} \int_0^L \left[\frac{d}{dz}\left(u - \frac{\gamma h}{2}\right)\right]^2 dz$$
$$= \frac{1}{2} \int_0^L \left[\left(\frac{du}{dz}\right)^2 - h\left(\frac{du}{dz}\right)\left(\frac{d\gamma}{dz}\right) + \frac{h^2}{4}\left(\frac{d\gamma}{dz}\right)^2\right] dz \quad (6.4.91)$$

Finally, substituting Eqs. (6.4.90) and (6.4.91) into Eq. (6.4.85), and then substituting the resulting expression for α into Eq. (6.4.84), leads to

$$V_2 = -M_x \int_0^L \left(\frac{du}{dz}\right)\left(\frac{d\gamma}{dz}\right) dz \quad (6.4.92)$$

6.4 Calculus of Variations

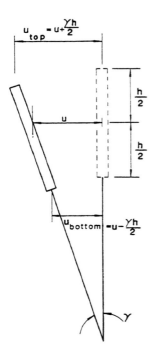

FIGURE 6.14 Lateral displacement at the top and bottom fibers of the beam

Combining Eqs. (6.4.83) and (6.4.92), the potential energy V of the beam is

$$V = V_1 + V_2$$
$$= M_x\left(\frac{dv}{dz}\right)\bigg|_0^L - M_x \int_0^L \left(\frac{du}{dz}\right)\left(\frac{d\gamma}{dz}\right) dz \qquad (6.4.93)$$

Total Potential Energy of the Beam

The total potential energy Π of the beam can be obtained by combining Eqs. (6.4.82) and (6.4.93)

$$\Pi = U + V$$
$$= \frac{1}{2}\int_0^L EI_x\left(\frac{d^2v}{dz^2}\right)^2 dz + \frac{1}{2}\int_0^L EI_y\left(\frac{d^2u}{dz^2}\right)^2 dz$$
$$+ \frac{1}{2}\int_0^L GJ\left(\frac{d\gamma}{dz}\right)^2 dz + M_x\left(\frac{dv}{dz}\right)\bigg|_0^L - M_x\int_0^L \left(\frac{du}{dz}\right)\left(\frac{d\gamma}{dz}\right) dz \qquad (6.4.94)$$

For equilibrium, the first variation of the total potential energy of the beam must vanish, that is

$$\delta\Pi = \int_0^L EI_x\left(\frac{d^2v}{dz^2}\right)\delta\left(\frac{d^2v}{dz^2}\right)dz + \int_0^L EI_y\left(\frac{d^2u}{dz^2}\right)\delta\left(\frac{d^2u}{dz^2}\right)dz$$

$$+ \int_0^L GJ\left(\frac{d\gamma}{dz}\right)\delta\left(\frac{d\gamma}{dz}\right)dz + M_x\delta\left(\frac{dv}{dz}\right)\bigg|_0^L$$

$$- M_x\int_0^L \left(\frac{du}{dz}\right)\delta\left(\frac{d\gamma}{dz}\right)dz - M_x\int_0^L \left(\frac{d\gamma}{dz}\right)\delta\left(\frac{du}{dz}\right)dz$$

$$= 0 \qquad (6.4.95)$$

or

$$\int_0^L EI_x\left(\frac{d^2v}{dz^2}\right)\left(\frac{d^2\,\delta v}{dz^2}\right)dz + \int_0^L EI_y\left(\frac{d^2u}{dz^2}\right)\left(\frac{d^2\,\delta u}{dz^2}\right)dz$$

$$+ \int_0^L GJ\left(\frac{d\gamma}{dz}\right)\left(\frac{d\,\delta\gamma}{dz}\right)dz + M_x\,\delta\left(\frac{dv}{dz}\right)\bigg|_0^L$$

$$M_x\int_0^L \left(\frac{du}{dz}\right)\left(\frac{d\,\delta\gamma}{dz}\right)dz - M_x\int_0^L \left(\frac{d\gamma}{dz}\right)\left(\frac{d\,\delta u}{dz}\right)dz = 0 \qquad (6.4.96)$$

Using integration by parts as done previously to eliminate derivatives of the variation, and realizing that $\delta u = \delta v = \delta\gamma = 0$ at $z = 0$ and $x = L$, it can be shown that Eq. (6.4.96) can be written as

$$\left(EI_x\frac{d^2v}{dz^2} + M_x\right)\left(\delta\frac{dv}{dz}\right)\bigg|_0^L + \int_0^L EI_x\frac{d^4v}{dz^4}\,dz\,\delta v$$

$$+ \left(EI_y\frac{d^2u}{dz^2}\right)\left(\delta\frac{du}{dz}\right)\bigg|_0^L + \int_0^L \left(EI_y\frac{d^4u}{dz^4} + M_x\frac{d^2\gamma}{dz^2}\right)dz\,\delta u$$

$$\int_0^L \left(GJ\frac{d^2\gamma}{dz^2} - M_x\frac{d^2u}{dz^2}\right)dz\,\delta\gamma = 0 \qquad (6.4.97)$$

To ascertain that Eq. (6.4.97) is satisfied for all arbitrary values of δv, δu, and $\delta\gamma$, each and every term of the equation must vanish, that is, for the first term, either

$$\left(EI_x\frac{d^2v}{dz^2} + M_x\right) = 0 \quad \text{at} \quad z = 0 \quad \text{and} \quad L \qquad (6.4.98)$$

or

$$\delta\left(\frac{dv}{dz}\right) = 0 \quad \text{at} \quad z = 0 \quad \text{and} \quad L \qquad (6.4.99)$$

and for the second term

$$\int_0^L EI_x\frac{d^4v}{dz^4}\,dz = 0$$

6.4 Calculus of Variations

or

$$EI_x \frac{d^4v}{dz^4} = 0 \qquad (6.4.100)$$

and for the third term

$$EI_y \frac{d^2u}{dz^2} = 0 \quad \text{at} \quad z = 0 \quad \text{and} \quad L \qquad (6.4.101)$$

or

$$\delta\left(\frac{du}{dz}\right) = 0 \quad \text{at} \quad z = 0 \quad \text{and} \quad L$$

and for the fourth term

$$\int_0^L \left(EI_y \frac{d^4u}{dz^4} + M_x \frac{d^2\gamma}{dz^2}\right) dz = 0$$

or

$$EI_y \frac{d^4u}{dz^4} + M_x \frac{d^2\gamma}{dz^2} = 0 \qquad (6.4.102)$$

and for the fifth term

$$\int_0^L \left(GJ \frac{d^2\gamma}{dz^2} - M_x \frac{d^2u}{dz^2}\right) dz = 0$$

or

$$GJ \frac{d^2\gamma}{dz^2} - M_x \frac{d^2u}{dz^2} = 0 \qquad (6.4.103)$$

The first two terms of Eq. (6.4.97), and, therefore, the conditions expressed in Eqs. (6.4.98) to (6.4.100) refer to the in-plane bending behavior of the beam. Because the member is simply supported, it is obvious that $\delta(dv/dz) \neq 0$ at $z = 0$ and L; we therefore have

$$-EI_x \frac{d^2v}{dz^2} = M_x \quad \text{at} \quad z = 0 \quad \text{and} \quad L \qquad (6.4.104)$$

which are the natural boundary conditions stating that the moments about the x axis at the ends are equal to the applied moment. The minus sign indicates that a positive moment corresponds to a negative curvature.

Equation (6.4.100) is simply the differential equation of equilibrium governing the in-plane bending behavior of the beam.

It should be noted that for small displacement analysis, this in-plane behavior is uncoupled with the out-of-plane behavior of the beam and, therefore, can be examined independently.

The out-of-plane or lateral torsional buckling behavior of the beam is expressed in the last three terms of Eq. (6.4.97), and so the conditions expressed in Eqs. (6.4.101) to (6.4.103) apply. Again, because the beam is simply supported, it is obvious that $\delta(du/dz) \neq 0$ and $z = 0$ and L. Therefore, Eq. (6.4.101) will apply. Equation (6.4.101) expresses the natural boundary conditions of the beam that state that the moment about the y axis vanishes at the simply supported ends.

Equations (6.4.102) and (6.4.103), which correspond, respectively, to Eqs. (5.4.8) and (5.4.9), express the lateral torsional buckling behavior of the beam. By eliminating the variable u, these two equations can be combined to give

$$\frac{d^4\gamma}{dz^4} + k^2 \frac{d^2\gamma}{dz^2} = 0 \qquad (6.4.105)$$

where

$$k^2 = \frac{M_x^2}{GJEI_y} \qquad (6.4.106)$$

Equation (6.4.105) corresponds to Eq. (5.4.11) in Chapter 5. Equation (5.4.11) was developed using the conventional free-body approach, whereas Eq. (6.4.105) was developed based on the variational principle, using the principle of stationary total potential energy of the system. Although comparable differential equations of equilibrium were developed using either approach, the energy approach has the added advantage that the natural boundary conditions were obtained as well in the solution process. Furthermore, as pointed out earlier, in using the energy approach we need only consider the energy of the system. The merit of using an energy criterion to characterize the conditions of equilibrium of an elastic system will be obvious for systems involving complicated geometries. For such systems, it is usually much simpler to establish the energy expressions and derive the governing differential equations of equilibrium using the calculus of variations rather than to develop the differential equations based on the free-body approach. Nevertheless, one should bear in mind that the stationary total potential energy approach can only be used for systems exhibiting elastic behavior. For systems exhibiting inelastic behavior, recourse to numerical methods is often necessary.

The application of the energy method to determine approximate elastic buckling loads of members will be presented in the next two sections. The two methods to be presented here are the *Rayleigh–Ritz method* and the *Galerkin method*. In addition, we will present two numerical methods, the *Newmark method* and *numerical integration method* for the analysis of inelastic members, in the later part of this chapter.

6.5 RAYLEIGH–RITZ METHOD

In the previous section, the extremum principle, using the calculus of variations operating on the total potential energy of a system, was

6.5 Rayleigh–Ritz Method

applied to establish the governing differential equations of a system. However, the solutions to the differential equations are not obtained in this extremization process. For practical purposes, it is also necessary to obtain the critical loads from these governing differential equations. Theoretically, the exact elastic critical loads can be solved from these equations using the method of eigenvalue analysis described in Chapter 2. For many practical problems, however, the exact solutions to the characteristic equations of the differential equations are often difficult, if not impossible, to obtain, and recourse must be had to approximate or numerical methods to obtain solutions. In the following, we shall present a simple method to obtain an approximate solution of the critical load. This method is called the *Rayleigh–Ritz method*. In this method, an assumed displacement function satisfying the *geometric boundary conditions* of the system is used in the expression for the total potential energy function Π. By using such an assumed displacement function, a structural system with an infinite degree of freedom is now reduced to a system of finite degrees of freedom. As a result of this simplification, the total potential energy function reduces from a functional to a function, and, so, instead of using the calculus of variations (which operates on functionals), we can now use ordinary calculus (which operates on functions) to obtain solutions directly from the total potential energy function.

The use of an assumed displacement function to approximate the true displacement of the system was first introduced by Rayleigh.[2,3] To illustrate this method, let us assume that the buckled shape of the member has the approximate form

$$\bar{v} = a\phi \qquad (6.5.1)$$

where

$\bar{v}$ = assumed lateral displacement of the column
ϕ = arbitrary function satisfying the boundary conditions of the column
a = undetermined coefficient

Upon substituting Eq. (6.5.1) into the strain energy and potential energy expressions, we obtain

$$\bar{U} = \bar{U}(a) \qquad (6.5.2)$$
$$\bar{V} = \bar{V}(P, a) \qquad (6.5.3)$$

The use of a bar above the strain energy U and the potential energy V expressions emphasizes that they are not the true strain energy and potential energy of the system but are approximate values as a result of using an approximate deflection curve.

Using the principle of stationary total potential energy, the equilibrium configuration of the member is identified if

$$\delta(\bar{U}+\bar{V})=0 \tag{6.5.4}$$

Since $\bar{U}$ and $\bar{V}$ are functions of a, we can write Eq. (6.5.4) as

$$\frac{\partial(\bar{U}+\bar{V})}{\partial a}\delta a=0 \tag{6.5.5}$$

or, because δa is arbitrary, we must have

$$\frac{\partial(\bar{U}+\bar{V})}{\partial a}=0 \tag{6.5.6}$$

The value of P satisfying Eq. (6.5.6) is the critical load of the member.

To obtain a better result, a series of arbitrary functions ϕ_i's can be used as the assumed deflected shape of the member. That is, we can let

$$\bar{v}=a_1\phi_1+a_2\phi_2+\cdots+a_n\phi_n=\sum_{i=1}^{n}a_i\phi_i \tag{6.5.7}$$

The use of this approach was due to Ritz,[4] which is essentially an extension of the Rayleigh method, in which only one arbitrary function is used.

Upon substituting Eq. (6.5.7) into the strain energy and the potential energy function, we have

$$\bar{U}=\bar{U}(a_1,a_2,\ldots,a_n) \tag{6.5.8}$$

$$\bar{V}=\bar{V}(P,a_1,a_2,\ldots,a_n) \tag{6.5.9}$$

In view of Eq. (6.5.4), we have

$$\frac{\partial(\bar{U}+\bar{V})}{\partial a_i}\delta a_i=0, \quad i=1,2,\ldots,n \tag{6.5.10}$$

or, since δa_i are arbitrary, it follows that

$$\frac{\partial(\bar{U}+\bar{V})}{\partial a_i}=0 \tag{6.5.11}$$

Equations (6.5.11) represent a system of n simultaneous homogeneous equations with the a's and the load P as unknowns. For nontrivial solution of a's, the determinant of the coefficient matrix of the system of equations (which contains P as a variable) must vanish. The lowest value of P that renders the determinant of the coefficient matrix zero is the critical load of the member.

The following examples are selected to illustrate the use of Rayleigh–Ritz method to obtain approximate elastic critical loads for columns and beams.

6.5 Rayleigh–Ritz Method

Example 6.1. Critical Load of a Cantilever Column Loaded at the Tip
Figure 6.15a shows a cantilever column loaded by an axial force P at the tip. The exact elastic buckling load, $P_{cr} = \pi^2 EI/4L^2 = 2.47 EI/L^2$, has been obtained in Chapter 2. We now want to obtain an approximate solution of the critical load using the Rayleigh–Ritz method.

SOLUTION: Assume the deflection shape of the cantilever column to be

$$\bar{v} = ax^2 \qquad (6.5.12)$$

Since $\bar{v}(0) = \bar{v}'(0) = 0$, the assumed deflection function satisfies the geometric boundary conditions of the member.

The total potential energy of the member is

$$\Pi = U + V$$

$$= \frac{1}{2}\int_0^L EI\left(\frac{d^2v}{dx^2}\right)^2 dx - \frac{P}{2}\int_0^L \left(\frac{dv}{dx}\right)^2 dx \qquad (6.5.13)$$

Substituting Eq. (6.5.12) into Eq. (6.5.13), we have

$$\bar{\Pi} = \bar{U} + \bar{V} = \frac{1}{2}\int_0^L EI(2a)^2 \, dx - \frac{P}{2}\int_0^L (2ax)^2 \, dx$$

$$= 2a^2 EIL - \tfrac{2}{3} Pa^2 L^3 \qquad (6.5.14)$$

Using Eq. (6.5.6), the critical load can be obtained as

$$\frac{\partial(\bar{U} + \bar{V})}{\partial a} = \frac{\partial(2a^2 EIL - 2Pa^2 L^3/3)}{\partial a} = 0$$

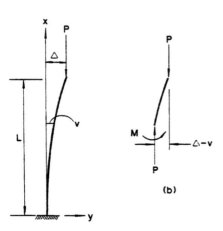

FIGURE 6.15 Tip-loaded cantilever column

or
$$4aEIL - 4PaL^3/3 = 0$$

from which we have

$$P_{cr} = 3\frac{EI}{L^2} \tag{6.5.15}$$

Comparing with the exact elastic critical load of $2.47EI/L^2$, we see that the approximate critical load is 21% in error.

The approximate solution can be improved if we use Eq. (6.4.14) rather than Eq. (6.4.15) for the strain energy of the column. The moment expression for the column can be written as

$$M = P(\Delta - v) \tag{6.5.16}$$

where Δ is the lateral tip deflection of the column (Fig. 6.15b).

From Eq. (6.5.12), we know

$$\Delta = aL^2 \tag{6.5.17}$$

Thus, the strain energy can be written as

$$\bar{U} = \frac{1}{2}\int_0^L \frac{[P(aL^2 - ax^2)]^2}{EI} dx = \frac{8}{30}\frac{P^2 a^2 L^5}{EI} \tag{6.5.18}$$

The total potential energy of the column becomes

$$\bar{\Pi} = \bar{U} + \bar{V} = \frac{4}{15}\frac{P^2 a^2 L^5}{EI} - \tfrac{2}{3}Pa^2 L^3 \tag{6.5.19}$$

from which the equation for the critical load is obtained

$$\frac{\partial(\bar{U}+\bar{V})}{\partial a} = \frac{\partial\left(\frac{4}{15}\frac{P^2 a^2 L^5}{EI} - \tfrac{2}{3}Pa^2 L^3\right)}{\partial a} = 0$$

or

$$\frac{8P^2 aL^5}{15EI} - \tfrac{4}{3}PaL^3 = 0$$

Solving the equation, we obtain

$$P_{cr} = 2.5\frac{EI}{L^2} \tag{6.5.20}$$

Compared to the exact solution of $2.47EI/L^2$, we see that the approximate solution for P_{cr} is just 1.2% in error. Since the assumed

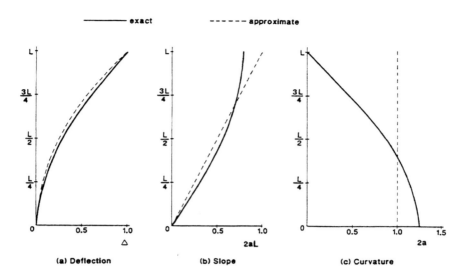

FIGURE 6.16 Comparison between exact and approximate deflection, slope, and curvature of the cantilever column

deflection shape [Eq. (6.5.12)] is not the true deflection curve of the column, the strain energy evaluated based on Eq. (6.4.14) and Eq. (6.4.15) will be different because the error involved by using the moment expression is different from the error involved in the curvature expression. In general, if a function is in error, the derivatives of the function will be in larger error. For instance, Fig. 6.16a–c shows a comparison between the actual and assumed deflection, slope and curvature of the cantilever column. As can be seen, the error becomes more and more noticeable for higher and higher derivatives. Since v'' is in greater error than v, the strain energy calculated using Eq. (6.4.15) will be less accurate than that of Eq. (6.4.14). Consequently, a better solution can be expected by using Eq. (6.4.14). Despite this shortcoming, Eq. (6.4.15) is much easier to manipulate in actual computation of Eq. (6.4.14). As a result, Eq. (6.4.15) is usually used in the calculation.

In Eq. (6.5.12), only one term is used for the assumed deflection curve. We can therefore improve the approximate solution by adding more terms to the deflection expression. For example, if we let

$$\bar{v} = a_1 x^2 + a_2 x^3 \tag{6.5.21}$$

the total potential energy can be written as

$$\bar{\Pi} = \bar{U} + \bar{V}$$

$$= \frac{1}{2}\int_0^L EI\left(\frac{d^2\bar{v}}{dx^2}\right)^2 dx - \frac{P}{2}\int_0^L \left(\frac{d\bar{v}}{dx}\right)^2 dx$$

$$= \frac{1}{2}\int_0^L EI(2a_1 + 6a_2x)^2 dx$$

$$- \frac{P}{2}\int_0^L (2a_1x + 3a_2x^2)^2 dx$$

$$= 2EIL(a_1^2 + 3a_1a_2L + 3a_2^2L^2)$$

$$- \frac{PL^3}{30}(20a_1^2 + 45a_1a_2L + 27a_2^2L^2) \qquad (6.5.22)$$

Note that the total potential energy function is now a function of two variables a_1 and a_2. Using Eq. (6.5.11) with $i = 1$ and 2, we have

$$\frac{\partial(\bar{U}+\bar{V})}{\partial a_1} = 2EIL(2a_1 + 3a_2L) - \frac{PL^3}{30}(40a_1 + 45a_2L) = 0 \qquad (6.5.23)$$

$$\frac{\partial(\bar{U}+\bar{V})}{\partial a_2} = 2EIL(3a_1L + 6a_2L^2) - \frac{PL^3}{30}(45a_1L + 54a_2L^2) = 0 \qquad (6.5.24)$$

If we denote

$$\lambda = \frac{PL^2}{EI} \qquad (6.5.25)$$

Equations (6.5.23) and (6.5.24) can be written as

$$(24 - 8\lambda)a_1 + (36 - 9\lambda)La_2 = 0 \qquad (6.5.26)$$
$$(20 - 5\lambda)a_1 + (40 - 6\lambda)La_2 = 0 \qquad (6.5.27)$$

or, in matrix form, we have

$$\begin{bmatrix}(24-8\lambda) & (36-9\lambda)L \\ (20-5\lambda) & (40-6\lambda)L\end{bmatrix}\begin{pmatrix}a_1 \\ a_2\end{pmatrix} = \begin{pmatrix}0 \\ 0\end{pmatrix} \qquad (6.5.28)$$

For a nontrivial solution, the determinant of the coefficient matrix of Eq. (6.5.28) must vanish

$$\det\begin{vmatrix}(24-8\lambda) & (36-9\lambda)L \\ (20-5\lambda) & (40-6\lambda)L\end{vmatrix} = 0 \qquad (6.5.29)$$

Expanding Eq. (6.5.29), we obtain

$$3\lambda^2 - 104\lambda + 240 = 0 \qquad (6.5.30)$$

6.5 Rayleigh–Ritz Method

The smallest positive root of this equation, called the *characteristic equation*, is $\lambda = 2.49$ and from Eq. (6.5.25), we obtain

$$P_{cr} = 2.49 \frac{EI}{L^2} \qquad (6.5.31)$$

which differs from the exact solution by only 0.81%. The solution can be further improved by using $U = \int_0^L (M^2/EI)\, dx$ in Eq. (6.5.22). This is not attempted here.

Example 6.2. Critical Load of a Cantilever Column Loaded at the Tip and Midheight

In the preceding example, the Rayleigh–Ritz method has been used to determine an approximate critical load of a relatively simple problem whose exact critical load is well known. In the following we shall consider a more difficult or rather more cumbersome example problem.

Figure 6.17 shows a cantilever column loaded by two axial forces: one at the tip and the other at midheight. The exact elastic buckling load as obtained previously in Chapter 2 has the value of $2.067 EI/L^2$.

SOLUTION: To solve the same problem using the Rayleigh–Ritz method, we assume the deflected shape of the column at buckling to be

$$\bar{v} = a\left(1 - \cos\frac{\pi x}{2L}\right) \qquad (6.5.32)$$

Equation (6.5.32) is the exact deflection curve for a centrally loaded column without the load at midheight. This shape becomes approximate

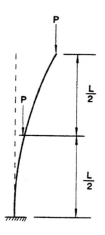

FIGURE 6.17 Cantilever column loaded at midheight and tip

for the column shown in Fig. 6.17 in which loads are present at both the tip and at midheight.

Using Eq. (6.4.15), the strain energy of the column corresponding to the assumed deflection shape is

$$\bar{U} = \frac{1}{2} \int_0^L EI \left(\frac{d^2\bar{v}}{dx^2}\right)^2 dx$$

$$= \frac{1}{2} \int_0^L EI \left[a\left(\frac{\pi}{2L}\right)^2 \cos\frac{\pi x}{2L}\right]^2 dx$$

$$= \frac{EI a^2 \pi^4}{64 L^3} \tag{6.5.33}$$

The potential energy is

$$\bar{V} = -\frac{2P}{2}\int_0^{L/2}\left(\frac{d\bar{v}}{dx}\right)^2 dx - \frac{P}{2}\int_{L/2}^L\left(\frac{d\bar{v}}{dx}\right)^2 dx$$

$$= -\frac{2P}{2}\int_0^{L/2}\left[a\left(\frac{\pi}{2L}\right)\sin\frac{\pi x}{2L}\right]^2 dx$$

$$\quad -\frac{P}{2}\int_{L/2}^L\left[a\left(\frac{\pi}{2L}\right)\sin\frac{\pi x}{2L}\right]^2 dx$$

$$= -\frac{Pa^2\pi}{32L}(3\pi - 2) \tag{6.5.34}$$

Using Eq. (6.5.6), we have at equilibrium

$$\frac{\partial(\bar{U}+\bar{V})}{\partial a} = \frac{EI a \pi^4}{32L^3} - \frac{Pa\pi}{16L}(3\pi - 2) = 0$$

from which

$$P_{cr} = \frac{\pi^3}{2(3\pi - 2)}\frac{EI}{L^2} = 2.09 \frac{EI}{L^2} \tag{6.5.35}$$

Compared with the exact solution of $2.067 EI/L^2$, the approximate solution is in error of just 1.1%.

It is worth mentioning here that the approximate critical load calculated using the Rayleigh–Ritz method is *always higher* than the exact critical load. This is because when we use an assumed displacement curve for the member, the member is mathematically constrained to displace according to the assumed shape rather than the natural shape, and so the apparent stiffness of the member increases, as does the buckling load. On the other hand, when we use more terms for the assumed deflected shape of the member, we are mathematically introducing more degrees of freedom to the member. As a result, the stiffening effect of the constraint brought by the assumed shape is reduced.

To obtain a reasonable critical load for a problem in the Rayleigh–Ritz method, the deflection functions chosen for any problem must be consistent with the geometric constraints of the problem. That is, the displacement functions used must satisfy the geometric boundary conditions of the problem. Furthermore, if the assumed displacement function consists of more than one term [Eq. (6.5.7)], it is advisable to use orthogonal functions for the ϕ_i's whenever possible. Functions ϕ_a and ϕ_b are said to be orthogonal if

$$\int_0^L \phi_a \phi_b \, dx = 0 \quad \text{for} \quad a \neq b \tag{6.5.36}$$

The merit of utilizing the orthogonality property of the functions will be apparent in the following example.

Example 6.3. Critical Load of a Bar on Elastic Foundation

Figure 6.18a shows a bar resting on an elastic foundation and subjected to a concentrated compressive force at the ends. As the bar buckles, the elastic foundation will exert a force of kiloNewtons per unit length per unit lateral deflection on the member (Fig. 6.18b).

FIGURE 6.18 Column on elastic foundation

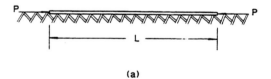

(a)

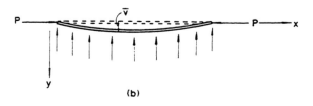

(b)

SOLUTION: Let the assumed deflection shape of the bar at buckling be approximated by

$$\bar{v} = \sum_{i=1}^{n} a_i \phi_i = \sum_{i=1}^{n} a_i \sin\left(\frac{i\pi x}{L}\right) \qquad (6.5.37)$$

The strain energy stored in the bar at buckling is then

$$\bar{U} = \frac{1}{2} \int_0^L EI \left(\frac{d^2 \bar{v}}{dx^2}\right)^2 dx$$

$$= \frac{1}{2} \int_0^L EI \left[\sum_{i=1}^{n} \frac{-a_i i^2 \pi^2}{L^2} \sin\frac{i\pi x}{L}\right]^2 dx$$

$$= \frac{EI\pi^4}{2L^4} \int_0^L \left(\sum_{i=1}^{n} a_i i^2 \sin\frac{i\pi x}{L}\right)\left(\sum_{i=1}^{n} a_i i^2 \sin\frac{i\pi x}{L}\right) dx \quad (6.5.38)$$

Equation (6.5.38) can be simplified if we use the *orthogonality* property of the sine function. It can easily be shown that

$$\int_0^L \sin ax \sin bx\, dx = 0 \quad \text{for} \quad a \neq b \qquad (6.5.39)$$

In view of Eq. (6.5.39), Eq. (6.5.38) can be simplified to

$$\bar{U} = \frac{EI\pi^4}{2L^4} \int_0^L \left(\sum_{i=1}^{n} a_i^2 i^4 \sin^2\frac{i\pi x}{L}\right) dx$$

$$= \sum_{i=1}^{n} \frac{EI\pi^4 a_i^2 i^4}{4L^3} \qquad (6.5.40)$$

The potential energy of the bar is

$$\bar{V} = -\frac{P}{2} \int_0^L \left(\frac{d\bar{v}}{dx}\right)^2 dx + \frac{1}{2} \int_0^L k\bar{v}^2\, dx$$

$$= -\frac{P}{2} \int_0^L \left(\sum_{i=1}^{n} \frac{a_i i\pi}{L} \cos\frac{i\pi x}{L}\right)^2 dx$$

$$+ \frac{1}{2} \int_0^L k\left(\sum_{i=1}^{n} a_i \sin\frac{i\pi x}{L}\right)^2 dx$$

$$= -\frac{P\pi^2}{2L^2} \int_0^L \left(\sum_{i=1}^{n} a_i i \cos\frac{i\pi x}{L}\right)\left(\sum_{i=1}^{n} a_i i \cos\frac{i\pi x}{L}\right) dx$$

$$+ \frac{k}{2} \int_0^L \left(\sum_{i=1}^{n} a_i \sin\frac{i\pi x}{L}\right)\left(\sum_{i=1}^{n} a_i \sin\frac{i\pi x}{L}\right) dx \quad (6.5.41)$$

The first term in Eq. (6.5.41) represents the familiar potential energy change resulting from bending shortening. It is negative because the corresponding external work done is positive as the directions of force

6.5 Rayleigh–Ritz Method

and displacement are the same. The second term represents the potential energy change due to the foundation's resistance to buckling of the bar. It is positive because the corresponding external work done is negative as the directions of the resistance force from the foundation and the lateral displacement of the bar are opposite.

To simplify Eqs. (6.5.41) and (6.5.24), we again use the orthogonality property of the terms in the assumed displacement function. For cosine function, it can easily be shown that

$$\int_0^L \cos ax \cos bx \, dx = 0 \quad \text{for} \quad a \neq b \tag{6.5.42}$$

and in view of Eq. (6.5.39), Eq. (6.5.41) can be simplified to

$$\bar{V} = -\frac{P\pi^2}{2L^2} \int_0^L \left(\sum_{i=1}^n a_i^2 i^2 \cos^2 \frac{i\pi x}{L} \right) dx$$

$$+ \frac{k}{2} \int_0^L \left(\sum_{i=1}^n a_i^2 \sin^2 \frac{i\pi x}{L} \right) dx$$

$$= -\sum_{i=1}^n \frac{P\pi^2 a_i^2 i^2}{4L} + \sum_{i=1}^n \frac{k a_i^2 L}{4} \tag{6.5.43}$$

If we substitute Eqs. (6.5.40) and (6.5.43) into Eq. (6.5.11), the critical load can be obtained from the condition

$$\frac{\partial(\bar{U}+\bar{V})}{\partial a_1} = \frac{EI\pi^4 a_1}{2L^3} - \frac{P\pi^2 a_1}{2L} + \frac{k a_1 L}{2} = 0$$

and

$$\frac{\partial(\bar{U}+\bar{V})}{\partial a_2} = \frac{EI\pi^4 a_2 2^4}{2L^3} - \frac{P\pi^2 a_2 2^2}{2L} + \frac{k a_2 L}{2} = 0$$

and

$$\vdots \qquad \vdots \qquad \vdots \qquad \vdots$$

$$\frac{\partial(\bar{U}+\bar{V})}{\partial a_n} = \frac{EI\pi^4 a_n n^4}{2L^3} - \frac{P\pi^2 a_n n^2}{2L} + \frac{k a_n L}{2} = 0 \tag{6.5.44}$$

from which, we obtain

$$P_{cr1} = P_e \left[1 + \frac{kL^2}{\pi^2 P_e} \right]$$

$$P_{cr2} = P_e \left[2^2 + \frac{kL^2}{2^2 \pi^2 P_e} \right]$$

$$\vdots \qquad \vdots \qquad \vdots$$

$$P_{crn} = P_e \left[n^2 + \frac{kL^2}{n^2 \pi^2 P_e} \right] \tag{6.5.45}$$

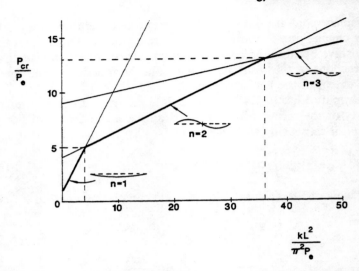

FIGURE 6.19 Critical loads and buckling modes of the column on elastic foundation

where

$$P_e = \frac{\pi^2 EI}{L^2}.$$

Any one of the above expressions can be the critical load of the bar. The one that governs depends on the stiffness of the foundation. Figure 6.19 shows a plot of the nondimensional critical load P_{cr}/P_e versus the parameter $kL^2/\pi^2 P_e$. As the stiffness k of the foundation increases, the buckling mode of the bar is seen to change to the higher mode and the corresponding buckling load is also increased.

Heretofore, the Rayleigh–Ritz method has been used to determine approximate in-plane buckling loads of columns or bars. In the following examples, the Rayleigh–Ritz method will be used to determine the approximate lateral buckling loads of beams.

Example 6.4. Lateral Buckling Load of a Cantilever Strip

In this example, the Rayleigh–Ritz method will be used to determine the critical tip load acting at the centroid of the cross section that will cause lateral buckling of the cantilever strip as shown in Fig. 6.20.

SOLUTION: The strain energy stored in the member during buckling is

$$U = \frac{1}{2}\int_0^L EI_y \left(\frac{d^2 u}{dz^2}\right)^2 dz + \frac{1}{2}\int_0^L GJ \left(\frac{d\gamma}{dz}\right)^2 dz \qquad (6.5.46)$$

6.5 Rayleigh–Ritz Method

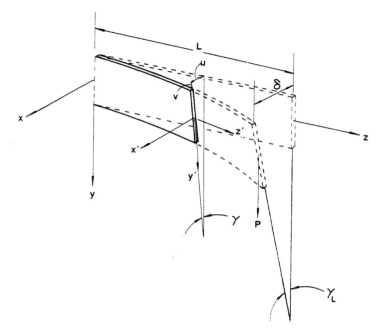

FIGURE 6.20 Tip-loaded cantilever strip

Note that Eq. (6.5.46) is similar to Eq. (6.4.82), except for the omission of the term $\frac{1}{2}\int_0^L EI_x(d^2v/dz^2)^2\,dz$. This term, described previously in Section 6.4, corresponds to the in-plane deformation of the member that exists before buckling. This in-plane action uncouples from the lateral buckling action in a small displacement analysis; hence, it has no effect on the lateral buckling load and therefore it is omitted in the analysis.

The strain energy expression in Eq. (6.5.46) contains two variables, u and γ. To facilitate computation, it is more desirable to express the strain energy in terms of only one variable γ. This can be done by reference to Fig. 6.21. From the figure, it can be seen that

$$M_{y'} = -\gamma M_x - \frac{dv}{dz}M_z = \gamma P(L-z) - \frac{dv}{dz}P(\delta - u)$$

$$\approx \gamma P(L-z) \qquad (6.5.47)$$

Since $M_{y'}$ can also be written as

$$M_{y'} = EI_y \frac{d^2u}{dz^2} \qquad (6.5.48)$$

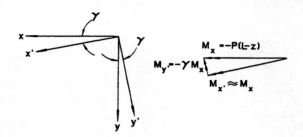

x-y plane

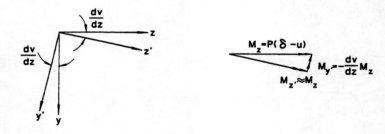

y-z plane

FIGURE 6.21 Moment components

Therefore, by equating the above two equations, we can obtain a relationship between d^2u/dz^2 and γ.

$$\frac{d^2u}{dz^2} = \frac{\gamma P}{EI_y}(L-z) \tag{6.5.49}$$

Substituting Eq. (6.5.49) into Eq. (6.5.46) gives

$$U = \frac{1}{2}\frac{P^2}{EI_y}\int_0^L \gamma^2(L-z)^2\,dz + \frac{1}{2}\int_0^L GJ\left(\frac{d\gamma}{dz}\right)^2 dz \tag{6.5.50}$$

The potential energy of the member at buckling is equal to the negative product of the applied force P and the distance it travels when the member buckles. In Fig. 6.22a if we can suppose a curvature d^2u/dz^2 exists between points a and b that are a distance of dz apart, then the change in angle from point a to point b is $(d^2u/dz^2)\,dz$. If the rest of the member remains straight, the tip of the cantilever will travel a distance $d\Delta$ in the x'-z' plane (Fig. 6.22c) by the amount

$$d\Delta = (L-z)\frac{d^2u}{dz^2}dz \tag{6.5.51}$$

6.5 Rayleigh–Ritz Method

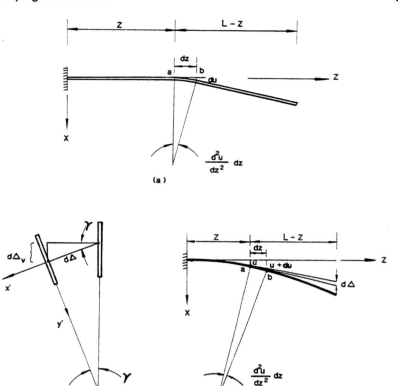

FIGURE 6.22 Vertical drop of load

The vertical component of this deflection is Figure 6.22b

$$d\Delta_v = \gamma(d\Delta) = \gamma(L-z)\frac{d^2u}{dz^2}dz \qquad (6.5.52)$$

Equation (6.5.52) represents the vertical drop of the centroid of the cross section as a result of a curvature existing between points a and b. If curvatures exist everywhere along the member (Fig. 6.22c), the total vertical displacement of the centroid of the tip cross section is

$$\Delta_v = \int_0^L d\Delta_v = \int_0^L \gamma(L-z)\frac{d^2u}{dz^2}dz \qquad (6.5.53)$$

Knowing this vertical displacement, the potential energy of the member due to the applied force P is

$$V = -P\Delta_v = -P\int_0^L \gamma(L-z)\frac{d^2u}{dz^2}\,dz \tag{6.5.54}$$

In view of Eq. (6.5.49), we can write Eq. (6.5.54) as

$$V = -\frac{P^2}{EI_y}\int_0^L \gamma^2(L-z)^2\,dz \tag{6.5.55}$$

Combining Eqs. (6.5.50) and (6.5.55), the total potential energy of the member at buckling is

$$\begin{aligned}\Pi = U + V &= \frac{1}{2}\frac{P^2}{EI_y}\int_0^L \gamma^2(L-z)^2\,dz + \frac{1}{2}\int_0^L GJ\left(\frac{d\gamma}{dz}\right)^2 dz \\ &\quad -\frac{P^2}{EI_y}\int_0^L \gamma^2(L-z)^2\,dz \\ &= -\frac{1}{2}\frac{P^2}{EI_y}\int_0^L \gamma^2(L-z)^2\,dz + \frac{1}{2}\int_0^L GJ\left(\frac{d\gamma}{dz}\right)^2 dz\end{aligned} \tag{6.5.56}$$

To obtain an approximate value for P by the Rayleigh–Ritz method, we need to assume a function for γ. The assumed function must satisfy the boundary conditions of

$$\gamma|_{z=0} = 0 \quad\text{and}\quad \frac{d\gamma}{dz}\bigg|_{z=L} = 0 \tag{6.5.57}$$

The first condition states that there is no rotation of the cross section at the fixed end and the second condition indicates that the St. Venant torsion $GJ(d\gamma/dz)$ vanishes at the free end.

If we use an assumed function for γ, denoted as $\bar{\gamma}$

$$\bar{\gamma} = \frac{\gamma_L}{L^2}(2Lz - z^2) \tag{6.5.58}$$

it can easily be shown that the boundary conditions in Eq. (6.5.57) are satisfied. In Eq. (6.5.58), γ_L, is the rotation of the cross section at the free end (Fig. 6.20).

Substituting the assumed function into Eq. (6.5.56), we have

$$\begin{aligned}\bar{\Pi} = \bar{U} + \bar{V} &= -\frac{1}{2}\frac{P^2}{EI_y}\int_0^L \left[\frac{\gamma_L}{L^2}(2Lz - z^2)\right]^2 (L-z)^2\,dz \\ &\quad + \frac{1}{2}\int_0^L GJ\left[\frac{\gamma_L}{L^2}(2L - 2z)\right]^2 dz\end{aligned} \tag{6.5.59}$$

6.5 Rayleigh–Ritz Method

Integrating, we have

$$\bar{\Pi} = \bar{U} + \bar{V} = -\frac{1}{2}\frac{P^2}{EI_y}\left(\frac{8\gamma_L^2 L^3}{105}\right) + \frac{1}{2}GJ\left(\frac{4}{3}\frac{\gamma_L^2}{L}\right)$$

$$= -\frac{4P^2\gamma_L^2 L^3}{105 EI_y} + \frac{2GJ\gamma_L^2}{3L} \quad (6.5.60)$$

For equilibrium, we must have

$$\delta\bar{\Pi} = \delta(\bar{U} + \bar{V}) = \frac{d(\bar{U} + \bar{V})}{d\gamma_L}\delta\gamma_L$$

$$= \left(-\frac{8P^2\gamma_L L^3}{105 EI_y} + \frac{4GJ\gamma_L}{3L}\right)\delta\gamma_L = 0 \quad (6.5.61)$$

Since $\delta\gamma_L$ is arbitrary and not necessarily zero, the quantity in parenthesis must vanish. By setting the quantity to zero, P_{cr} can be evaluated as

$$P_{cr} = \frac{4.183}{L^2}\sqrt{EI_y GJ} \quad (6.5.62)$$

The exact solution for this problem is given in the book by Timoshenko and Gere.[5] It has the value $(4.013/L^2)\sqrt{EI_y GJ}$. Upon comparison, the approximate solution overpredicts the critical load by about 4%.

We will now present, as a last example to demonstrate the use of the Rayleigh–Ritz method, the approximate critical lateral buckling load of a simply supported I-beam loaded by a concentrated force at midspan.

Example 6.5. Lateral Buckling Load of a Simply Supported I-Beam Loaded at Midspan

Figure 6.23a shows a simply supported I-beam subjected to a concentrated force P acting at the centroid of the midspan cross section. As P is increased gradually, it will reach a stage at which the deflected shape in the y-z plane ceases to be stable and lateral buckling occurs, as shown in Fig. 6.23b. Because of symmetry, we need consider only half of the beam. Note that at buckling, a twisting moment of $Pu_m/2$ will develop at the simply supported end of the beam to maintain equilibrium.

SOLUTION: To apply the Rayleigh–Ritz method to obtain the critical buckling load of the beam, we need to evaluate the strain energy and the potential energy of the beam at buckling.

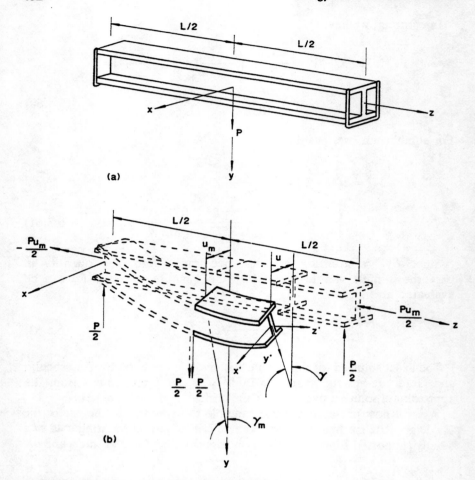

FIGURE 6.23 Lateral buckling of an I-shaped beam

The strain energy of the beam at buckling is given by

$$U = \frac{1}{2}\int_0^L EI_y\left(\frac{d^2u}{dz^2}\right)^2 dz + \frac{1}{2}\int_0^L GJ\left(\frac{d\gamma}{dz}\right)^2 dz$$
$$+ \frac{1}{2}\int_0^L EC_w\left(\frac{d^2\gamma}{dz^2}\right)^2 dz \qquad (6.5.63)$$

The first term in Eq. (6.5.63) represents the strain energy result from the out-of-plane bending of the cross section during buckling. The second term represents the strain energy due to St. Venant torsion. The third

6.5 Rayleigh–Ritz Method

term represents the strain energy due to warping restraint torsion. As indicated in Chapter 5, for all cross sections except circular sections or cross sections that are made up of thin-walled elements for which all elements intersect at a common point, the cross sections will warp. that is, plane sections will not remain plane as the member is twisted. As a result of this warping deformation, warping restraint torsion will be developed in the cross section that is not allowed to warp freely. For the case of an I-section, the strain energy stored in the member as a result of warping restraint torsion is due primarily to the coupled bending energy of the two flanges. The bending energy of the flanges is equal to

$$U_f = 2\left[\frac{1}{2}\int_0^L EI_f\left(\frac{d^2u_f}{dz^2}\right)^2 dz\right] \tag{6.5.64}$$

in which I_f is the moment of inertia of one flange about the weak axis of the cross section. The variable x is indicated in Fig. 6.24. The factor two in front of the square bracket accounts for the fact that there are two flanges.

From Fig. 6.24, we can write

$$u_f = \frac{\gamma h}{2} \tag{6.5.65}$$

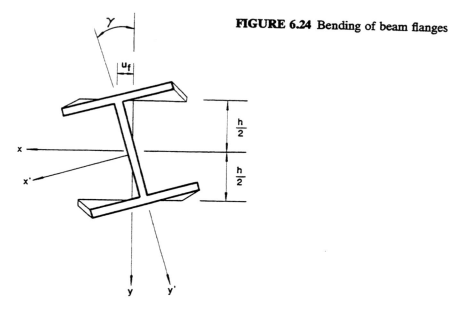

FIGURE 6.24 Bending of beam flanges

Substituting Eq. (6.5.65) into Eq. (6.5.64), we obtain

$$U_f = \int_0^L EI_f\left(\frac{h^2}{4}\right)\left(\frac{d^2\gamma}{dz^2}\right)^2 dz$$

$$= \frac{1}{2}\int_0^L EC_w\left(\frac{d^2\gamma}{dz^2}\right)^2 dz \qquad (6.5.66)$$

where

$$C_w = \frac{I_f h^2}{2} \qquad (6.5.67)$$

is the warping constant for the section.

Note that Eq. (6.5.66), which is the third term in Eq. (6.5.63), is not present in the previous case of narrow rectangular section [see Eq. (6.5.46)] because warping restraint torsion is negligible for such cross sections.

Because of symmetry, the strain energy expression of Eq. (6.5.63) can be written as

$$U = \int_0^{L/2} EI_y\left(\frac{d^2u}{dz^2}\right)^2 dz + \int_0^{L/2} GJ\left(\frac{d\gamma}{dz}\right)^2 dz$$

$$+ \int_0^{L/2} EC_w\left(\frac{d^2\gamma}{dz^2}\right)^2 dz \qquad (6.5.68)$$

Here, as in the above example, it is desirable to express the strain energy of Eq. (6.5.68) in terms of only one variable, γ. Referring to Fig. 6.25, we see that the out-of-plane bending moment $M_{y'}$ can be expressed as

$$M_{y'} = -\gamma M_x - \frac{dv}{dz} M_z$$

$$= -\gamma \frac{P}{2}\left(\frac{L}{2} - z\right) + \frac{dv}{dz}\frac{P}{2}(u_m - u)$$

$$\approx -\frac{\gamma P}{2}\left(\frac{L}{2} - z\right) \qquad (6.5.69)$$

The second term is dropped because it is of higher order compared to the first term.

Since $M_{y'}$ can also be expressed as

$$M_{y'} = EI_y \frac{d^2u}{dz^2} \qquad (6.5.70)$$

6.5 Rayleigh–Ritz Method

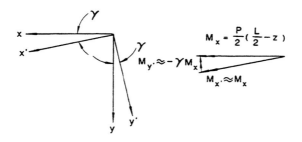

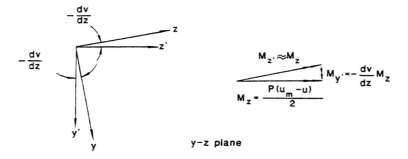

FIGURE 6.25 Moment components

therefore, by equating Eq. (6.5.69) with Eq. (6.5.70), we have

$$\frac{d^2u}{dz^2} = \frac{-\gamma P}{2EI_y}\left(\frac{L}{2} - z\right) \quad (6.5.71)$$

Upon substituting Eq. (6.5.71) into Eq. (6.5.68), we have

$$U = \int_0^{L/2} \frac{\gamma^2 P^2}{4EI_y}\left(\frac{L}{2} - z\right)^2 dz + \int_0^{L/2} GJ\left(\frac{d\gamma}{dz}\right)^2 dz \\ + \int_0^{L/2} EC_w\left(\frac{d^2\gamma}{dz^2}\right)^2 dz \quad (6.5.72)$$

The potential energy of the member is equal to the negative product of the applied force and the vertical drop of the force. The vertical drop is given by

$$\Delta_v = \int_0^{L/2} \gamma\left(\frac{L}{2} - z\right)\left(-\frac{d^2u}{dz^2}\right) dz \quad (6.5.73)$$

Equation (6.5.73) was attained from Eq. (6.5.53) by considering the simply supported beam to be fixed at midspan (because of symmetry) and loaded at the end by a force of $P/2$. The minus sign for the term d^2u/dz^2 accounts for the fact that the quantity d^2u/dz^2 is negative.

Upon substituting Eq. (6.5.71) into Eq. (6.5.73), the potential energy of the member at buckling can be written as

$$V = -P\,\Delta_v = -\int_0^{L/2} \frac{P^2 \gamma^2}{2EI_y}\left(\frac{L}{2} - z\right)^2 dz \qquad (6.5.74)$$

Combining Eqs. (6.5.72) and (6.5.74), the total potential energy for the member at buckling is

$$\Pi = U + V = \int_0^{L/2} \frac{\gamma^2 P^2}{4EI_y}\left(\frac{L}{2} - z\right)^2 dz + \int_0^{L/2} GJ\left(\frac{d\gamma}{dz}\right)^2 dz$$

$$+ \int_0^{L/2} EC_w\left(\frac{d^2\gamma}{dz^2}\right)^2 dz - \int_0^{L/2} \frac{P^2 \gamma^2}{2EI_y}\left(\frac{L}{2} - z\right)^2 dz$$

$$= -\int_0^{L/2} \frac{P^2 \gamma^2}{4EI_y}\left(\frac{L}{2} - z\right)^2 dz + \int_0^{L/2} GJ\left(\frac{d\gamma}{dz}\right)^2 dz$$

$$+ \int_0^{L/2} EC_w\left(\frac{d^2\gamma}{dz^2}\right)^2 dz \qquad (6.5.75)$$

In using the Rayleigh–Ritz method, we need to assume a function for γ. The assumed function must satisfy the boundary conditions of zero torsion at $z = 0$ (because of symmetry) and zero angle of twist at $z = L/2$ (because of the support)

$$\left.\frac{d\gamma}{dz}\right|_{z=0} = 0 \quad \text{and} \quad \gamma|_{z=L/2} = 0 \qquad (6.5.76)$$

If we choose the function

$$\bar{\gamma} = \gamma_m \cos\frac{\pi z}{L} \qquad (6.5.77)$$

where γ_m is the rotation of the cross section at midspan (Fig. 6.23b), the boundary conditions of Eq. (6.5.76) are satisfied.

Substituting Eq. (6.5.77) into Eq. (6.5.75) gives

$$\bar{\Pi} = \bar{U} + \bar{V} = \frac{-P^2}{4EI_y}\int_0^{L/2}\left(\gamma_m^2 \cos^2\frac{\pi z}{L}\right)\left(\frac{L}{2} - z\right)^2 dz$$

$$+ GJ\int_0^{L/2}\left(\frac{\gamma_m \pi}{L}\right)^2 \sin^2\frac{\pi z}{L}\, dz$$

$$+ EC_w \int_0^{L/2} \gamma_m^2\left(\frac{\pi}{L}\right)^4 \cos^2\frac{\pi z}{L}\, dz \qquad (6.5.78)$$

6.5 Rayleigh–Ritz Method

By integrating, Eq. (6.5.78) can be simplified to

$$\bar{\Pi} = \bar{U} + \bar{V} = -\frac{P^2}{4EI_y}\gamma_m^2 L^3\left(\frac{1}{48} + \frac{1}{8\pi^2}\right)$$
$$+ GJ\gamma_m^2\left(\frac{\pi}{L}\right)^2\left(\frac{L}{4}\right) + EC_w\gamma_m^2\left(\frac{\pi}{L}\right)^4\left(\frac{L}{4}\right) \quad (6.5.79)$$

For equilibrium, the first variation of the total potential energy must vanish

$$\delta\bar{\Pi} = \delta(\bar{U}+\bar{V}) = \frac{d(\bar{U}+\bar{V})}{d\gamma_m}\delta\gamma_m$$
$$= \left[-\frac{P^2}{2EI_y}\gamma_m L^3\left(\frac{1}{48}+\frac{1}{8\pi^2}\right)\right.$$
$$+ 2GJ\gamma_m\left(\frac{\pi}{L}\right)^2\left(\frac{L}{4}\right)$$
$$\left. + 2EC_w\gamma_m\left(\frac{\pi}{L}\right)^4\left(\frac{L}{4}\right)\right]\delta\gamma_m = 0 \quad (6.5.80)$$

Since $\delta\gamma_m$ is arbitrary and not necessarily zero, the quantity in brackets must vanish. By setting the quantity in brackets equal zero, P_{cr} is found to be

$$P_{cr} = 5.464\frac{\pi}{L^2}\sqrt{\left(EI_yGJ + \frac{\pi^2}{L^2}EI_yEC_w\right)} \quad (6.5.81)$$

Compared with the exact answer given in the book by Timoshenko and Gere,[5] we see that the approximate critical load expressed in Eq. (6.5.81) is in error by only 0.5%. Here, as in other examples, the approximate critical load that is obtained this way is always higher than the exact one because of the stiffening effect imposed on the member by using an assumed deflection curve.

In summary, in applying the Rayleigh–Ritz method, we are approximating the true deflected shape of the member by an assumed deflected shape [Eq. (6.5.7)] with the functions ϕ_i's satisfying the kinematic boundary conditions and the constants a_i's as unknown coefficients. The task is to adjust these coefficients such that the true deflected shape can be best approximated. The best approximated shape is the one in which the total potential energy of the member is minimized with respect to each and every unknown coefficient [Eq. (6.5.11)].

6.6 GALERKIN'S METHOD

In the preceding section, the Rayleigh–Ritz method has been applied to obtain approximate values of critical loads. A somewhat similar technique that uses approximate deflected shapes to obtain approximate critical loads is a result of work done by B. G. Galerkin.[6] Unlike the Rayleigh–Ritz method, in which an expression for the total potential energy is required, Galerkin's method requires that the differential equation of equilibrium be known.

For illustration purposes, it has been shown in Section 6.4 that the total potential energy of a hinged-hinged column will assume a stationary value if the following condition is satisfied

$$\left(EI\frac{d^2v}{dx^2}\right)\left(\delta\frac{dv}{dx}\right)\bigg|_{x=L} - \left(EI\frac{d^2v}{dx^2}\right)\left(\delta\frac{dv}{dx}\right)\bigg|_{x=0}$$
$$+ \int_0^L \left(EI\frac{d^4v}{dx^4} + P\frac{d^2v}{dx^2}\right) dx\, \delta v = 0 \qquad (6.6.1)$$

Equation (6.6.1) corresponds to Eq. (6.4.46), except that the first three terms in Eq. (6.4.46) that describe the axial behavior of the column before buckling are neglected. This is because we are only concerned here with the buckling behavior of the column.

Assume the deflection function has the form

$$\bar{v}_1 = a_1\phi_1 + a_2\phi_2 + \cdots + a_n\phi_n = \sum_{i=1}^n a_i\phi_i \qquad (6.6.2)$$

in which the ϕ_i's are independent continuous functions satisfying *both the kinematic and natural boundary conditions* of the member and a_i's are undetermined coefficients.

Substituting the assumed deflection function (6.6.2) into Eq. (6.6.1), we obtain

$$\int_0^L \left(EI\frac{d^4\bar{v}}{dx^4} + P\frac{d^2\bar{v}}{dx^2}\right) dx\, \delta\bar{v} = 0 \qquad (6.6.3)$$

The first two terms of Eq. (6.6.1) vanish because the assumed deflection function $\bar{v}$ satisfies the natural boundary conditions of zero moments at the ends. That is, the conditions of

$$(d^2\bar{v}/dx^2)|_{x=L} = (d^2\bar{v}/dx^2)|_{x=0} = 0$$

are satisfied.

Equation (6.6.3) can be further simplified if we denote an operator Q as

$$Q = EI\frac{d^4}{dx^4} + P\frac{d^2}{dx^2} \qquad (6.6.4)$$

6.6 Galerkin's Method

and recognize that

$$\delta \bar{v} = \delta\left(\sum_{i=1}^{n} a_i \phi_i\right) = \sum_{i=1}^{n} \phi_i \delta a_i \tag{6.6.5}$$

Thus, using Eqs. (6.6.4) and (6.6.5), Eq. (6.6.3) can be written as

$$\int_0^L Q(\bar{v}) \sum_{i=1}^{n} \phi_i \delta a_i \, dx = 0 \tag{6.6.6}$$

Since all the ϕ_i's are independent of one another, the only way that Eq. (6.6.6) can be satisfied is when each and every term vanishes individually, that is

$$\int_0^L Q(\bar{v}) \phi_i \delta a_i \, dx = 0, \quad i = 1, 2, \ldots, n \tag{6.6.7}$$

Since δa_i are arbitrary, we must have

$$\int_0^L Q(\bar{v}) \phi_i \, dx = 0, \quad i = 1, 2, \ldots, n \tag{6.6.8}$$

Equation (6.6.8) is called the Galerkin's Integral.

If n terms are used in the deflection function [Eq. (6.6.2)], Eq. (6.6.8) represents n simultaneous equations. For an eigenvalue problem, the approximate critical load can be obtained by setting the determinant of the coefficient matrix of the n equations equal to zero.

Although Eq. (6.6.8) has been derived using the differential equation for a hinged-hinged column, the same expression can be used for columns with other end conditions. This is because the differential equation

$$EI \frac{d^4 v}{dx^4} + P \frac{d^2 v}{dx^2} = Q(v) = 0 \tag{6.6.9}$$

is valid for columns with other end conditions (fixed, guided) other than hinged (see Section 2.3). The only restriction is that the assumed deflected shape must conform to the kinematic and natural boundary conditions of the specific problem.

Galerkin's method can also be used to evaluate approximate lateral buckling loads for beams. For example, for a simply supported beam of narrow rectangular cross section under uniform moment, the differential equation of equilibrium is

$$\frac{d^4 \gamma}{dz^4} + k^2 \frac{d^2 \gamma}{dz^2} = 0 \tag{6.6.10}$$

where

$$k^2 = \frac{M_x^2}{GJEI_y}$$

Thus, by redefining the operator Q as

$$Q = \frac{d^4}{dz^4} + k^2 \frac{d^2}{dz^2} \qquad (6.6.11)$$

Eq. (6.6.8) can be used to obtain approximate value for the critical moment M_{cr}.

The following examples illustrate the use of Galerkin's method to perform approximate buckling analyses.

Example 6.6. Buckling Load of a Hinged-Fixed Column

In this example, the approximate buckling load of a hinged-fixed column as shown in Fig. 6.26a will be determined using Galerkin's method.

SOLUTION: The first step is to assume a deflection function for the buckled configuration of the column that satisfies both the kinematic and natural boundary conditions of the member, that is, the conditions

$$y|_{x=0} = 0, \quad \frac{d^2y}{dx^2}\bigg|_{x=0} = 0, \quad y|_{x=L} = 0, \quad \frac{dy}{dx}\bigg|_{x=L} = 0 \qquad (6.6.12)$$

must be satisfied.

By using the deflected shape of a hinged-fixed beam under uniform distributed load (Fig. 6.26b) to approximate the buckled shape of the

FIGURE 6.26 Critical load of a hinged-fixed column by Galerkin's method

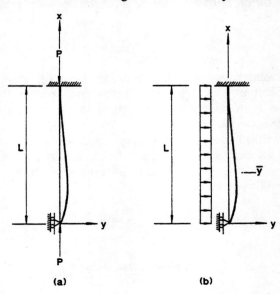

(a)   (b)

6.6 Galerkin's Method

column

$$\bar{y} = a(L^3 x - 3Lx^3 + 2x^4) \tag{6.6.13}$$

it can easily be shown that the conditions expressed in Eq. (6.6.12) are satisfied.

Substituting the assumed deflection curve of Eq. (6.6.13) into Eq. (6.6.8) with $n = 1$, $\bar{v} = \bar{y}$, $\phi = L^3 x - 3Lx^3 + 2x^4$ and

$$Q = EI \frac{d^4}{dx^4} + P \frac{d^2}{dx^2}$$

gives

$$\int_0^L a[48EI + P(-18Lx + 24x^2)][xL^3 - 3Lx^3 + 2x^4]\,dx = 0 \tag{6.6.14}$$

By integrating, Eq. (6.6.14) can be reduced to

$$a\left[\frac{36EIL^5}{5} - \frac{12PL^7}{35}\right] = 0 \tag{6.6.15}$$

from which

$$P_{cr} = 21\frac{EI}{L^2} \tag{6.6.16}$$

which, when compared with the exact solution of $20.2EI/L^2$, the approximate critical load is in error by only 4%.

Example 6.7. Lateral Buckling Load of a Simply Supported Rectangular Beam under Uniform Bending Moment

To further demonstrate the use of Galerkin's method, the approximate M_{cr} for the simply supported rectangular beam as shown in Fig. 6.11 will be determined.

SOLUTION: The differential equation for this beam is, from Eq. (6.6.10),

$$\frac{d^4\gamma}{dz^4} + k^2 \frac{d^2\gamma}{dz^2} = 0 \tag{6.6.17}$$

where

$$k^2 = \frac{M_x^2}{GJEI_y}$$

The boundary conditions are

$$\gamma|_{x=0} = \gamma|_{x=L} = 0, \quad \left.\frac{d^2\gamma}{dz^2}\right|_{x=0} = \left.\frac{d^2\gamma}{dz^2}\right|_{x=L} = 0 \tag{6.6.18}$$

The first two conditions are kinematic boundary conditions that state that the angle of twist at the ends are zero. The last two conditions are natural boundary conditions that state that the cross sections at the ends are free to warp.

By using the polynomial solution of the deflected shape of a simply supported beam under uniform distributed load

$$\bar{\gamma} = a(L^3 z - 2Lz^3 + z^4) \qquad (6.6.19)$$

to approximate the angle of twist γ, it can easily be shown that the conditions expressed in Eq. (6.6.18) are satisfied.

Substituting the assumed function for the angle of twist [Eq. (6.6.19)] into Eq. (6.6.8) with $n = 1$, $\bar{v} = \bar{\gamma}$, $\phi = L^3 z - 2Lz^3 + z^4$, and $Q = (d^4/dz^4) + k^2(d^2/dz^2)$ gives

$$\int_0^L a[24 + k^2(-12Lz + 12z^2)][L^3 z - 2Lz^3 + z^4] \, dz = 0 \qquad (6.6.20)$$

which, when simplified, becomes

$$a\left[\frac{24}{5} L^5 - \frac{17}{35} k^2 L^7\right] = 0 \qquad (6.6.21)$$

from which

$$k^2 = \frac{9.88}{L^2} \qquad (6.6.22)$$

And since $k^2 = M_x^2/GJEI_y$, we have

$$M_{cr} = k\sqrt{GJEI_y} = \frac{3.144}{L}\sqrt{GJEI_y} \qquad (6.6.23)$$

which, when compared to the exact solution of $M_{cr} = \pi/L\sqrt{GJEI_y}$, the approximate solution is in error by less than 0.1%.

Before leaving the subject, it should be mentioned that if the true deflection curves such as

$$y = a\left(\frac{x}{L} - \frac{\sin kx}{kL \cos kL}\right)$$

for the first problem, and

$$\gamma = a \sin kz$$

for the second problem had been used, we would have obtained the exact critical loads.

In the above examples we used approximate rather than true deflection curves to demonstrate that the critical loads so obtained will be *larger* than the exact ones. For more complicated problems for which the true deflection curves are not known, the use of Galerkin's method in conjunction with an approximate deflection curve will furnish an *upper bound value* for the critical load.

In comparing the Rayleigh–Ritz method with Galerkin's method, two important differences can be identified. First, the Rayleigh–Ritz method deals with the total potential energy of the system, whereas Galerkin's method deals with the differential equation of equilibrium of the system. Second, the assumed deflection curve used in the Rayleigh–Ritz method need satisfy only the kinematic boundary conditions, whereas the assumed deflection curve used in Galerkin's method must satisfy both the geometric and natural boundary conditions. The nature of the problem will decide which method will be more appropriate to use. For example, if it is easier to evaluate the expression for the total potential energy function and to find an approximate deflection curve that satisfies only the kinematic boundary conditions, the Rayleigh–Ritz method will be simpler to use. On the other hand, if it is easier to set up the differential equations of equilibrium and to find an approximate deflection curve that satisfies both the kinematic and natural boundary conditions, Galerkin's method will be more appropriate. Regardless of which method one uses, the approximate value for the critical load obtained will always be larger than the true critical load.

6.7 NEWMARK'S METHOD

Newmark's method[7] is a numerical iterative procedure that involves the use of successive approximations to obtain solutions when the load on (or deflection of) the member is increased in steps. Unlike the Rayleigh–Ritz and Galerkin's methods, in which the assumed deflection curve is expressed as a continuous function, Newmark's method divides the member into several equal segments and a numerical value of initial deflection is assumed for each division point along the member. Each of the division points is referred to as a *station*. Since Newmark's method is valid for both the elastic perfect member and the inelastic imperfect member, it follows that the moment-curvature relationship of an inelastic member must be either known beforehand or derived first before one can proceed with the numerical procedure. If the member remains fully elastic, the moment-curvature relationship has the simple form

$$M = EI\Phi \tag{6.7.1}$$

In the determination of the critical load using the Newmark's method, numerical values for the deflections at stations $x_0, x_1, x_2, \ldots, x_{n-1}, x_n$

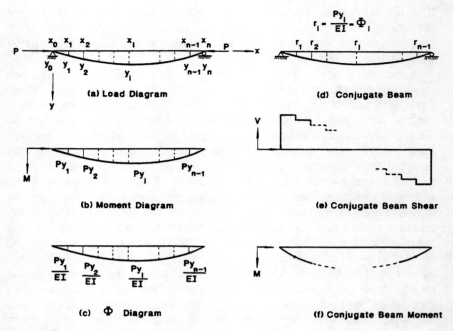

FIGURE 6.27 Newmark's method

along the member are first assumed. They are denoted as y_0, y_1, $y_2, \ldots, y_{n-1}, y_n$ in Fig. 6.27a, where a hinged-hinged column is shown for illustration. The bending moments at stations 0 to n are computed for each division point or station from the condition of equilibrium (Fig. 6.27b). For an elastic member, the curvatures at each station is then computed by using the elastic moment-curvature relationship of Eq. (6.7.1) (Fig. 6.27c). Knowing the curvatures at all the stations, the slopes and deflections of the member can be determined by the *conjugate beam method*.

Since the moment diagram is usually irregular in nature, the conjugate beam loading, which uses the M/EI diagram as the distributed loadings, will also be irregular. This is particularly the case if the flexural rigidity EI changes along the member. To simplify the computation, it is convenient to express the distributed loading by *equivalent concentrated loads* acting at the stations. The formulas for computing the equivalent concentrated loads for a linear and parabolic distributions of loadings are shown in Fig. 6.28a–d. Figure 6.28a gives the formulas for the equivalent concentrated loads at stations x_{i-1} and x_i for a linearly varying distributed load between the two stations. These formulas should be used if there is a

6.7 Newmark's Method

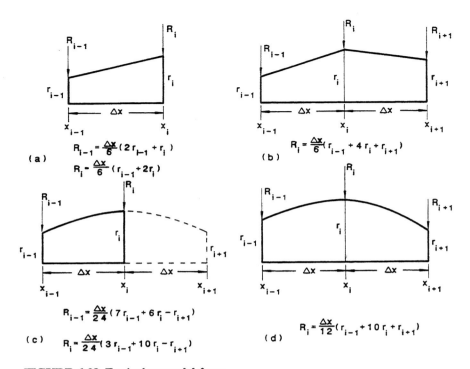

FIGURE 6.28 Equivalent nodal force

jump in magnitude in the M/EI diagram at either of the stations. Figure 6.28b gives the formula for the equivalent concentrated loads for stations x_{i-1}, x_i, and x_{i+1}. This formula is for the case in which the actual loading is continuous over the station x_i. Figure 6.28c gives the formulas for the equivalent concentrated loads for stations x_{i-1} and x_i for a parabolic varying distributed load. Note that station x_{i+1} in the formula is fictitious because three stations are needed to define a parabola. In using the formulas, the value r_{i+1} at station x_{i+1} must be known. The value r_{i+1} can be taken as equal to r_{i-1} if a plane of symmetry can be established at station x_i. Otherwise, the value of r_{i+1} can be taken as zero. The formulas in Fig. 6.28c are for the cases if station x_{i-1} or x_i is an endpoint or if there is a jump in the M/EI diagram at either of the stations. Figure 6.28d gives the formula for the equivalent concentrated loads for stations x_{i-1}, x_i, and x_{i+1} if the distributed loading is parabolic over the three stations and is continuous over station x_i.

In many cases, the actual distribution of loading is neither linear nor parabolic. Nevertheless, the parabolic load distributions can still be used to approximate the actual loading for most practical purposes.

Having replaced the actual distributed loading, namely, the M/EI diagram, by equivalent concentrated loads at the stations of the conjugate beam (Fig. 6.27d), the slope of the real member can then be obtained as the shear force of the conjugate beam (Fig. 6.27e) and the deflection as the moment of the conjugate beam (Fig. 6.27f). Since equivalent concentrated loads are used, rather than the actual distributed load, the slopes evaluated represent only the *average* slopes of two adjacent stations.

The deflections thus obtained contain P as an unknown (see the example problem). Thus, by equating these deflections to the assumed deflections at each station, numerical values for P can be calculated. The lowest value of P obtained this way will represent a lower bound to the critical load, and the largest value of P will represent an upper bound to the critical load. Consequently, by using Newmark's method, both the upper and lower bound solutions to the actual critical load are obtained. This represents an obvious advantage over the Rayleigh–Ritz or Galerkin's method, in which only an upper bound solution can be obtained. To estimate the critical load, the average value of the calculated deflections at all stations is equated to the average value of the assumed deflections; thus

$$P_{cr} = \frac{(y_{\text{assumed}})_{\text{avg}}}{(y_{\text{calculated}})_{\text{avg}}} = \frac{\sum_{i=0}^{n} (y_{\text{assumed}})_i}{\sum_{i=0}^{n} (y_{\text{calculated}})_i} \qquad (6.7.2)$$

A better approximation of the critical load can be obtained by using the calculated deflections as the new assumed deflections and repeat the same procedure outlined above. The process can be repeated as many times as needed until the lower and upper bounds for P_{cr} are not too far apart.

Example 6.8. Critical Load of a Cantilever Column Loaded at the Tip and at Midheight

SOLUTION: Table 6.1 shows Newmark's numerical procedure for the determination of the critical load of the column. The column is divided into 10 stations. The initial deflections of the stations are assumed to follow the function

$$y_{\text{assumed}} = \delta\left(1 - \cos\frac{\pi x}{2L}\right)$$

The moments M are then obtained from the deflections by consideration of equilibrium. For example, the moment at Station 9 is obtained

6.7 Newmark's Method

from

$$M_9 = P(y_{10} - y_9) = P(100 - 84.4)\frac{\delta}{100} = 15.6\frac{P\delta}{100}$$

and the moment at Station 4 is obtained from

$$M_4 = P(y_{10} - y_4) + P(y_5 - y_4)$$
$$= P(100 - 19.1)\frac{\delta}{100} + P(29.3 - 19.1)\frac{\delta}{100}$$
$$= 91.1\frac{P\delta}{100}$$

The curvature Φ can be obtained from the moment by using the elastic moment-curvature relationship given in Eq. (6.7.1). The values of the curvatures are then used in conjunction with Fig. 6.28 to calculate the equivalent concentrated loads R at the stations. For instance, the equivalent concentrated load at Station 9 is (using Fig. 6.28d)

$$R_9 = \frac{L/10}{12}[30.9 + 10(15.6) + 0]\frac{P\delta}{100EI} = 1.56\frac{P\delta L}{100EI}$$

and at Station 4

$$R_4 = \frac{L/10}{12}[108 + 10(91.1) + 70.7]\frac{P\delta}{100EI} = 9.08\frac{P\delta L}{100EI}$$

The average slope θ of the real column between two stations is obtained as the shear of the conjugate beam and the deflection $y_{\text{calculated}}$ of the real column as the moment of the conjugate beam (Fig. 6.29).

By taking the ratio of y_{assumed} to $y_{\text{calculated}}$ for all the stations, a bound for the critical load can be obtained:

$$1.91\frac{EI}{L^2} \leq P_{\text{cr}} \leq 2.07\frac{EI}{L^2}$$

Using Eq. (6.7.2), an approximate value for the critical load is obtained as $2.02(EI/L^2)$. Compare this with the exact value of $2.067EI/L^2$, the error is about 2.3%.

To improve the accuracy of the solution, we will now start a second cycle of calculation. The assumed deflection (y_{assumed}) for the second cycle is now taken as proportional to the calculated deflection ($y_{\text{calculated}}$) of the first cycle. To adjust the common factor, a factor of 100/48.4 is used to multiply every term of ($y_{\text{calculated}}$) in Cycle 1 to obtain (y_{assumed}) in Cycle 2. This factor is used so that Station 10 will have a deflection equal to δ. The results of the second cycle indicates that a better bound for P_{cr}

Table 6.1 Critical Load by Newmark's Method

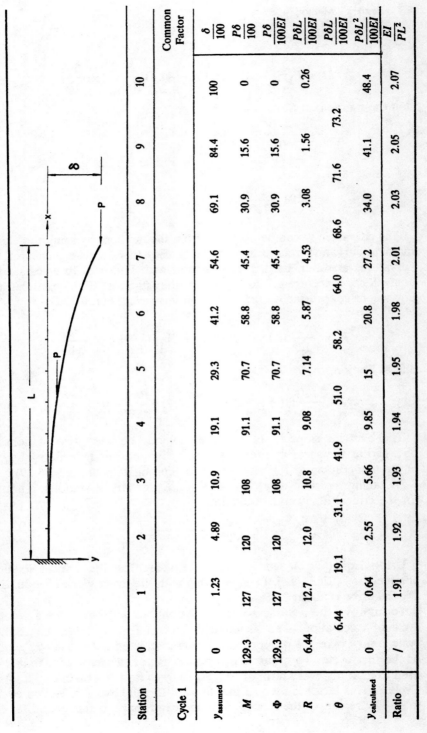

Station	0	1	2	3	4	5	6	7	8	9	10	Common Factor
Cycle 1												
$y_{assumed}$	0	1.23	4.89	10.9	19.1	29.3	41.2	54.6	69.1	84.4	100	$\dfrac{\delta}{100}$
M	129.3	127	120	108	91.1	70.7	58.8	45.4	30.9	15.6	0	$\dfrac{P\delta}{100}$
Φ	129.3	127	120	108	91.1	70.7	58.8	45.4	30.9	15.6	0	$\dfrac{P\delta}{100EI}$
R	6.44	12.7	12.0	10.8	9.08	7.14	5.87	4.53	3.08	1.56	0.26	$\dfrac{P\delta L}{100EI}$
θ		6.44	19.1	31.1	41.9	51.0	58.2	64.0	68.6	71.6	73.2	$\dfrac{P\delta L}{100EI}$
$y_{calculated}$	0	0.64	2.55	5.66	9.85	15	20.8	27.2	34.0	41.1	48.4	$\dfrac{P\delta L^2}{100EI}$
Ratio	/	1.91	1.92	1.93	1.94	1.95	1.98	2.01	2.03	2.05	2.07	$\dfrac{EI}{PL^2}$

Station	0	1	2	3	4	5	6	7	8	9	10	Common Factor
Cycle 2												
$y_{assumed}$	0	1.33	5.26	11.7	20.3	30.9	42.9	56.1	70.1	84.9	100	$\dfrac{\delta}{100}$
M	131	128	120	108	90.3	69.1	57.1	43.9	29.9	15.1	0	$\dfrac{P\delta}{100}$
Φ	131	128	120	108	90.3	69.1	57.1	43.9	29.9	15.1	0	$\dfrac{P\delta}{100EI}$
R	6.53	12.8	12.0	10.8	9.0	6.99	5.7	4.38	2.98	1.51	0.253	$\dfrac{P\delta L}{100EI}$
θ	6.53	19.3	31.3	42.1	51.1	58.1	63.8	68.2	71.2	72.7		$\dfrac{P\delta L}{100EI}$
$y_{calculated}$	0	0.653	2.58	5.71	9.92	15.0	20.8	27.2	34.0	41.1	48.4	$\dfrac{P\delta L^2}{100EI}$
Ratio	/	2.04	2.04	2.05	2.05	2.06	2.06	2.06	2.06	2.06	2.07	$\dfrac{EI}{PL^2}$

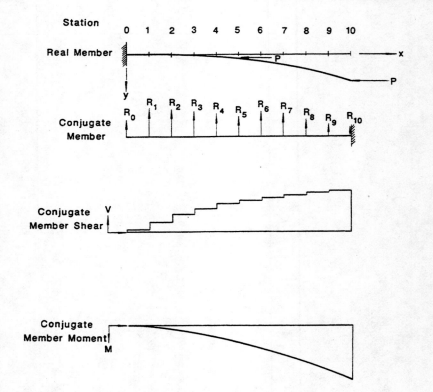

FIGURE 6.29 Critical load by Newmark's method

is obtained

$$2.04\frac{EI}{L^2} \leq P_{cr} \leq 2.07\frac{EI}{L^2}$$

with an estimated P_{cr} [Eq. (6.7.2)] of $2.06EI/L^2$, which is only 0.3% in error.

Newmark's numerical procedure to obtain critical loads for elastic columns can be extended to the determination of maximum load-carrying capacity of inelastic beam-columns. The procedure for the analysis of inelastic beam-column is similar to that for the analysis of elastic column. However, instead of using the conjugate beam method to determine the average slopes and deflections of the stations, we will now present an alternative method to obtain these quantities. If we denote θ_i and y_i as the average slope and deflection of station i, respectively, their values can

6.7 Newmark's Method

be obtained by numerically integrating the curvature and slope.

$$\theta_i = \sum_{k=0}^{i} \left(\frac{\Delta^2 y}{\Delta x^2}\right)_k \Delta x = -\sum_{k=0}^{i} \Phi_k \Delta x, \qquad i = 0, 1, 2, \ldots, n \quad (6.7.3)$$

$$y_i = \sum_{k=0}^{i} \left(\frac{\Delta y}{\Delta x}\right)_k \Delta x = \sum_{k=0}^{i} \theta_k \Delta x, \qquad i = 0, 1, 2, \ldots, n \quad (6.7.4)$$

In these formulas, Δx is the interval between the stations. The minus sign in Eq. (6.7.3) indicates that the curvature Φ is related to d^2y/dx^2 negatively as a result of the manner the coordinate axes are set up. In writing Eq. (6.7.4), it is tacitly assumed that the initial slope θ_0 is equal to zero. If θ_0 is not zero, the deflections calculated from Eq. (6.7.4) must be adjusted. The adjustment can be made by realizing that if θ_0 is not equal to zero, the deflections calculated in Eq. (6.7.4) will differ from the corrected deflections by a proportionate amount equal to $(i/n)y_n$ (Fig. 6.30). Thus the corrected deflections can be written as

$$y_{ic} = y_i - \left(\frac{i}{n}\right) y_n \quad (6.7.5)$$

We should mention that the conjugate beam method used in the preceding example for an elastic member works equally well for an inelastic member. In other words, instead of using Eqs. (6.7.3) to (6.7.5), one can construct a conjugate beam with the curvature diagram as the distributed loadings. The values of the shear and moment of this conjugate beam at the i station will represent the values of θ_i and y_i of the real member, respectively. The use of Eqs. (6.7.3) and (6.7.4) simply

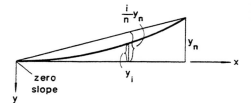

FIGURE 6.30 Correction to deflection

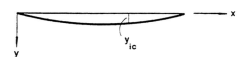

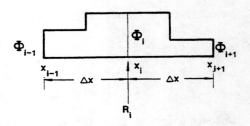

FIGURE 6.31 Equivalent concentrated force for constant curvature

means that the curvature Φ_i is assumed constant over the i station, as shown in Fig. 6.31, whereas by using the conjugate beam method, one can assume a linear or parabolic variation of curvature across the station (Fig. 6.28).

In using Newmark's method to analyze an *inelastic* beam-column, the moment-curvature-thrust (M-Φ-P) relationship must be known in advance. The moment-curvature-thrust relationship for a rectangular cross section has been developed analytically in detail in Chapter 3. For other cross sections, the moment-curvature-thrust relationship are more complicated and are usually obtained numerically[8,9] with the aid of a computer. With a known moment-curvature-thrust relationship, the steps for analyzing an inelastic beam-column by Newmark's method can be summarized as follows:

1. Divide the member into n segments.
2. Assume a numerical value for the initial deflection at each station $(y_{\text{assumed}})_i$, $i = 0, 1, \ldots, n$.
3. Compute the bending moment M_i at each station by considering equilibrium.
4. Using the moment-curvature-thrust relationship, compute the curvature Φ_i at each station.
5. Use Eq. (6.7.3) to evaluate the average slope θ_i.
6. Use Eq. (6.7.4) to evaluate the deflection y_i of each station.
7. Adjust the deflections calculated in Step 6 by Eq. (6.7.5).
8. Compare the deflections calculated in Step 7 to the assumed deflections in Step 2.
9. If the deflections of Step 7 and Step 2 are comparable, a solution is obtained. If not, use the deflections of Step 7 as the assumed deflections and repeat Steps 3 through 8 until convergence.

The following example illustrates Newmark's numerical procedure for the analysis of an inelastic beam-column.

6.7 Newmark's Method

Example 6.9. Inelastic Beam-Column Analysis by Newmark's Method
Figure 6.32 shows a simply supported member of rectangular cross section subjected to two end moments M_0 and $0.5M_0$ and an axial force P equal to $0.5P_y$, where P_y is the yield load of the member. Now we want to find the maximum end moment M_0 that will cause failure of the member.

SOLUTION: Before proceeding with the analysis, it is convenient to determine the following quantities from the given cross-sectional and material properties (Fig. 6.32)

$$P_y = A\sigma_y = bd\sigma_y$$

$$M_y = S\sigma_y = \frac{bd^2}{6}\sigma_y$$

Thus $M_y = (d/6)P_y = (0.06L/6)P_y(P_yL/100)$

$$\Phi_y = \frac{M_y}{EI} = \frac{P_yL}{100EI} = \frac{Lbd\sigma_y}{100E\left(\frac{bd^3}{12}\right)} = \frac{12L\sigma_y}{100Ed^2} = \frac{1}{30L}$$

$$\left(\frac{L}{4}\right)^2 \Phi_y = \frac{L}{480}$$

FIGURE 6.32 Member under combined moment and axial thrust

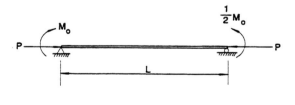

Cross-Section Geometry: $d=0.06L$

Material Property: $\sigma_y = 0.001E$

The moment-curvature-thrust relationship used is from Eqs. (3.9.30a–c) with $p = P/P_y = 0.5$

$$m = \phi, \qquad \phi \leq 0.5$$

$$m = 1.5 - \sqrt{\frac{1}{2\phi}} \qquad 0.5 \leq \phi \leq 2.0$$

$$m = 1.125 - \frac{1}{2\phi^2} \qquad \phi \geq 2.0$$

from which

$$\phi = m, \qquad m \leq 0.5$$

$$\phi = \frac{1}{2(1.5 - m)^2} \qquad 0.5 \leq m \leq 1.0$$

$$\phi = \frac{1}{\sqrt{(2.25 - 2m)}} \qquad m \geq 1.0$$

As a start, assume $M_0 = 0.4 M_y$. Table 6.2a shows all the necessary steps in determining the equilibrium configuration of the member subjected to applied axial force combined with end moments.

The member is divided into four equal segments. The primary moment (moment caused by the end moments) at the stations are then evaluated.

A deflected shape of the member is then assumed. For this problem, the numerical values of the stations are assumed to be that of the deflected shape of a beam under the action of the end moments only.

With these assumed deflection values, the secondary or Py moment (moment caused by the axial force acting through the deflection of the member relative to its chord) of the stations can be calculated by multiplying the values of the assumed deflection by the axial force of $0.5 P_y$.

The next step is simply a change of multiplier from $P_y L$ to M_y by making use of the relationship $M_y = P_y L/100$.

The total moment can then be obtained by adding the values of primary moment to that of the secondary moment.

Using the moment-curvature-thrust relationship, the curvatures at the stations can be evaluated.

Using Eq. (6.7.3), the average slopes between two adjacent stations can be obtained by numerically integrating the curvatures.

Using Eq. (6.7.4), the deflections at the stations can be calculated by numerically integrating the slopes.

As explained previously, the values of these deflections must be adjusted due to the assumption of zero slope at Station 0. The adjusted deflections can be evaluated using Eq. (6.7.5) and are reported as

6.7 Newmark's Method

Table 6.2a Determination of Equilibrium Configuration by Newmark's Method ($M_0 = 0.4M_y$)

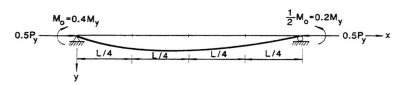

Station	0	1	2	3	4	Common Factor
Primary Moment M_0	0.4	0.35	0.30	0.25	0.2	M_y
Cycle 1						
$y_{assumed}$	0	0.00099	0.00125	0.000885	0	L
Secondary Moment $Py_{assumed}$	0	0.000495	0.000625	0.000443	0	$P_y L$
Change Multiplier	0	0.0495	0.0625	0.0443	0	M_y
Total Moment $M_0 + Py_{assumed}$	0.4	0.4	0.363	0.294	0.2	M_y
Curvature Φ_i	0.4	0.4	0.363	0.294	0.2	Φ_y
Average Slope θ_i		-0.4	-0.8	-1.163	-1.456	$\left(\frac{L}{4}\right)\Phi_y$
Deflection y_i	0	-0.4	-1.2	-2.363	-3.819	$\left(\frac{L}{4}\right)^2 \Phi_y$
Corrected Deflection y_{ic}	0	0.555	0.710	0.501	0	$\left(\frac{L}{4}\right)^2 \Phi_y$
Change Multiplier $y_{calculated}$	0	0.0012	0.0015	0.001	0	L

"Corrected Deflection" in Table 6.2a. As shown in the table, the corrected deflection values have a common factor of $(L/4)^2\Phi_y$. To correlate this calculated deflection with the assumed deflection, a change in multiplier is necessary. This can be done by using the relationship $(L/4)^2\Phi_y = L/480$. Once the multiplier is changed, one can make a direct comparison between the assumed and calculated deflections.

Table 6.2a Determination of Equilibrium Configuration by Newmark's Method ($M_0 = 0.4M_y$) *(continued)*

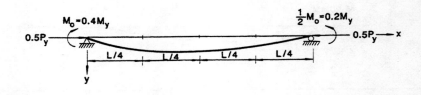

Station	0	1	2	3	4	Common Factor
Cycle 2						
$y_{assumed}$	0	0.0012	0.0015	0.001	0	L
Secondary Moment $Py_{assumed}$	0	0.0006	0.00075	0.0005	0	P_yL
Change Multiplier	0	0.06	0.075	0.05	0	M_y
Total Moment $M_0 + Py_{assumed}$	0.4	0.41	0.375	0.30	0.2	M_y
Curvature Φ_i	0.4	0.41	0.375	0.30	0.2	Φ_y
Average Slope θ_i		−0.4	−0.81	−1.185	−1.485	$\left(\dfrac{L}{4}\right)\Phi_y$
Deflection y_i	0	−0.4	−1.21	−2.395	−3.88	$\left(\dfrac{L}{4}\right)^2\Phi_y$
Change Multiplier $y_{calculated}$	0	0.0012	0.0015	0.0011	0	L

Since $y_{calculated} \approx y_{assumed}$, solution has converged.

If the calculated deflection is comparable to the assumed deflection, an equilibrium configuration of the member is said to have found. If the calculated deflection is not comparable to the assumed deflection, the calculated deflection is used as the assumed deflection and the calculation is repeated.

A second cycle of calculation is shown in Table 6.2a. As can be seen, convergence is achieved at the second cycle of calculation. Thus, the values of the deflection at the end of the second cycle will represent the equilibrium configuration of the member corresponding to an axial force of $0.5P_y$ and $M_0 = 0.4M_y$.

6.7 Newmark's Method

Table 6.2b Determination of Equilibrium Configuration by Newmark's Method ($M_0 = 0.90M_y$)

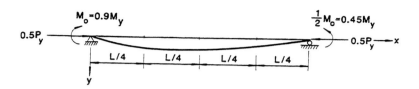

Station	0	1	2	3	4	Common Factor
Primary Moment M_0	0.90	0.788	0.675	0.563	0.45	M_y
Cycle 1						
$y_{assumed}$	0	0.0022	0.0028	0.002	0	L
Secondary Moment $Py_{assumed}$	0	0.0011	0.0014	0.001	0	$P_y L$
Change Multiplier	0	0.11	0.14	0.1	0	M_y
Total Moment $M_0 + Py_{assumed}$	0.90	0.898	0.815	0.663	0.45	M_y
Curvature Φ_i	1.389	1.380	1.066	0.714	0.45	Φ_y
Average Slope θ_i		-1.389	-2.769	-3.835	-4.549	$\left(\frac{L}{4}\right)\Phi_y$
Deflection y_i	0	-1.389	-4.158	-7.993	-12.54	$\left(\frac{L}{4}\right)^2 \Phi_y$
Corrected Deflection y_{ic}	0	1.747	2.113	1.414	0	$\left(\frac{L}{4}\right)^2 \Phi_y$
Change Multiplier $y_{calculated}$	0	0.0036	0.0044	0.0029	0	L

For any given loading, an equilibrium configuration of the member can be found using the procedure outlined above so long as the member remains stable. Therefore, by increasing the magnitude of end moments from $M_0 = 0.4M_y$ to $M_0 = 0.5M_y$, one can find a new equilibrium configuration of the member. Hence, by increasing the value of M_0 and

Table 6.2b Determination of Equilibrium Configuration by Newmark's Method ($M_0 = 0.90M_y$) (*continued*)

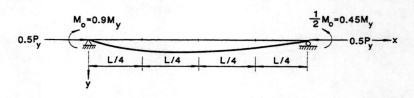

Station	0	1	2	3	4	Common Factor
Cycle 2						
$y_{assumed}$	0	0.0036	0.0044	0.0029	0	L
Secondary Moment $Py_{assumed}$	0	0.0018	0.0022	0.00145	0	$P_y L$
Change Multiplier	0	0.18	0.22	0.145	0	M_y
Total Moment $M_0 + Py_{assumed}$	0.9	0.968	0.895	0.708	0.450	M_y
Curvature Φ_i	1.389	1.767	1.366	0.797	0.45	Φ_y
Average Slope θ_i		−1.389	−3.156	−4.522	−5.319	$\left(\frac{L}{4}\right)^2 \Phi_y$
Deflection y_i	0	−1.389	−4.545	−9.067	−14.386	$\left(\frac{L}{4}\right)^2 \Phi_y$
Corrected Deflection y_i	0	2.208	2.648	1.723	0	$\left(\frac{L}{4}\right)^2 \Phi_y$
Change Multiplier $y_{calculated}$	0	0.0046	0.0055	0.0036	0	L

Since $y_{calculated} \gg y_{assumed}$, solution diverges.

repeating the procedure, a nondimensional load-deflection ($M_0/M_y - y_2/L$) curve can be plotted. Such a curve is shown in Fig. 6.33. The curve is nonlinear for $M_0/M_y > 0.5$ because the effects of geometrical and material nonlinearities become significant in the analysis. The effect of geometrical nonlinearity is accounted for by introducing the secondary (Py) moment

6.7 Newmark's Method

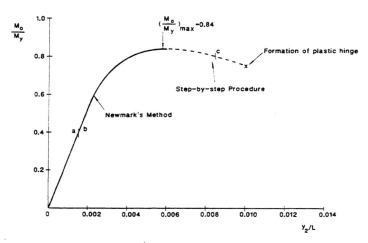

FIGURE 6.33 Load-deformation behavior of an elastic beam-column

in the calculation and the effect of material nonlinearity is accounted for by using the nonlinear M-P-Φ relationship.

As seen from the figure, as the load (M_0/M_y) increases, the deflection (y_2/L) increases. However, the rate of increase in deflection is not directly proportioned to the rate of increase in load for $M_0/M_y > 0.5$. The rate of increase in deflection exceeds that of the load when $M_0/M_y > 0.5$ and at $M_0/M_y > 0.8$, a slight increase in load will bring about a tremendous increase in deformation. The peak point of the curve, that is, $M_0/M_y \approx 0.84$, represents the maximum load the member can carry. Thus, for the given member subjected to an axial force of $0.5P_y$, the maximum end moments that the member can carry are $0.84M_y$ and $0.42M_y$ at the left and right ends, respectively.

If a value greater than $M_0 = 0.84M_y$ is used in the calculation, the values of the deflection will diverge rather than converge, indicating that the assigned value of M_0 exceeds the load-carrying capacity of the member. To illustrate this, Newmark's method is used in an attempt to establish the equilibrium configuration of the member for $M_0 = 0.90M_y$. The detailed calculations are shown in Table 6.2b. As can be seen, the values of the deflections diverge in subsequent cycles of calculations.

Before leaving the subject, it is important to note that Newmark's method can only be used to generate the *ascending* branch of the load-deflection curve. In other words, this method is applicable only for tracing the load-deflection behavior of the member from the start of loading to the peak point of the curve. If we are also interested in the descending branch of the curve, that is, the load-deflection behavior of the member beyond the peak point, Newmark's method cannot be used directly. Instead, another numerical procedure known as the *numerical integration procedure* should be used. This procedure is discussed in the following section.

6.8 NUMERICAL INTEGRATION PROCEDURE

In the step-by-step numerical integration procedure, the member is divided into segments, and, just as in Newmark's method, the moment-curvature-thrust relationship must be known or calculated in advance. However, unlike Newmark's method, in which an assumed deflection is assigned at each and every station, only the deflection of the first station is specified in the step-by-step numerical integration procedure. The deflections at subsequent stations are calculated from station to station in a systematic, forward-marching manner (which we will describe in a moment). Another difference between Newmark's method and the step-by-step numerical integration procedure is that in the former the external applied forces are kept constant during the iteration and the equilibrium configuration of the member that corresponds to these applied forces is sought. In the step-by-step numerical integration procedure, the assumed deflection of the first station is kept constant during the iteration and the forces that correspond to that deflection are sought. Newmark's method is essentially a *load control* iterative procedure, while the step-by-step numerical integration procedure is a *displacement control* iteration procedure.

Consider Fig. 6.34, in which a pinned-pinned member is shown. The member is subjected to an axial force P and end moments M_0 and βM_0. β is a constant, representing the ratio of the moment applied at the right end to the moment applied at the left end of the member. The solution procedure is outlined as follows.

1. Specify a deflection y_1 at Station 1 and assume a value for M_0.
2. Calculate the secondary moment at Station 1 by multiplying y_1 by the axial force P.
3. Calculate the total moment at Station 1 by adding the secondary moment to the primary moment.
4. Obtain the curvature Φ_1 of Station 1 from the moment-curvature-thrust relationship.
5. Calculate the deflection of Station 2, y_2, from

$$y_2 = \left(\frac{\Delta^2 y}{\Delta x^2}\right)_1 (\Delta x)^2 + 2y_1 - 0_0$$
$$= -\Phi_1 (\Delta x)^2 + 2y_1 \qquad (6.8.1)$$

6.8 Numerical Integration Procedure

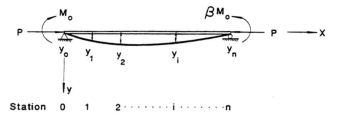

FIGURE 6.34 Step-by-step procedure

Equation (6.8.1) is the second-order central difference equation. Δx is the length of the segment.

6. Knowing y_2, the secondary moment at Station 2 can be obtained by multiplying y_2 by P.
7. The total moment is obtained by adding the secondary moment to the primary moment.
8. The curvature Φ_2 at Station 2 can then be obtained from the moment-curvature-thrust relationship.
9. The deflection at Section 3 is obtained from the second-order central difference equation

$$y_3 = \left(\frac{\Delta^2 y}{\Delta x^2}\right)_2 (\Delta x)^2 + 2y_2 - y_1$$
$$= -\Phi_2 (\Delta x)^2 + 2y_2 - y_1 \qquad (6.8.2)$$

Steps 6 to 9 are repeated for successive stations until the end of the member is reached. If the displacement y_n at the end of the member differs from zero, a correction is made to M_0 by

$$M_{oc} = \left(1 + \frac{1}{n}\frac{y_n}{y_1}\right) M_0 \qquad (6.8.3)$$

where n is the number of segments in the model. The process is then repeated starting from Step 2 until y_n becomes zero.

Equation (6.8.3) was developed on the postulation that

$$\frac{\text{error in } M_0}{M_0} = \frac{\text{error in } y_1}{y_1} \qquad (6.8.4)$$

that is

$$\frac{M_{oc} - M_0}{M_0} = \frac{y_n/n}{y_1} \qquad (6.8.5)$$

Solving for M_{oc} in Eq. (6.8.5) gives Eq. (6.8.3).

To illustrate the procedure, the problem shown in Fig. 6.32 is reworked using the step-by-step numerical integration procedure. The

Table 6.3a Determination of Equilibrium Configuration by the Step-by-Step Numerical Integration Procedure ($y_1 = 0.0012L$)

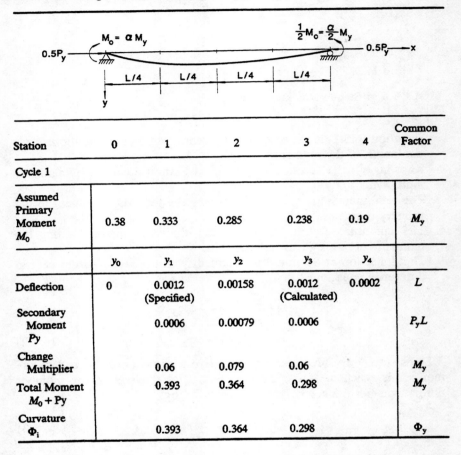

Station	0	1	2	3	4	Common Factor
Cycle 1						
Assumed Primary Moment M_0	0.38	0.333	0.285	0.238	0.19	M_y
	y_0	y_1	y_2	y_3	y_4	
Deflection	0	0.0012 (Specified)	0.00158	0.0012 (Calculated)	0.0002	L
Secondary Moment Py		0.0006	0.00079	0.0006		$P_y L$
Change Multiplier		0.06	0.079	0.06		M_y
Total Moment $M_0 + Py$		0.393	0.364	0.298		M_y
Curvature Φ_i		0.393	0.364	0.298		Φ_y

detailed calculations are shown in Table 6.3a. The solution procedure begins with a value of y_1 equal to $0.0012L$ and an assumed moment M_0 equal to $0.38M_y$. After that, Steps 2 through 5 are followed to calculate y_2. Steps 6 through 9 are then followed to calculate y_3, and, finally, by repeating Steps 6 to 9, y_4 can be calculated. The calculated value of y_4 is $0.0002L$, which differs from the expected value of zero. Therefore, a second cycle of calculation is necessary. This time the modified value for M_0 is calculated from Eq. (6.8.3) to be $0.4M_y$. By following through the same procedure, the value of y_4 is found to be $0.00003L$, which, for practical purposes, can be taken as zero, and so the solution process is stopped.

It is important to mention here that unlike Newmark's method, in

6.8 Numerical Integration Procedure

Table 6.3a (continued)

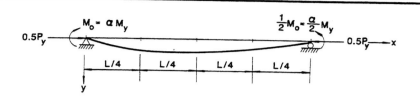

Station	0	1	2	3	4	Common Factor
Cycle 2						
Assumed Primary Moment M_0	0.40	0.35	0.30	0.25	0.20	M_y
	y_0	y_1	y_2	y_3	y_4	
Deflection	0	0.0012 (Specified)	0.00155	0.00111 (Calculated)	0.00003	L
Secondary Moment Py		0.0006	0.000773	0.000557		$P_y L$
Change Multiplier		0.06	0.0773	0.0557		M_y
Total Moment $M_0 + Py$		0.41	0.377	0.306		M_y
Curvature Φ_i		0.41	0.377	0.306		Φ_y

Since $y_4 \approx 0$, therefore stop.

which the solution process proceeds from row to row, the solution process for the step-by-step numerical integration procedure proceeds from column to column in the tabulated form. In addition, the numerical integration procedure can be used to generate the descending branch of the load-deflection curve. This can be achieved by assuming a somewhat larger starting value for y_1. Table 6.3b shows one such calculation and the complete load-deflection curve, including the descending branch, is plotted in Fig. 6.33 (dotted line). Points a, b, and c on the curve correspond to the values calculated in Table 6.2a and 6.3a,b, respectively. Note that for the ascending branch, the Newmark's and the numerical integration methods give almost identical results.

Table 6.3b Determination of Equilibrium Configuration by the Step-by-Step Numerical Integration Procedure ($y_1 = 0.007L$)

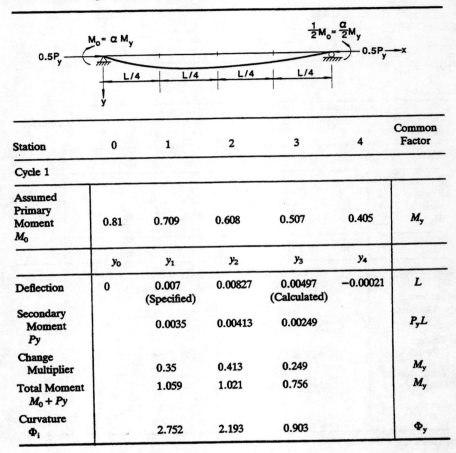

Station	0	1	2	3	4	Common Factor
Cycle 1						
Assumed Primary Moment M_0	0.81	0.709	0.608	0.507	0.405	M_y
	y_0	y_1	y_2	y_3	y_4	
Deflection	0	0.007 (Specified)	0.00827	0.00497 (Calculated)	−0.00021	L
Secondary Moment Py		0.0035	0.00413	0.00249		$P_y L$
Change Multiplier		0.35	0.413	0.249		M_y
Total Moment $M_0 + Py$		1.059	1.021	0.756		M_y
Curvature Φ_i		2.752	2.193	0.903		Φ_y

6.9 SUMMARY

For compression members whose elastic buckling loads are difficult to obtain by analytical means, approximate methods can be used to estimate the critical loads. In this chapter, we introduced two energy methods (Rayleigh–Ritz and Galerkin's) and two numerical methods (Newmark's and numerical integration). In the energy methods, we investigated the energy of the member, in particular, formulating the total potential energy function. By using the technique of variational calculus, we showed that the equilibrium conditions of the member can be established.

The Rayleigh–Ritz method is the first energy method that deals directly with the total potential energy function. By assuming a deflected

6.9 Summary

Table 6.3b (continued)

Diagram: Beam with $0.5P_y$ load at left end, $M_o = \alpha M_y$ applied, four segments of $L/4$, $\frac{1}{2}M_o = \frac{\alpha}{2}M_y$ at right, $0.5P_y \rightarrow x$

Station	0	1	2	3	4	Common Factor
Cycle 2						
Assumed Primary Moment M_o	0.804	0.704	0.603	0.503	0.402	M_y
	y_0	y_1	y_2	y_3	y_4	
Deflection	0	0.007 (Specified)	0.00847	0.00523 (Calculated)	0.00006	L
Secondary Moment Py		0.0035	0.00424	0.00262		$P_y L$
Change Multiplier		0.35	0.424	0.262		M_y
Total Moment $M_o + Py$		1.054	1.027	0.765		M_y
Curvature Φ_i		2.654	2.259	0.926		Φ_y

Since $y_4 \approx 0$, therefore stop.

shape that satisfies the kinematic boundary conditions of the member, the total potential energy function is reduced from a functional to a function. As a result, ordinary calculus, rather than variational calculus, is sufficient to carry out the energy minimization process. The final product of this procedure is the elastic critical load.

Galerkin's method, on the other hand, is derived from energy formulation, but deals with the differential equation of equilibrium in the solution procedure. By assuming a deflected shape that satisfies both the kinematic and natural boundary conditions of the member, the elastic critical load can be obtained directly from the Galerkin's integral [Eq. (6.6.8)].

The approximate critical loads evaluated using the Rayleigh–Ritz or Galerkin's method are always higher than the true critical loads, unless the assumed deflection is the true deflection of the member, in which case the energy method will lead to the true buckling load of the problem. In general, the approximate critical loads represent upper bound solutions to the true critical loads.

An alternative procedure that furnishes both an upper and a lower bound to the true critical load was also presented: Newmark's method. Here the member is divided into segments and numerical values for the deflections at the stations are assumed. Through an iteration process, the true values of the deflections at the stations can be calculated. This procedure is extremely useful to evaluate the critical loads of members with a variable moment of inertia. In addition, Newmark's method is also applicable to the analysis of *inelastic* members so long as the moment-curvature-thrust relationship is known or available.

Another numerical solution scheme, presented in this chapter, that is suitable for inelastic analysis is the step-by-step numerical integration procedure. In this procedure, the numerical calculation proceeds from station to station. This procedure has an advantage over Newmark's method in that both the ascending and descending branches of the load-deflection curve can be traced. Thus, a complete load-deflection analysis can be performed.

Because of the systematic nature of Newmark's method and the numerical integration procedure, both can be carried out conveniently in a computer-based analysis.

PROBLEMS

6.1 Find the approximate elastic critical loads of the columns shown in Fig. P6.1a–c by the Rayleigh–Ritz method.

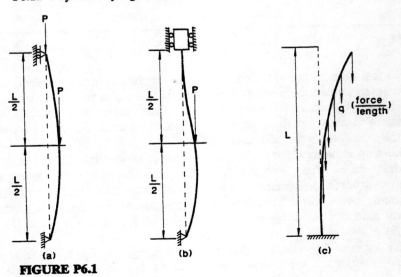

FIGURE P6.1

Problems

6.2 Determine the approximate elastic critical load for the spring-supported column shown in Fig. P6.2 by the Rayleigh–Ritz method. k_s is the spring stiffness.

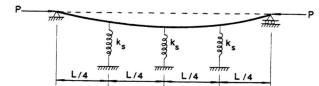

FIGURE P6.2

6.3 Using the Rayleigh–Ritz method, determine the approximate elastic critical load for the column with variable EI shown in Fig. P6.3.

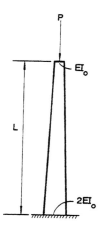

FIGURE P6.3

6.4 Using Galerkin's method, determine the approximate elastic critical load for the pinned-pinned column with variable EI shown in Fig. P6.4.

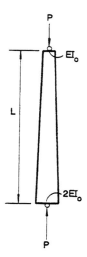

FIGURE P6.4

6.5 Determine an approximate value for the elastic critical moment of the fixed-ended beam shown in Fig. P6.5 using Galerkin's method. The cross section of the beam is rectangular.

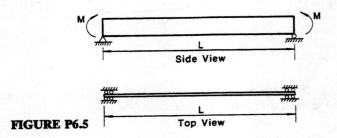

FIGURE P6.5

6.6 Using Newmark's method, determine the approximate elastic critical loads of the stepped columns shown in Fig. P6.6a–b. Establish bounds to the critical loads after each cycle of calculation.

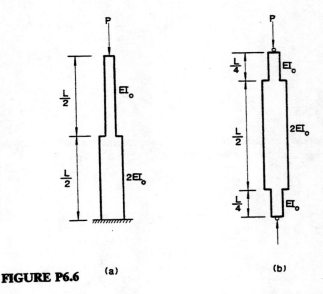

FIGURE P6.6

6.7 Using Newmark's method, trace the moment-rotation $(M - \theta)$ curve for the beam shown in Fig. P6.7, taking into consideration inelastic action. The beam has a rectangular cross section; and the nondimensional moment-curvature relationship is given by the following: for

$$m \leq 1,$$
$$m = \phi$$

Problems

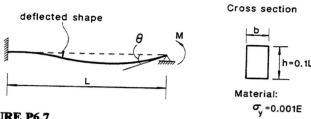

FIGURE P6.7

and for

$$m > 1,$$

$$m = \tfrac{3}{2}\left(1 - \frac{1}{3\phi^2}\right)$$

where $m = M/M_y$, $\phi = \Phi/\Phi_y$, in which M_y is the yield moment and Φ_y is the curvature that corresponds to first yield of the cross section. What is the value for M_{max}?

6.8 Using Newmark's method and the moment-curvature-thrust relationship shown in Section 3.9.1 of Chapter 3, plot the load-deflection (moment-centerline deflection) curve of the initially crooked beam-column shown in Fig. P6.8. Determine the maximum moment that the member can carry for $P = 0.5P_y$. Material: $\sigma_y = 0.001E$.

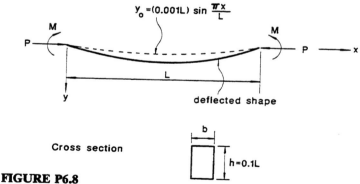

FIGURE P6.8

6.9 Repeat Problem 6.8 using the step-by-step numerical integration procedure. Also trace the descending branch of the load-deflection curve.

6.10 By using the calculus of variations in conjunction with the Principle of Stationary Total Potential Energy, derive the governing differential equation and the corresponding boundary conditions (kinematic or natural) describing the buckling behavior of a tip-loaded cantilever column. Explain why Galerkin's method is difficult to implement for members with a free end.

REFERENCES

1. Hoff, N. J. The Analysis of Structures. John Wiley & Sons, New York, 1956.
2. Strutt, J. W. On the theory of resonance. Transactions of the Royal Society (London). A161, 1870, 77–118.
3. Strutt, J. W., and Rayleigh, B. The Theory of Sound: Vols. 1 and 2. Second edition. Dover, New York, 1945.
4. Ritz, W. Über eine neue methode zur Lösung gewisser Variations. Probleme der Mathematischen Physik. J. für Reine und Angewandte Mathematik. 135: 1–61, 1909.
5. Timoshenko, S. P., and Gere, J. M. Theory of Elastic Stability. Second edition. McGraw-Hill, New York, 1961.
6. Galerkin, B. G. Series solutions of some problems of elastic equilibrium of rods and plates. Vestnik Inzheneron. 1: 879–908, 1915. (In Russian.)
7. Newmark, N. M. Numerical procedure for computing deflections, moments, and buckling loads. Transaction. ASCE, 108: 1161, 1943.
8. Chen, W. F., and Atsuta, T. Theory of Beam-Columns. Vol. 1: In-Plane Behavior and Design; Vol. 2: Space Behavior and Design. McGraw-Hill, New York, 1976, 1977.
9. Chen, W. F., and Han, D. J. Tubular Members in Offshore Structures. Pitman, London, 1985.

General References

Lanczos, C. The Variational Principles of Mechanics. Fourth edition. University of Toronto Press, 1970.

Langhaar, H. L. Energy Methods in Applied Mechanics. John Wiley, 1962.

Richards, T. H. Energy Methods in Stress Analysis. Ellis Horwood, England, 1977.

Washizu, K. Variational Methods in Elasticity and Plasticity. Third edition. Pergamon Press, England, 1982.

ANSWERS TO SOME SELECTED PROBLEMS

CHAPTER 1 GENERAL PRINCIPLES

1.1 **a.** $P_{cr} = \dfrac{k_{s1}}{L} + k_{s2}L$

   b. $P_{cr} = \dfrac{13k_{s1}}{15L} + \dfrac{3k_{s2}L}{5}$

   c. $P_{cr} = \dfrac{3}{20}k_{s1}L + \dfrac{1}{15}k_{s2}L$

1.2 **a.** $P_{cr} = \dfrac{k_{s1}}{L} + k_{s2}L$

   b. $P_{cr} = \dfrac{13k_{s1}}{15L} + \dfrac{3k_{s2}L}{5}$

   c. $P_{cr} = \dfrac{3}{20}k_{s1}L + \dfrac{1}{15}k_{s2}L$

1.3 $P_{cr} = \dfrac{k_s L}{2}$

$P = \dfrac{k_s L}{2 \sin \theta}(\cos 2\theta - \cos \theta + \sin \theta)$

$90° > \theta > 68.53°$,   stable

$68.53° > \theta > 0°$,   unstable

$0° > \theta > -68.53°$,   stable

$-68.53° > \theta > -90°$, unstable

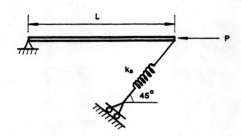

FIGURE P1.3

1.4 $P = k_s L[\cos(\alpha - \theta) - \cos\alpha]\tan(\alpha - \theta)$

$P_{cr} = 0$

$\theta < \alpha - \cos^{-1}(\cos\alpha)^{1/3}$, stable

or

$\theta > \alpha + \cos^{-1}(\cos\alpha)^{1/3}$, stable

$\alpha - \cos^{-1}(\cos\alpha)^{1/3} < \theta < \alpha + \cos^{-1}(\cos\alpha)^{1/3}$, unstable

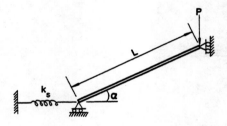

FIGURE P1.4

CHAPTER 2   COLUMNS

2.1 a. $n = 1$, $P_{cr} = \dfrac{\pi^2 EI}{L^2}$

$n = 2$, $P_{cr} = \dfrac{4\pi^2 EI}{L^2}$

$\dfrac{P_{cr(n=1)}}{P_{cr(n=2)}} = \dfrac{1}{4}$

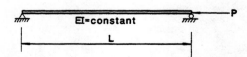

FIGURE P2.1a

b. $n = 1,$ $\quad P = P_{cr} = \dfrac{\pi^2 EI}{4L^2}$

$\qquad n = 3,$ $\quad P = P_{cr} = \dfrac{9\pi^2 EI}{4L^2}$

$$\dfrac{P_{cr(n=1)}}{P_{cr(n=3)}} = \dfrac{1}{9}$$

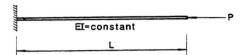

FIGURE P2.1b

2.2 $P_{cr} = \dfrac{\pi^2 EB^4}{8L^2} = 1.234 \dfrac{EB^4}{L^2}$

2.3 $\tan kL \tan\left(\dfrac{kL}{\sqrt{2}}\right) = \sqrt{2}$

$P_{cr} = \dfrac{1.0336 EI_0}{L^2}$

2.4 a. $\tan k_2 L = -k_2 L \qquad k_2^2 = \dfrac{P}{EI_0}$

$\qquad P_{cr} = \dfrac{4.1159 EI_0}{L^2} = \dfrac{\pi^2 EI_0}{(1.55L)^2}$

b. $P_{cr} = \dfrac{4.223 EI}{L^2}$

c. $P_{cr} = \dfrac{3.7185 EI}{L^2}$

2.5 $A_F = \dfrac{1 + \dfrac{1}{4}\left(\dfrac{P}{P_e}\right)}{1 - \left(\dfrac{P}{P_e}\right)} = \dfrac{a}{1 - \left(\dfrac{P}{P_e}\right)}$

$a = 1 + 0.25\left(\dfrac{P}{P_e}\right)$

2.8 a. $\left(\dfrac{P_I}{P_t}\right)_{\lambda_0 = 0.4} = 1.18$

$\qquad \left(\dfrac{P_I}{P_t}\right)_{\lambda_0 = 1.2} = 1.046$

2.11 a. Use W10 × 30 ASD
      b. Use W10 × 30 PD
      c. Use W8 × 24 LRFD

2.12 $E_r = \dfrac{2EE_t}{E + E_t}$

2.13 $P_{cr} = \dfrac{3.1EI_0}{L^2}$

2.14 $P_{cr} = \dfrac{0.1245EI}{h^2}$

CHAPTER 3   BEAM-COLUMNS

3.2 $A = \dfrac{Q \sin k(L - a)}{EIk^3 \sin kL}$

$B = 0$

$C = \dfrac{-Q \sin ka}{EIk^3 \tan kL}$

$D = \dfrac{Q \sin ka}{EIk^3}$

3.4 a.

$$R_A = \dfrac{w\left(\cos kL - \dfrac{k^2 L^2}{2} \cos kL + kL \sin kL - 1\right)}{k(\sin kL - kL \cos kL)}$$

$$y_{max} = \left[\dfrac{w\left(1 - \cos kL - kL \sin kL + \dfrac{k^2 L^2}{2}\right)}{EIk^4(\sin kL - kL \cos kL)}\right] \sin kx_1$$

$$+ \left(\dfrac{w}{EIk^4}\right) \cos kx_1 + \left(\dfrac{w}{2EIk^2}\right) x_1^2 - \dfrac{w}{EIk^4}$$

$$- \left[\dfrac{w\left(\cos kL - \dfrac{k^2 L^2}{2} \cos kL + kL \sin kL - 1\right)}{EIk^3(\sin kL - kL \cos kL)}\right] x_1$$

$$M_B = \dfrac{wL(2 - 2\cos kL - kL \sin kL)}{2k(\sin kL - kL \cos kL)}$$

$$= \left(\dfrac{wL^2}{8}\right) \dfrac{4(kL \sin kL + 2 \cos kL - 2)}{kL(kL \cos kL - \sin kL)}$$

Chapter 3   Beam-Columns

$$\bar{M}_{max} = \left(\frac{w}{k^2}\right)$$

$$\times \left\{ \frac{\sqrt{\left[\frac{k^4L^4}{4} + 2k^2L^2 + 2 - (2+k^2L^2)(kL \sin kL + \cos kL)\right]}}{\sin kL - kL \cos kL} - 1 \right\}$$

$$M_{max} = \text{the larger of } (M_B, \bar{M}_{max}), \qquad M_0 = \frac{wL^2}{8}$$

$$M_B \approx \frac{1 - 0.4 \dfrac{P}{P_{ek}}}{1 - \dfrac{P}{P_{ek}}} \left(\frac{wL^2}{8}\right)$$

b.

$$M_B = \frac{\frac{1}{2}\sin kL - \sin \frac{kL}{2}}{kL \cos kL - \sin kL} QL$$

$$= \left(\frac{3QL}{16}\right) \frac{16\left(\sin \frac{kL}{2} - \frac{1}{2}\sin kL\right)}{3(\sin kL - kL \cos kL)}$$

$$\bar{M}_{max} = \frac{Q \sin\left(\dfrac{kL}{2}\right)}{k \sin kL} - \frac{M_B}{\sin kL}$$

$$M_{max} = \text{the larger of } (M_B, \bar{M}_{max})$$

$$M_B \approx \frac{1 - 0.3 \dfrac{P}{P_{ek}}}{1 - \dfrac{P}{P_{ek}}} \left(\frac{3QL}{16}\right)$$

3.5   **a.** For Fig. P3.5a

$$y = \left(\frac{Q}{EIk^3}\right)\left[\frac{\sin k(L-a)}{\sin kL}\right] \sin kx - \left(\frac{Q(L-a)}{LEIk^2}\right)x$$

$$\text{for } 0 \leqslant x \leqslant a$$

$$y = -\left(\frac{Q}{EIk^3}\right)\left[\frac{\sin ka}{\tan kL}\right] \sin kx + \left(\frac{Q \sin ka}{EIk^3}\right) \cos kx - \frac{Qa(L-x)}{LEIk^2}$$

$$\text{for } 0 \leqslant x \leqslant \frac{L}{3}$$

3.5 b. For Fig. P3.5b

$$y = \left(\frac{Q}{EIk^3}\right)\left[\frac{\cos\left(\frac{kL}{6}\right)}{\cos\left(\frac{kL}{2}\right)}\right]\sin kx - \left(\frac{Q}{EIk^2}\right)x \quad \text{for} \quad 0 \leq x \leq \frac{L}{3}$$

$$y = \left(\frac{Q}{EIk^3}\right)\sin\left(\frac{kL}{3}\right)\left[\tan\frac{kL}{2}\sin kx + \cos kx\right]$$
$$-\frac{QL}{3EIk^2} \quad \text{for} \quad \frac{L}{3} \leq x \leq \frac{2}{3}L$$

$$y = -\left(\frac{Q}{EIk^3}\right)\left[\frac{\left(\sin\frac{kL}{3} + \sin\frac{2kL}{3}\right)}{\tan kL}\right]\sin kx$$
$$+ \left[\frac{Q\left(\sin\frac{kL}{3} + \sin\frac{2kL}{3}\right)}{EIk^3}\right]\cos kx$$
$$-\frac{Q(L-x)}{EIk^2} \quad \text{for} \quad \frac{2}{3}L \leq x \leq L$$

c. For Fig. P3.5c

$$y = \left(\frac{Q}{EIk^3}\right)\left[\frac{\sin\left(\frac{kL}{3}\right)}{\sin kL}\right]\sin kx - \frac{Qx}{3EIk^2} + \frac{w}{EIk^2}\left(\frac{x^2}{2} - \frac{5xL}{18}\right)$$
$$+ \left(\frac{w}{EIk^4 \sin kL}\right)\left[\sin kx \cos\frac{2kL}{3} + \sin k(L-x) - \sin kL\right]$$
$$\text{for} \quad 0 \leq x \leq \frac{L}{3}$$

$$y = \left(\frac{Q}{EIk^3}\right)\left[\frac{\sin\left(\frac{kL}{3}\right)}{\sin kL}\right]\sin kx - \frac{Qx}{3EIk^2} + \frac{w(xL - L^2)}{18EIk^2}$$
$$+ \frac{w}{EIk^4 \sin kL}\left[\left(1 - \cos\frac{kL}{3}\right)\sin k(L-x)\right]$$
$$\text{for} \quad \frac{L}{3} \leq x \leq \frac{2L}{3}$$

$$y = -\left(\frac{Q}{EIk^3}\right)\frac{\sin\left(\frac{2kL}{3}\right)}{\tan kL}\sin kx + \frac{Q\sin\left(\frac{2kL}{3}\right)}{EIk^3}\cos kx - \frac{2Q(L-x)}{3EIk^2}$$
$$+ \left(\frac{w}{EIk^4 \sin kL}\right)\left[\left(1 - \cos\frac{kL}{3}\right)\sin k(L-x)\right] + \frac{w(xL - L^2)}{18EIk^2}$$
$$\text{for} \quad \frac{2L}{3} \leq x \leq L$$

Chapter 3 Beam-Columns

d. For Fig. P3.5d

$$y = \left(\frac{Q}{EIk^3}\right)\left[\frac{\sin\frac{kL}{3}}{\sin kL}\right]\sin kx - \frac{Qx}{3EIk^2} + \left(\frac{wx}{EIk^2L}\right)\left(\frac{x^2}{2} - \frac{7L^2}{54}\right)$$

$$+ \left(\frac{3w}{EIk^4 L \sin kL}\right)\left(\frac{L}{3}\sin kx \cos\frac{2kL}{3}\right.$$

$$\left. + \frac{1}{k}\sin kx \sin\frac{2kL}{3} - x \sin kL\right) \quad \text{for} \quad 0 \le x \le \frac{L}{3}$$

$$y = \left(\frac{Q}{EIk^3}\right)\left[\frac{\sin\frac{kL}{3}}{\sin kL}\right]\sin kx - \frac{Qx}{3EIk^2} - \frac{wL(L-x)}{27EIk^2}$$

$$\times \left(\frac{3w}{LEIk^4}\right)\left(\frac{\sin k(L-x)}{\sin kL}\right)\left[\frac{1}{k}\sin\frac{kL}{3} - \frac{L}{3}\cos\frac{kL}{3}\right]$$

$$\text{for} \quad \frac{L}{3} \le x \le \frac{2}{3}L$$

$$y = \left[\frac{Q \sin\left(\frac{2kL}{3}\right)}{EIk^3 \sin kL}\right]\sin k(L-x) - \frac{2Q(L-x)}{3EIk^2} - \frac{wL(L-x)}{27EIk^2}$$

$$+ \left(\frac{3w}{LEIk^4}\right)\left[\frac{\sin k(L-x)}{\sin kL}\right]\left[\frac{1}{k}\sin\frac{kL}{3} - \frac{L}{3}\cos\frac{kL}{3}\right]$$

$$\text{for} \quad \frac{2}{3}L \le x \le L$$

3.6 a. $\dfrac{M_A}{M_B} = 0.4$

$k\bar{x} = 1.982$

$\bar{x} = 0.998L \approx L$

$M_{max} = M_B$

b. $\dfrac{M_A}{M_B} = 0$

$\bar{x} = 0.79L$

$M_{max} = -1.0933 M_B$

c. $\dfrac{M_A}{M_B} = -0.4$

$\bar{x} = 0.637L$

$M_{max} = 1.3316 M_B$

3.7

$$M_{FA} = M_{FB} = \frac{P\Delta(1 - \cos kL)}{kL \sin kL + 2 \cos kL - 2}$$

$$= \frac{P\Delta}{kL \cot \dfrac{kL}{2} - 2}$$

$$k^2 = \frac{P}{EI}$$

3.13

$$M_u = \frac{PL}{1 - \dfrac{16P}{P_e}}$$

$$P_e = \frac{\pi^2 EI}{L^2}$$

CHAPTER 4 RIGID FRAMES

4.1 $P_{cr} = \dfrac{30.5656 EI}{L^2}$

4.2 **a.** For Fig. P4.2(a)

(i) Sway-prevented case

$P_{cr} = 2.407 P_e$

(ii) Sway-permitted case

$P_{cr} = 0.6694 P_e$

b. For Fig. P4.2(b)

(i) Sway-prevented case

$P_{cr} = 1.314 P_e$

(ii) Sway-permitted case

$P_{cr} = 0.661 P_e$

4.3 **b.**

Upper bound $P_{cr} = \dfrac{\pi^2 EI}{(0.5L)^2}$

Lower bound $P_{cr} = \dfrac{\pi^2 EI}{L^2}$

c. $a = 1,$ $P_{cr} = \dfrac{\pi^2 EI}{(0.58L)^2}$

4.4 $P_p = \dfrac{3.2 M_p}{L}$

4.5 a. $P_{cr} = \dfrac{\pi^2 EI}{(0.555L)^2}$

$K = 0.555$

b. $G_T = 0.25$

$G_B = 0$

$K = 0.555$

4.6 a. Nonsway

(i) Pinned end $G'_A = \dfrac{\sum\limits_A \left(\dfrac{EI}{L}\right)_c}{\tfrac{3}{2}\sum\limits_A \left(\dfrac{EI}{L}\right)_b}$

(ii) Fixed end $G'_A = \dfrac{\sum\limits_A \left(\dfrac{EI}{L}\right)_c}{2\sum\limits_A \left(\dfrac{EI}{L}\right)_b}$

b. Sway

(i) Pinned end $G'_A = \dfrac{\sum\limits_A \left(\dfrac{EI}{L}\right)_c}{\tfrac{1}{2}\sum\limits_A \left(\dfrac{EI}{L}\right)_b}$

(ii) Fixed end $G'_A = \dfrac{\sum\limits_A \left(\dfrac{EI}{L}\right)_c}{\tfrac{2}{3}\sum\limits_A \left(\dfrac{EI}{L}\right)_b}$

4.7 $P_{cr} = 225$ kips

CHAPTER 5 BEAMS

5.1 $EC_w \dfrac{d^4\gamma}{dz^4} - GJ \dfrac{d^2\gamma}{dz^2} - \dfrac{1}{EI_y}\left[\dfrac{M(1+\beta)z - L}{L}\right]^2 \gamma = 0$

5.2 a, b, and c

$$\frac{d^4\gamma}{dz^4} - \frac{GJ}{EC_w}\frac{d^2\gamma}{dz^2} - \frac{P^2(L-z)^2}{EC_wEI_y}\gamma = 0$$

5.3 $M_{ocr} = \frac{\pi}{L}\left(\frac{bt^3}{6}\right)\sqrt{29 \times 12 \times 10^6 c}$ k-in.²/ft

$= 67.8 \frac{bt^3}{L}\sqrt{C}$ k-ft (L in ft).

Solution 1, 2t × t

$$M_{ocr1} = \frac{110.379}{L}$$

Solution 2, 4t × t

$$M_{ocr2} = \frac{247.633}{L}$$

Solution 3, 8t × t

$$M_{ocr3} = \frac{515.427}{L}$$

5.4 (1) $C_b = 1.0$ (5) $C_b = 1.143$
 (2) $C_b = 1.714$ (6) $C_b = 1.0$
 (3) $C_b = 2.4$ (7) $C_b = 1.385$
 (4) $C_b = 1.333$

5.7 $M_n = 3645$ in-kip

5.9 Method 1. Use Fig. 5.20,

$$P_{cr} = 28.24 \text{ kips}$$

Method 2. Use Table 5.9, Case 11 and Eqs. (5.6.16) and (5.6.17),

$$P_{cr} = 49.75 \text{ kips}$$

Method 3. Use Table 5.4 and Eqs. (5.5.16) and (5.5.19),

$$P_{cr} = 32.1 \text{ kips}$$

5.10 a. Solution 1: $M_{cr} = 8680$ k-in.

$(M_A/M_B = 0, \quad C_b = 1.75, \quad M_{ocr} = 4960$ in-kip)

Solution 2: $M_n = 6484.5$ k-in.

$M_{cr} = 9219, \quad (C_{bs} = 5.115, \quad M_{ocr} = 1802$ in-kip)

5.11 $EC_w\frac{d^4\gamma}{dz^4} - GJ\frac{d^2\gamma}{dz^2} - \frac{\gamma}{EI_y}\left[\frac{w}{2}\left(\frac{L^2}{4} - z^2\right)\right]^2 = 0$

CHAPTER 6 ENERGY AND NUMERICAL METHODS

6.1 a. $\bar{v} = a \sin \dfrac{\pi x}{L}$

$$P_{cr} = \dfrac{2}{3}\dfrac{\pi^2 EI}{L^2} = \dfrac{\pi^2 EI}{(1.2247L)^2}$$

b. $\bar{v} = a(L^3 x - 3Lx^2 + 2x^4)$

$$P_{cr} = \dfrac{23.344\pi^2 EI}{L^2} = \dfrac{\pi^2 EI}{(0.207L)^2}$$

c. $\bar{v} = a\left(1 - \cos\dfrac{\pi x}{2L}\right)$

$$(qL)_{cr} = 0.7992 \dfrac{\pi^2 EI}{L^2}.$$

6.2 $\bar{v} = \sum\limits_{i=1}^{n} a_i \sin \dfrac{i\pi x}{L}$

$$i = 1, \quad P_{cr} = P_e\left[1 + \dfrac{2k_s L}{a_1 \pi^2 P_e}\left\{\sqrt{2}\left(\sum_{i=1}^{n} a_i \sin \dfrac{i\pi}{4}\right)\right.\right.$$
$$\left.\left. + \sum_{i=1}^{n} a_i \sin \dfrac{i\pi}{2}\right\}\right]$$

$$i = 2, \quad P_{cr} = P_e\left[2^2 + \dfrac{2k_s L}{2^2 a_2 \pi^2 P_e}\left\{2\left(\sum_{i=1}^{n} a_i \sin \dfrac{i\pi}{4}\right)\right\}\right]$$

$$i = 3, \quad P_{cr} = P_e\left[3^2 + \dfrac{2k_s L}{3^2 a_3 \pi^2 P_e}\left\{\sqrt{2}\left(\sum_{i=1}^{n} a_i \sin \dfrac{i\pi}{4}\right)\right.\right.$$
$$\left.\left. - \sum_{i=1}^{n} a_i \sin \dfrac{i\pi}{2}\right\}\right]$$

$$i = n, \quad P_{cr} = P_e\left[n^2 + \dfrac{2k_s L}{n^2 a_n \pi^2 P_e}\left\{2\left(\sum_{i=1}^{n} a_i \sin \dfrac{i\pi}{4}\right)\sin \dfrac{n\pi}{4}\right.\right.$$
$$\left.\left. + \left(\sum_{i=1}^{n} a_i \sin \dfrac{i\pi}{2}\right)\sin \dfrac{n\pi}{2}\right\}\right]$$

6.3 $\bar{v} = a\left(1 - \cos\dfrac{\pi x}{2L}\right)$

$$P_{cr} = 4.2011 \dfrac{EI_0}{L^2} = 0.4257 \dfrac{EI_0 \pi^2}{L^2}$$

6.4 a. $\bar{y} = a(\tfrac{3}{2}L^3 x - 3Lx^3 + \tfrac{3}{2}x^4)$

$$P_{cr} = \dfrac{14.8235 EI_0}{L^2}$$

b. $\bar{y} = a \sin \dfrac{\pi x}{L}$

$$P_{cr} = \dfrac{14.8 EI_0}{L^2}$$

6.5 $\bar{\gamma} = a\left(1 - \cos \dfrac{2\pi z}{L}\right)$

$$M_{cr} = \dfrac{2\pi}{L}\sqrt{GJEI_y}$$

6.6 a. Assume the deflection shape

$$y = \delta\left(1 - \cos \dfrac{\pi x}{2L}\right)$$

and divide the member into 10 segments.
Cycle 1.

$$4.442 \dfrac{EI_0}{L^2} \leq P_{cr} \leq 4.94 \dfrac{EI_0}{L^2}$$

$$P_{cr} = 4.6626 \dfrac{EI_0}{L^2}$$

Cycle 2.

$$4.334 \dfrac{EI_0}{L^2} \leq P_{cr} \leq 4.424 \dfrac{EI_0}{L^2}$$

$$P_{cr} = 4.37 \dfrac{EI_0}{L^2}$$

b. Assume the deflected shape $y = \delta \sin \dfrac{\pi x}{L}$ and divide the member into 10 segments

Cycle 1

$$13.836 \dfrac{EI_0}{L^2} \leq P_{cr} \leq 15.456 \dfrac{EI_0}{L^2}$$

$$P_{cr} = 14.763 \dfrac{EI_0}{L^2}$$

Cycle 2

$$14.582 \dfrac{EI_0}{L^2} \leq P_{cr} \leq 14.802 \dfrac{EI_0}{L^2}$$

$$P_{cr} = 14.714 \dfrac{EI_0}{L^2}$$

6.8 Divide the member into 4 segments

$$\frac{M}{M_y} = 0.4, \quad y_{\text{midspan}} = 0.00114L$$

$$\frac{M}{M_y} = 0.8, \quad y_{\text{midspan}} = 0.00372L$$

b. $M_n = 6484.5$ k-in.

INDEX

A

Alignment chart (nomograph)
  sway permitted case, 287
  sway prevented case, 283
Allowable bending stress, 357–363
ASD (Allowable Stress Design), 37, 41, 169,
  beam, 355–363
  beam-column, 211–213, 218, 230
  column, 123, 211, 213
Amplification factor, 55, 56, 94, 153, 351
  design, 154, 156–157, 160, 161, 231
  deflection, 155, 160
  moment, 94, 155, 156, 160, 161, 170, 206
  P-δ moment (B_1), 215, 218, 230
  P-Δ moment (B_2), 216, 218, 230
  theoretical, 154, 157, 160, 161
Analysis
  large deflection, 24–31, 59, 60
  small deflection (deformation), 12–24, 404
  small displacement, 42, 413
Angle of twist, 307–380, 408–414, 426–437
Asymmetric
  bifurcation, 4, 6, 7
  postbuckling behavior, 7
Axial shortening, 392–395, 401, 403

B

B_1 factor, 215, 218, 230
B_2 factor, 215, 216, 218, 230
Beams, 307–380
  amplification factor, 351
  cantilever, 336, 426
  continuous, 346
  design curves, (*see* Design curves for steel beams)
  effective length, 338–339, 340, 342–343
  fixed-end, 339
  inelastic, 351
  initially crooked, 348
  lateral buckling, 317–324
  on elastic foundation, 423–426
  other end conditions, 343
  simply supported, 317–333, 431–437
  spring system, 14
  stocky, 354
Beam-column, 48, 147–235
  basic differential equation, 175–177
  design format (AISC), 211–219
  fixed-ended, 172–175, 178–182, 192, 229
    AISC/ASD, 170, 211–213, 218, 230
    AISC/LRFD, 170, 215–219, 230
    AISC/PD, 214, 218, 230
    interaction equation, 205–219
      stability (control), 211, 214
      strength (yielding control), 212, 214
  inelastic analysis, 193–205
  simply supported, 148, 149, 156, 170–172, 453–460
  slope-deflection equation, 182–193
  superposition solutions, 170–175
Bending
  curvature, 392, 401
  shortening, 395–396, 401, 403, 410

485

Index

Bifurcation, 4, 5, 9, 11, 13, 16, 18, 20, 46, 401
  analysis, (*see* eigenvalue analysis)
  approach, 11, 13, 16, 18, 20
  buckling, 5, 9
  equilibrium, 4, 46, 401
  instability, 4
  load, 11
  stable symmetric, 6
  symmetric, 4, 6
  unstable symmetric, 7

Buckling
  analysis, 4
  elastic, 4
  inelastic, 96
  lateral, 164, 307, 364, 367, 373, 390, 391, 403–414, 426–437, 441–443
  local, 355, 364, 365
  nonsway, 148, 170, 215, 229, 239–242, 248
  stress (load), 212
  sway, 148, 215, 217, 230, 243–247, 250, 260

C

Calculus of variations, 390–415
C_b factor, 326, 327, 332–335
C_m factor, 206–210, 229–230
  Austin expression, 169
  Massonnet expression, 169

Codes,
  AISC/ASD, (*see* ASD)
  AISC/LRFD, (*see* LRFD)

Coefficient of variation
  load effects, 39
  resistance, 39

Collapse
  load, 238, 270–276
  mechanism, 238, 354

Columns, 45–146
  aluminum, 108–110
  cantilever, 65, 83, 85, 86, 417–423, 446–449
  curve, 73, 75, 112
  design curves
    AISC/ASD, 123, 211, 213
    AISC/LRFD, 128, 206
    AISC/PD, 124
    CRC, 122, 206, 213
    SSRC, 125, 206
  eccentrically loaded, 53
  end-restrained, 61, 74
  fixed-fixed, 62, 82, 402, 403
  fixed-pinned, 67, 402, 403, 440–442

  fixed and guided, 69
  fourth order differential equation, 80
  hinged and guided, 71
  hinged-hinged, 48, 81, 91, 390–403, 438, 439, 444
  inelastic buckling, 96
  large displacement analysis, 58
  long, 47
  medium length, 47
  Perry–Roberston formula, 95
  reduced modulus, 48, 97, 100–105, 381
  slender, 47
  stub, 111
  stocky, 47
  selection table
    AISC/LRFD curve, 127
  Shanley's inelastic theory, 105

Conjugate beam method, 444, 450, 451
Compact section, 355, 357–359
Conservative forces, 11, 12, 42, 390
Contragradient law, 263
Critical load, 4, 8, 11, 42, 46, 239, 381, 382, 401, 415, 417
  elastic buckling stress (load), 212
  lower bound, 446
  upper bound, 443, 446
Crookedness, 47, 91

D

Deflection
  primary or fundamental, 4, 25, 26, 29, 30
  problem, 147
  secondary or postbuckling, 4, 25, 26, 29, 30

Design curves for steel beams, 355–371
  AISC/ASD, 355–363
  AISC/LRFD, 365–370
  AISC/PD, 363–365
    lateral instability, 365
    local buckling, 365
    plastic yielding, 365
  ECCS, 371
  SSRC, 371

Design curves for column
  AISC/ASD, 123, 211, 213
  AISC/LRFD, 128, 206
  AISC/PD, 124
  CRC, 122, 206, 213
  mathematical form
    Lui–Chen, 131
    Rondal–Maquoi, 130
  multiple curves, 125, 129
  SSRC, 125, 206

Index

Design format, 129
  AISC/ASD, 37, 124
  AISC/LRFD, 39, 128
  AISC/PD, 38, 125
Design format for beam-column
  AISC/ASD, 211–213, 218, 230
  AISC/LRFD, 215–218, 230
  AISC/PD, 214, 218, 230
Design
  interaction formulas, 205–219
  philosophies, 37–42, 206
Displacement control iteration procedure, 460–463
Double (reverse) curvature (bending), 162, 164, 167, 193, 211
Double modulus theory, 100–105
Ductility, 271
Dynamic approach, 12

E

Eccentricity factor, 58
Effective length, 61
  alignment charts, 148, 225, 251, 283
  factor, 64, 67, 69, 71, 73, 74, 148, 230, 282–287, 338
Effective modulus, 97, 100, 381
Eigenvalue analysis, 11, 24, 49, 239, 381, 401, 415
Elastic
  buckling analysis, 4
  buckling stress (load), 212
  frameworks, 382
  instability, 237
  lateral buckling, 367
  lateral torsional, 307
  linear system, 388, 389, 392, 404
  modulus, 381
  nonlinear system, 388, 389
  -perfectly plastic, 194
  -plastic analysis, 194, 238, 280
  restrained ends, 74
Element stiffness formulation, 253
End-restrained column, 61
Energy method, 4, 11, 14, 17, 19, 22–36, 381–443
Equilibrium
  effect of imperfection on, 37
  bifurcation, 4, 46
  neutral, 3, 4, 46, 381, 401
  postbuckling, 37
  stable, 2, 15, 45, 46, 390
  unstable, 2, 4, 15, 28, 37, 46, 390
Equivalent concentrated loads, 444, 445, 447

Equivalent moment
  concept, 168, 326, 374
  factor C_b, 326, 327, 332–335
  factor C_m, 168, 169, 170, 211
Euler load, 46, 52, 152, 225
European Committee on Constructional Steelwork, 371
Extremum principle, 414

F

Factor of safety, 38
Finite-disturbance buckling, 8, 9
First-order
  elastic analysis, 215
  moment, 94, 218
  probability theory, 39
  second moment probability analysis, 39
  structural analysis, 1
First variation of the total potential energy, 390, 397, 412
Fixed-end
  beams, 339
  beam-column, 172–182, 229
  column, 62, 82, 402, 403
  moment, 154, 172, 173, 191, 222
Fourier sine series, 95
Fourth-order differential equation, 79, 228
Frame, 219–218, 236–306
  nonsway buckling, 239–242, 248
  simple portal, 237, 277
  sway buckling, 243–247, 250, 260
Free-body approach, 388, 414
Functionals, 391, 415
Fundamental path, 4, 26, 29

G

Galerkin's method, 414, 438–443, 446, 463, 465
Generalized
  displacements, 11, 12, 386
  forces, 386
Geometrical imperfection, 11, 381
Guyed tower, 7

H

Hinge-by-hinge analysis, 272–275
Hooke's Law, 50, 96, 149

I

Imperfection, 8, 11, 31–37, 381
  column, 47, 91–108
  factor, 58, 134
  system, 31–37

Incremental load approach, 266
Inelastic
   analysis, 193–205, 351, 381
   beams, 351
   beam-columns, 193–205, 450, 452, 453–460
   buckling, 96
   column, 96–108
   lateral buckling, 367
Initially crooked
   beams, 348–350
   columns, 91–95
In-plane
   bending, 317–319, 404–409, 413
   buckling, 390, 391–403, 426
Instability, (see stability)
Interaction equations, (see beam-column)

K

Kinematic
   admissible displacement, 20, 383, 385, 398
   assumption, 196
   boundary condition, 402, 438, 439, 443, 464, 465
   constraints, 383, 385

L

Large deflection analysis, 24–31, 59, 60
Lateral (torsional) buckling (instability), 164, 208, 307, 317–324, 364, 367, 373, 390, 391, 403–414, 426–437, 441–443
Lateral unbraced length, 356
L'Hospital's rule, 187, 224
Limit load, 4, 8
Limit states, 1, 39, 47
   design, 1
   strength, 1
   serviceability, 1
Linear
   eigenvalue analysis, 24
   translational spring, 16
   rotational spring, 12, 20
LRFD (Load and Resistance Factor Design), 37, 39, 41, 169, 170
   beam, 365–370
   beam-column, 215–219, 230
   column, 128, 206
   frame, 283–287
Load
   combinations, 38, 39, 40
   control iterative procedure, 460
   effects, 38, 39
   factor, 38, 39, 40, 125, 214
Load-deflection
   problem, 11, 31, 47, 459
   analysis, 266–270, 381, 382
Local buckling, 355, 364, 365

M

Magnification factor, (see Amplification factor)
Matrix stiffness method, 253–266
Mechanism method, 277
   beam mechanism, 277
   combined mechanism, 277
   sway mechanism, 277
Merchant–Rankine equation, 280
Method of
   neutral equilibrium, 46, 238
   variation of parameters, 54, 150
   undetermined coefficient, 54, 150
Moment-curvature relationship, 49, 148, 197, 200, 443, 444, 447
M-Φ-P, 195–200, 452, 454, 459, 460, 461, 466
Moment reduction factor, 211
   mathematical form
      Lui–Chen, 131
      Rondal–Maquoi, 130
Monosymmetric section, 353
Multiple-design curves, 125, 129

N

Newark's method, 204, 382, 414, 443–460, 462, 463, 466
   beam-column, 453–458
   column, 446–450
Nomographs, (see alignment charts)
Noncompact section, 355–358
Nonsway (model), 148, 170, 215, 229, 230, 239–242, 248
Non-uniform torsion, (see warping restraint torsion)
Numerical method, 148, 382, 414, 415, 443–446
   beam-column, 453–458
   column, 446–450
   frame, 266–269
   integration, 204, 382, 414, 460–463, 464

O

Orthogonal
   functions, 423
   property, 423, 424, 425

Index

Out-(of)-plane
  bending (buckling), 317, 319, 404, 407, 409, 410, 432, 434
  displacement, 318, 322

P

P-delta effect, (see second order)
Perry–Roberston formula, 95
Plastic analysis
  buckling load, 381
  collapse loads, 238, 270–276
  collapse mechanism, 1, 270
  design (PD), 37, 38, 41, 124
    beam, 363–365
    beam-column, 214, 218, 230
    column, 124
  hinge, 270–272, 354
  limit moment, 135
  moment, 271, 355
  section modulus, 214
  strength, 39
Plastification, 271
Positive definite, 3
Postbuckling
  behavior, 6, 24, 42
  paths, 4, 6, 25, 27, 29, 30
Potential energy, 4, 11, 12, 14, 15, 17, 25, 390–443
Primary
  bending moments, 147, 155, 161, 208, 460
  deflections, 147
  path, 4, 25
  plastic, 197, 203
Principle of
  stationary total potential energy, 388–391, 402–404, 414, 415
  superposition, 154, 170, 172
  virtual work, 382–388, 389
Proportional limit, 96

R

Random parameters, 40
Rayleigh–Ritz method, 414–437, 443, 446, 463–465
Reduced modulus theory, (see double modulus theory)
Reliability (or safety) index, 39, 40, 41
Residual stresses, 112–117, 204
Resistance
  factor, 39, 40
  nominal, 38, 39
Rigid constraints, 385
Rigid frames, 236–306

Right-handed screw rule, 320

S

Safety index, 39, 40, 41
Secant formula, 58
Secondary
  deflection, 148
  moment, 94, 148, 155, 211, 230, 454, 460
  buckling path, 4
  path, 4, 6, 25, 27, 29, 30
  plastic, 198–200, 204
Second-order
  analysis, 2, 266–269
  elastic, 266–269
  p-δ effect, 130, 139, 215, 218, 230, 266
  p-Δ effect, 211, 216, 218, 230, 231, 266
Semicompact section, 357
Serviceability
  limit states, 1
  requirements, 42
Shanley's inelastic theory, (see column)
Shape factor, 133
Sidesway, 211, 230
Single curvature, 164, 165, 167, 192, 193, 211
Slenderness
  modified, 354
  parameter, 73, 121–136, 354
  ratio, 203, 204, 351
Slope-deflection equations, 182–187, 192, 193, 222, 248
  modified, 187–193
Snap-through buckling, 8, 9
Specification
  AISC/ASD (see ASD)
  AISC/LRFD (see LRFD)
Spring-bar system, 12, 24
Stability
  analysis, 1, 11, 42, 126
  behavior, theory, 42
  functions, 184
  interaction equation, (see beam-column)
  limit point, 4
  problem, 147
Stable equilibrium, 2, 15, 28, 45, 390
Stable equilibrium path
  fundamental, 28
  postbuckling, 28, 37
Standard deviation, 39, 40
Stationary (extremal, maximum, or minimum) value, 14, 391
Step-by-step numerical integration procedure, 460–463, 466

Stiffened elements, 355
Stiffness matrix, 3
   linear, 260
   geometric, 260
   structure, 264
Strain
   energy (function), 14, 17, 19, 22, 26, 28, 31, 35, 389–443
   hardening, 271
   reversal, 105
Strength (yielding control) interaction equation, (*see* beam-column design format)
Strength limit state, 1
Stress-strain curve (relation), 110, 111, 196, 270
Strong axis bending, 119
Structural Stability Research Council Curves
   beam, 371
   column, 125–126
Stub column, 111
St. Venant
   shear stresses, 309
   torsion, 310, 430, 432
Superposition
   principle of, 154, 170, 172
   solution of beam-column by, 170–175
Sway
   model (case), 148, 243
   permitted (unbraced) case, 243–252
   prevented (braced) case, 239–242, 248, 282
Symmetric bifurcation
   stable, 6, 28
   unstable, 7, 31

T

Tangent modulus
   load, 47, 100, 107, 381
   theory, 48, 97
Tangent stiffness, 11
Taylor series, 27, 30
Torsion
   narrow rectangular section, 317–321
   non-uniform, 311
   open cross section, 307, 309, 311
   thin-walled open section, 309, 311
   uniform (pure), 309, 311
   warping restraint, 311, 313, 433, 434
Torsional constant, 310
Total potential energy (function), 4, 11, 390–438, 464
   minimum, 4
   first variation, 390
   second variation, 390
   stationary, 388–391, 402–404, 414, 415
Total strain energy, 395, 408
Transition moment, 368
Twisting (deformation), 307–380, 404, 405, 408, 410

U

Ultimate
   moment capacity, 208
   plastic strength limit state, 39
   strength interaction diagrams (curves), 204, 206, 207
Uniform torsion, (*see* St. Venant torsion)
Unstable
   equilibrium, 2, 46, 390
   equilibrium path, 8, 15
   postbuckling equilibrium paths, 37
   symmetric bifurcation, 7
Unstiffened projecting elements, 355
Upper bound theorem, 276

V

Variational calculus, 391, 464
Virtual work, 382–388
   displacement, 382–389, 398
   external, 386, 389
   internal, 386, 388
   strain energy, 389, 390

W

Warping
   restraint torsion, 311, 313, 433, 434
   constant, 315, 434
   deformation, 433
Weak axis bending, 119

Y

Yield load, 47, 73
Yield moment, 271